Introduction to Modeling and Control of Internal Combustion Engine Systems

Lino Guzzella and Christopher H. Onder

Introduction to Modeling and Control of Internal Combustion Engine Systems

 Springer

Prof. Dr. Lino Guzzella
ETH Zürich
Institute for Dynamic Systems & Control
Sonneggstr. 3
8092 Zürich
ETH-Zentrum
Switzerland
E-mail: lguzzella@ethz.ch

Dr. Christopher H. Onder
ETH Zürich
Institute for Dynamic Systems & Control
Sonneggstr. 3
8092 Zürich
ETH-Zentrum
Switzerland
E-mail: onder@ethz.ch

ISBN 978-3-642-10774-0   e-ISBN 978-3-642-10775-7

DOI 10.1007/978-3-642-10775-7

Library of Congress Control Number: 2009940323

© 2010 Springer-Verlag Berlin Heidelberg

Typesetting: Data supplied by the authors

Production: Scientific Publishing Services Pvt. Ltd., Chennai, India

Cover Design: WMX Design, Heidelberg, Germany

Printed in acid-free paper

9 8 7 6 5 4 3 2 1

springer.com

Preface

Who should read this text?

This text is intended for students interested in the design of classical and novel IC engine control systems. Its focus lies on the control-oriented mathematical description of the physical processes involved and on the model-based control system design and optimization.

This text has evolved from a lecture series held during the last several years in the mechanical engineering (ME) department at ETH Zurich. The target readers are graduate ME students with a thorough understanding of basic thermodynamic and fluid dynamics processes in internal combustion engines (ICE). Other prerequisites are knowledge of general ME topics (calculus, mechanics, etc.) and a first course in control systems. Students with little preparation in basic ICE modeling and design are referred to [64], [97], [194], and [206].

Why has this text been written?

Internal combustion engines represent one of the most important technological success stories in the last 100 years. These systems have become the most frequently used sources of propulsion energy in passenger cars. One of the main reasons that this has occurred is the very high energy density of liquid hydrocarbon fuels. As long as fossil fuel resources are used to fuel cars, there are no foreseeable alternatives that offer the same benefits in terms of cost, safety, pollutant emission and fuel economy (always in a total cycle, or "well-to-wheel" sense, see e.g., [5] and [68]).

Internal combustion engines still have a substantial potential for improvements; Diesel (compression ignition) engines can be made much cleaner and Otto (spark ignition) engines still can be made much more fuel efficient. Each goal can be achieved only with the help of control systems. Moreover, with the systems becoming increasingly complex, systematic and efficient system design procedures have become technological and commercial necessities. This text addresses these issues by offering an introduction to model-based control system design for ICE.

What can be learned from this text?

The primary emphasis is put on the ICE (torque production, pollutant formation, etc.) and its auxiliary devices (air-charge control, mixture formation, pollutant abatement systems, etc.). Mathematical models for some of these processes will be developed below. Using these models, selected feedforward and feedback control problems will then be discussed.

A model-based approach is chosen because, even though more cumbersome in the beginning, it after proves to be the most cost-effective in the long run. Especially the control system development and calibration processes benefit greatly from mathematical models at early project stages.

The appendix contains a brief summary of the most important controller analysis and design methods, and a case study that analyzes a simplified idle-speed control problem. This includes some aspects of experimental parameter identification and model validation.

What cannot be learned from this text?

This text treats ICE systems, i.e., the load torque acting on the engine is assumed to be known and no drive-train or chassis problems will be discussed.

Moreover, this text does not attempt to describe *all* control loops present in engine systems. The focus is on those problem areas in which the authors have had the opportunity to work during earlier projects.

Acknowledgments

Many people have implicitly helped us to prepare this text. Specifically our teachers, colleagues and students have helped to bring us to the point where we felt ready to write this text. Several people have helped us more explicitly in preparing this manuscript: Alois Amstutz, with whom we work especially in the area of Diesel engines, several of our doctoral students whose dissertations have been used as the nucleus of several sections (we reference their work at the appropriate places), Simon Frei, Marzio Locatelli and David Germann who worked on the idle-speed case study and helped streamlining the manuscript, and, finally, Brigitte Rohrbach and Darla Peelle, who translated our manuscripts from "Germlish" to English.

Zurich,   *Lino Guzzella*
May 2004   *Christopher H. Onder*

Preface to the Second Edition

Why a second edition?

The discussions concerning pollutant emissions and fuel economy of passenger cars constantly intensified since the first edition of this book was published. Concerns about the air quality, the limited resources of fossil fuels and the detrimental effects of greenhouse gases further spurred the interest of both the industry and academia to work towards improved internal-combustion engines for automotive applications. Not surprisingly, the first edition of this monograph rapidly sold out. When the publisher inquired about a second edition, we decided to seize this opportunity for revising the text, correcting several errors, and adding some new material. The following list outlines the most important changes and additions included in this second edition:

- restructured and slightly extended section on superchargers, increasing the comprehensibility;
- short subsection on rotational oscillations and their treatment on engine test-benches, being a safety-relevant aspect;
- improved physical and chemical model for the three-way catalyst, simplifying the conception and realization of downstream air-to-fuel ratio control;
- complete section on modeling, detection, and control of engine knock;
- new methodology for the design of an air-to-fuel ratio controller exhibiting several advantages over the traditional H_∞ approach;
- short introduction to thermodynamic engine-cycle calculation and some corresponding control-oriented aspects.

As in the first edition, the text is focused on those problems we were (or still are) working on in our group at ETH. Many exciting new ideas (HCCI combustion, variable-compression engines, engines for high-octane fuels, etc.) have been proposed by other groups. However, simply reporting those concepts without being able to round them off by first-hand experience would not add any benefit to the existing literature. Therefore, they are not included in this book, which should remain an introductory reference for students and engineers new to the topic of internal-combustion engines.

Acknowledgements

We want to express our gratitude to the many colleagues and students who reported to us errors and omissions in the first edition of this text.

Several people have helped us improving this monograph, in particular Daniel Rupp, Roman Möller and Jonas Asprion who helped preparing the manuscript.

Zurich,   *Lino Guzzella*
September 2009   *Christopher H. Onder*

Contents

1
Introduction

In this chapter, first the notation used throughout this text is defined. It further contains some general remarks on electronic engine control systems and introduces the most common control problems encountered in spark ignition (Otto or gasoline) and compression ignition (Diesel) engine systems. The intention is to show the general motivation for using control systems and to give the reader an idea of the problems that can be tackled by feedforward and feedback control systems for both SI and CI engines.

The emphasis in this chapter is on qualitative arguments. The mathematically precise formulation is deferred to subsequent chapters. Those readers not familiar with modern electronic sensors, actuators, and control hardware for automotive applications may want to consult either [7], [108], or [125].

1.1 Notation

The notation used in this text is fairly standard. The *derivative* of a variable $x(t)$, with respect to its independent variable t, is denoted by

$$\frac{d}{dt}x(t)$$

while the notation

$$\dot{x}(t)$$

is used to indicate a *flow* of mass, energy, etc. Both variables $\frac{d}{dt}x(t)$ and $\dot{x}(t)$ have the same units, but they are different objects. No special distinction is made between scalars, vectors and matrices. The dimensions of a variable, if not a scalar, are explicitly defined in the context. Input signals are usually denoted by $u_{...}$ and output signals by $y_{...}$, whereas the index ... specifies what physical quantity is actuated or measured.

Concentrations of chemical species C are denoted by $[C]$, with units mol/mol, with respect to the reference substance. The concentrations are therefore limited to the interval $[0, 1]$. The concentration of pollutant species are often shown in plots or tables using *ppm* units (part per million), i.e., by using a amplification factor of 10^6. For mass storage and transportation models it is advantageous to use mass fractions, which are denoted by ξ having units [kg/kg].

In general, all variables are defined at that place in the text where they are used for the first time. To facilitate the reading, some symbols have been reserved for special physical quantities:

α	$[\frac{W}{m^2 K}]$	heat-transfer coefficient
A	$[m^2]$	area
c_x	$[\frac{J}{kg K}, \text{-}]$	specific heat capacities ($x = p, v$), concentration of x
ε	[-]	compression ratio, volume fraction
η	[-]	efficiency
ϕ	$[°, rad]$	crank angle
γ	[-]	gear ratio
H	$[J]$	enthalpy
κ	[-]	ratio of specific heats
λ	[-]	air-to-fuel ratio, volumetric efficiency, Lagrange multiplier
m	$[kg]$	mass
M	$[\frac{kg}{mol}]$	molar mass
N	[-]	number of engine revolutions per cycle
		(1 for two-stroke, 2 for four-stroke engines)
ν	[-]	stoichiometric coefficient
p	$[Pa, bar]$	pressure
P	$[W]$	power
Π	[-]	pressure ratio
Q	$[J]$	heat
r	$[m, \frac{mol}{s}]$	radius, reaction rate
ρ	$[\frac{kg}{m^3}]$	density
R	$[\frac{J}{kg K}]$	specific gas constant
$\mathcal{R}$	$[\frac{J}{mol K}]$	universal gas constant
σ_0	[-]	stoichiometric air-to-fuel ratio
t	$[s]$	time (independent variable)
τ	$[s]$	time (interval or constant)
ϑ	$[K, °C]$	temperature
θ	[-]	occupancy
T	$[Nm]$	torque
Θ	$[m^2 kg]$	rotational inertia
u, y	[-]	control input, system output (both normalized)
V	$[m^3, l]$	volume
ζ	$[°]$	ignition angle
ω	$[\frac{rad}{s}]$	rotational speed or angular frequency
ξ	[-]	mass fraction

Similarly, some indices have been reserved for special use. The following list shows what each of them stands for:

α, a, β ambient air
c compressor or cylinder
e engine
eg exhaust gas
egr, ε exhaust-gas recirculation
f, φ, ψ fuel
γ engine outlet
l load
m manifold or mean value
seg segment
t turbine
ξ combustion
ζ timing (e.g. of ignition or injection)

In a turbocharged engine system, the four most important locations are designated by the indices 1 for "before compressor," 2 for "after compressor," 3 for "after engine," and 4 for "after turbine."

In general, all numerical values listed in this text are shown in SI units. A few exceptions are made where non-SI units are widely accepted. These few cases are explicitly mentioned in the text.

The most commonly used acronyms are:

BDC (TDC)   bottom (top) dead center (piston at lowest (topmost) position)
BMEP or p_{me} (brake) mean-effective pressure
bsfc   brake specific fuel-consumption
CA   crank angle
CI   compression ignition (in Diesel engines)
CNG   compressed natural gas
COM   control-oriented model
DEM   discrete-event model
DPF   Diesel particulate-filter
ECU   electronic (or engine) control unit
IEG   induction-to-exhaust delay
IPS   induction-to-powerstroke delay
IVC (IVO)   inlet-valve closing (opening)
EVC (EVO)   exhaust-valve closing (opening)
MBT   maximum brake torque (ignition or injection timing)
OC   oxidation catalyst
ODE   ordinary differential equation
ON   octane number
PDE   partial differential equation
PM   particulate matter

SCR	selective catalytic reduction
SI	spark ignition (in Otto/gasoline/gas engines)
TPU	time-processing unit
TWC	three-way catalytic converter
VNT	variable-nozzle turbine
WOT	wide-open throttle

1.2 Control Systems for IC Engines

1.2.1 Relevance of Engine Control Systems

Future cars are expected to incorporate approximately one third of their parts value in electric and electronic components. These devices help to reduce the fuel consumption and the emission of pollutant species, to increase safety, and to improve the drivability and comfort of passenger cars. As the electronic control systems become more complex and powerful, an ever increasing number of mechanical functions are being replaced by electric and electronic devices. An example of such an advanced vehicle is shown in Fig. 1.1.

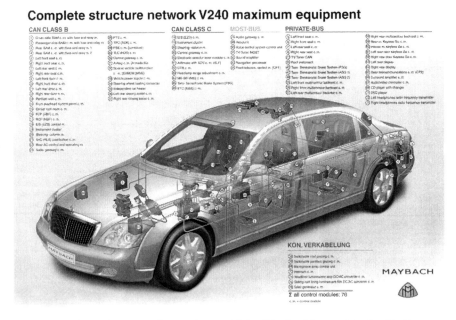

Fig. 1.1. Wiring harness of a modern vehicle (Maybach), reprinted with the permission of Daimler AG.

In such a system, the engine is only one part within a larger structure. Its main input and output signals are the commands issued by the electronic control unit (ECU) or directly by the driver, and the load torque transmitted through the clutch onto the engine's flywheel. Figure 1.2 shows a possible substructure of the vehicle control system. In this text, only the "ICE" (i.e., the engine and the corresponding hardware and software needed to control the engine) will be discussed.

Control systems were introduced in ICE on a larger scale with the advent of three-way catalytic converters for the pollutant reduction of SI engines. Good experiences with these systems and substantial progress in microelectronic components (performance and cost) have opened up a path for the application of electronic control systems in many other ICE problem areas. It is clear that the realization of the future, more complex, engine systems, e.g., hybrid power trains or homogeneous charge compression ignition engines, will not be possible without sophisticated control systems.

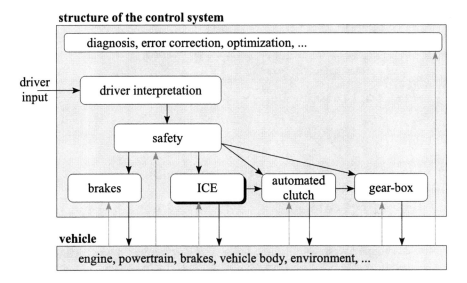

Fig. 1.2. Substructure of a complete vehicle control system.

1.2.2 Electronic Engine Control Hardware and Software

Typically, an electronic engine control unit (ECU) includes standard micro-controller hardware (process interfaces, RAM/ROM, CPU, etc.) and at least one additional piece of hardware, which is often designated as a *time processing unit* (TPU), see Fig. 1.3. This TPU synchronizes the engine control commands with the reciprocating action of the engine. The synchronization of the ECU with the engine is analyzed in more detail in Sec. 3.1.3.[1] Notice also that clock rates of ECU microprocessors are typically much lower than those of desktop computers due to electromagnetic compatibility considerations.

ECU software has typically been written in assembler code, with proprietary real-time kernels. In the last few years there has been a strong tendency towards standardized high-level programming interfaces. Interestingly, the software is structured to reflect the primary physical connections of the plant to be controlled [70].

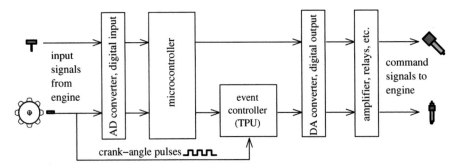

Fig. 1.3. Internal structure of an electronic engine control unit.

1.3 Overview of SI Engine Control Problems

1.3.1 General Remarks

The majority of modern passenger cars are still equipped with port (indirect) injection spark-ignited gasoline engines. The premixed and stoichiometric combustion of the Otto process permits an extremely efficient exhaust gas purification with three-way catalytic converters and produces very little particulate matter (PM). A standard configuration of such an engine is shown in Fig. 1.4.

The torque of a stoichiometric SI engine is controlled by the quantity of air/fuel mixture in the cylinder during each stroke (the quality, i.e., the air/fuel ratio, remains constant). Typically, this quantity is varied by changing the intake pressure and, hence, the density of the air/fuel mixture. Thus, a throttle plate is used upstream of the combustion process in the intake system. This solution is relatively simple and reliable, but it produces substantial "pumping losses" that negatively affect the part-load efficiency of the

---

[1] The reciprocating or *event-based* behavior of all ICE also has important consequences for the controller design process. These problems will be addressed in Chapters 3 and 4.

engine. Novel approaches, such as electronic throttle control, variable valve timing, etc., which offer improved fuel economy and pollutant emission, will be discussed below.

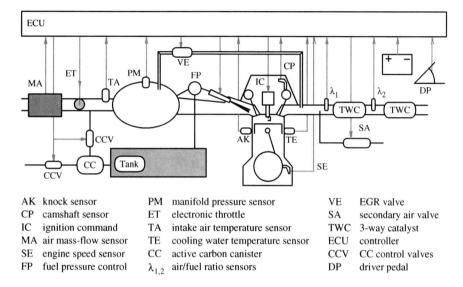

AK	knock sensor	PM	manifold pressure sensor	VE	EGR valve
CP	camshaft sensor	ET	electronic throttle	SA	secondary air valve
IC	ignition command	TA	intake air temperature sensor	TWC	3-way catalyst
MA	air mass-flow sensor	TE	cooling water temperature sensor	ECU	controller
SE	engine speed sensor	CC	active carbon canister	CCV	CC control valves
FP	fuel pressure control	$\lambda_{1,2}$	air/fuel ratio sensors	DP	driver pedal

Fig. 1.4. Overview over a typical SI engine system structure.

A simplified control-oriented substructure of an SI engine is shown in Fig. 1.5. The main blocks are the fuel path P_φ and the air path P_α, which define the mixture entering the cylinder, and the combustion block P_χ that determines the amount of torque produced by the engine.

Other engine outputs are the knock signal y_ζ (as measured by a knock sensor P_ζ) and the engine-out air/fuel ratio y_λ (as measured by a λ sensor P_λ mounted as close as possible to the exhaust valves). The engine speed ω_e is the output of the block P_Θ, taking into account the rotational inertia of the engine, whose inputs are the engine torque T_e and the load torque T_l.

The four most important control loops are indicated in Fig. 1.5 as well:

- the fuel-injection feedforward loop;
- the air/fuel ratio feedback loop;
- the ignition angle feedforward[2] loop; and
- the knock feedback loop.

In addition, the following feedforward or feedback loops are present in many engine systems:[3]

---

[2] Closed-loop control has been proposed in [60] using the spark plug as an ion current sensor.

[3] Modern SI engines can include several other control loops.

- idle and cruise speed control;
- exhaust gas recirculation (for reducing emission during cold-start or for lean-burn engines);
- secondary air injection (for faster catalyst light-off); and
- canister purge management (to avoid hydrocarbon evaporation).

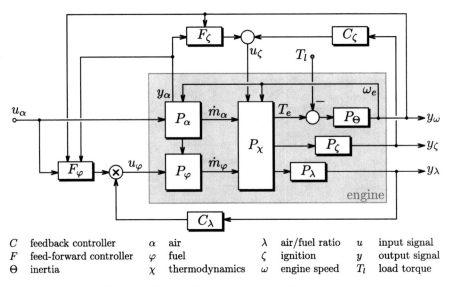

Fig. 1.5. Basic SI engine control substructure.

C	feedback controller	α	air	λ	air/fuel ratio	u	input signal
F	feed-forward controller	φ	fuel	ζ	ignition	y	output signal
Θ	inertia	χ	thermodynamics	ω	engine speed	T_l	load torque

1.3.2 Main Control Loops in SI Engines

Air/Fuel Ratio Control

The air/fuel ratio control problem has been instrumental in paving the road for the introduction of several sophisticated automotive control systems. For this reason, it is described in some detail.

The pollutant emissions of SI engines (mainly hydrocarbon (HC), carbon monoxide (CO), and nitrogen oxide (NO_x)) greatly exceed the limits imposed by most regulatory boards, and future emission legislation will require substantial additional reductions of pollutant emission levels. These requirements can only be satisfied if appropriate exhaust gas after-treatment systems are used.

The key to clean SI engines is a three-way catalytic converter (TWC) system whose stationary conversion efficiency is depicted in Fig. 1.6. Only for a very narrow air/fuel ratio "window," whose mean value is slightly below the stoichiometric level, can all three pollutant species present in the exhaust

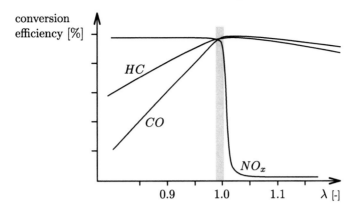

Fig. 1.6. Conversion efficiency of a TWC (after light-off, stationary behavior).

gas be almost completely converted to the innocuous components water and carbon dioxide. In particular, when the engine runs under lean conditions, the reduction of nitrogen oxide stops almost completely, because the now abundant free oxygen in the exhaust gas is used to oxidize the unburned hydrocarbon and the carbon monoxide. Only when the engine runs under rich conditions do the unburned hydrocarbon (HC) and the carbon monoxide (CO) act as agents reducing the nitrogen oxide on the catalyst, thereby causing the desired TWC behavior.

The mean air/fuel ratio can be kept within this narrow band only if electronic control systems and appropriate sensors and actuators are used. The air/fuel ratio sensor (λ sensor) is a very important component in this loop. A precise fuel injection system also is necessary. This is currently realized using "sequential multiport injectors." Each intake port has its own injector, which injects fuel sequentially, i.e., only when the corresponding intake valves are closed.

Finally, appropriate control algorithms have to be implemented in the ECU. The fuel-injection feedforward controller F_φ tries to quickly realize a suitable injection timing based only on the measured air-path input information (either intake air mass flow, intake manifold pressure, or throttle plate angle and engine speed). The air/fuel ratio feedback control system C_λ compensates the unavoidable errors in the feedforward loop. While it guarantees the mean value of the air/fuel ratio to be at the stoichiometric level, it cannot prevent transient excursions in the air/fuel ratio.

Ignition Control

Another important example of a control system in SI engines is the spark angle control system. This example shows how control systems can help improve fuel economy as well.

In fact, the efficiency of SI engines is limited, among other factors, by the knock phenomenon. Knock (although still not fully understood) results from an unwanted self-ignition process that leads to locally very high pressure peaks that can destroy the rim of the piston and other parts in the cylinder. In order to prevent knocking, the compression ratio must be kept below a safe value and ignition timing must be optimized off-line and on-line.

A first optimization takes place during the calibration phase (experiments on engine or chassis dynamometers) of the engine development process. The nominal spark timing data obtained are stored in the ECU. An on-line spark timing control system is required to handle changing fuel qualities and engine characteristics. The key to this component is a knock sensor and the corresponding signal processing unit that monitors the combustion process and signals the onset of knocking.

The feedforward controller F_ζ, introduced in Fig. 1.5, computes the nominal ignition angles (realizing maximum brake torque while avoiding knock and excessive engine-out pollution levels) depending on the engine speed and load (as measured by manifold pressure or other related signals). This correlation is static and is only optimal for that engine from which the ignition data was obtained during the calibration of the ECU. The feedback control system C_ζ utilizes the output of the knock detection system to adapt the ignition angle to a safe and fuel efficient value despite variations in environmental conditions, fuel quality, etc.

1.3.3 Future Developments

Pollutant emission levels of stoichiometric SI engines are or soon will be a "problem solved" such that the focus of research and development efforts can be redirected towards the improvement of the fuel economy. The most severe drawbacks of current SI engines are evident in part-load operating conditions. As Fig. 1.7 shows, the average efficiency even of modern SI engines remains substantially below their best bsfc[4] values. This is a problem because most passenger cars on the average (and also on the governmental test cycles) utilize less than 10% of the maximum engine power.[5] Not surprisingly, cycle-averaged "tank-to-wheel" efficiency data of actual passenger cars are between 12% and 18% only. The next step in the development of SI engines therefore will be a substantial improvement of their part-load efficiency.

Several ideas have been proposed to improve the fuel efficiency of SI engines, all of which include some control actions, e.g.,

- variable valve timing systems (electromagnetic or electrohydraulic);
- downsizing and supercharging systems;
- homogeneous and stratified lean combustion SI engines;

---

[4] Brake-specific fuel consumption (usually in g fuel/kWh mechanical work).
[5] Maximum engine power is mainly determined by the customer's expectation of acceleration performance and is, therefore, very much dependent on vehicle mass.

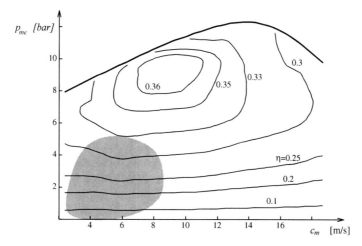

Fig. 1.7. Engine map (mean effective pressure versus mean piston speed) of a modern SI engine, gray area: part-load zone, η =const: iso-efficiency curves. For the definition of p_{me} and c_m see Sect. 2.5.1.

- variable compression ratio engines; and
- engines with improved thermal management.

These systems reduce the pumping work required in the gas exchange part of the Otto cycle, reduce mechanical friction, or improve the thermodynamic efficiency in part-load conditions.

Another approach to improving part-load efficiency is to include novel power train components, such as starter-generator[6] devices, CVTs[7], etc. As mentioned in the Introduction, these approaches will not be analyzed in this text. Interested readers are referred to the textbook [81].

1.4 Overview of Control Problems in CI Engines

1.4.1 General Remarks

Diesel engines are inherently more fuel-efficient than gasoline engines (see Appendix C), but they cannot use the pollutant abatement systems that have proved to be so successful in gasoline engines. In fact, the torque output of Diesel engines is controlled by changing the air/fuel ratio in the combustion

---

[6] These advanced electric motors and generators typically have around 5 kW mechanical power and permit several improvements like idle-load shut-off strategies or even "mild hybrid" concepts.

[7] Continuously variable transmissions allow for the operation of the engine at the lowest possible speed and highest possible load, thus partially avoiding the low efficiency points in the engine map.

chamber. This approach is not compatible with the TWC working principle introduced above.

In naturally aspirated Diesel engines, the amount of air available is approximately the same for all loads, and only the amount of fuel injected changes in accordance with the driver's torque request. In modern CI engines the situation is more complex since almost all engines are turbocharged. Turbochargers introduce additional feedback paths, considerably complicating the dynamic behavior of the entire engine system. Additionally, pre-chamber injection has been replaced by direct-injection systems. The injection is thereby realized using either integrated-pump injectors or so-called common-rail systems, of which particularly the latter introduces several additional degrees of freedom.

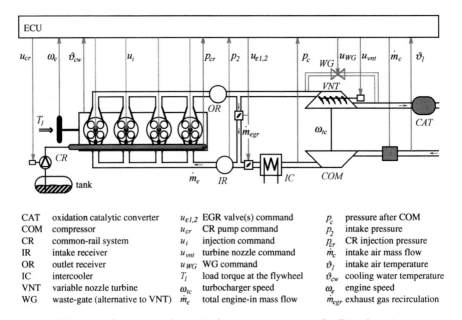

CAT	oxidation catalytic converter	$u_{\varepsilon1,2}$	EGR valve(s) command	P_c	pressure after COM
COM	compressor	u_{cr}	CR pump command	P_2	intake pressure
CR	common-rail system	u_i	injection command	P_{cr}	CR injection pressure
IR	intake receiver	u_{vnt}	turbine nozzle command	$\dot{m}_c$	intake air mass flow
OR	outlet receiver	u_{WG}	WG command	ϑ_l	intake air temperature
IC	intercooler	T_l	load torque at the flywheel	ϑ_{cw}	cooling water temperature
VNT	variable nozzle turbine	ω_{tc}	turbocharger speed	ω_e	engine speed
WG	waste-gate (alternative to VNT)	$\dot{m}_e$	total engine-in mass flow	$\dot{m}_{egr}$	exhaust gas recirculation

Fig. 1.8. Overview of a typical system structure of a Diesel engine.

Compression ignition, or Diesel engines, have been traditionally less advanced in electronic controller utilization due to cost, reliability, and image problems in the past. However this situation has changed, and today, electronic control systems help to substantially improve the total system behavior (especially the pollutant emission) of Diesel engines [79].

Figure 1.8 shows an overview of a typical modern Diesel engine as used in passenger cars. The main objective for electronic Diesel-engine control-systems is to provide the required engine torque while minimizing fuel consumption and complying with exhaust-gas emissions and noise level regulations. This requires an optimal coordination of injection, turbocharger and exhaust-gas recirculation (EGR) systems in stationary and transient operating conditions.

From a control-engineering point of view, there are three important paths which have to be considered: fuel, air and EGR. Figure 1.9 shows a schematic overview of the basic structure of a typical Diesel-engine control-system, clearly pointing out these three paths (for more details on the inner structure of the Diesel engine, see Sect. 2.1). Notice that a speed controller is standard in Diesel engines: The top speed must be limited in order to prevent engine damage whereas the lower limit is imposed by the desired running smoothness when idling.

The fuel path with the outputs torque, speed, and exhaust-gas emissions obtains its inputs from the injection controller. The control inputs to the fuel path are start of injection, injection duration, and injection pressure. With common-rail systems, new degrees of freedom, such as the choice of a pilot injection, main and after-injection quantities with different dwell times in-between, are added. The injected fuel mass is, if necessary, adjusted by the speed controller and has an upper boundary often called the smoke limit: Using the measurement of the air mass-flow into the engine, the maximum quantity of injected fuel is calculated such that the air/fuel ratio does not fall below a certain (constant or operating-point dependent) value. This prevents the engine from producing visible smoke as often seen on older vehicles during heavy acceleration.

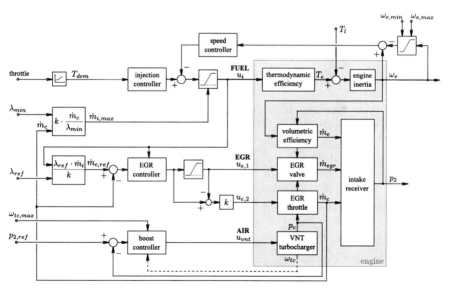

Fig. 1.9. Basic Diesel-engine control-system structure, variables as defined in Fig. 1.8.

The turbocharger dominates the air path. Especially in applications with heavy transient operations, turbocharger designs with small A/R ratios (nozzle area over diameter of the turbine wheel) are chosen to get a good acceleration performance of the supercharging device. Unfortunately, at high loads the small-sized turbocharger works inefficiently and creates high back pressure increasing the pumping work of the engine. Therefore, a substantial fraction of the exhaust gas has to bypass the turbine through a waste gate and at a certain point, not enough enthalpy can be extracted from the exhaust gas, leading to a lack in boost pressure and thus constraining the entire engine system. Besides this, the turbocharger speed has to be limited in order to prevent mechanical damage, further restricting the maximum power output of small-sized turbines.

With the expectations of good acceleration performance and high turbocharger efficiency over the whole range of operation, instead of waste-gate systems variable nozzle turbochargers (VNT) are used. They overcome the trade-off between acceleration performance and sufficient power output at high loads by adjusting the nozzle area as needed. Another approach is two-stage charging, combining a small and a large turbocharger in serial configuration. The former accelerates quickly and ensures good driveability, while the latter provides high boost pressures at high mass flows when needed.

Turbochargers are typically controlled using a closed-loop approach where the measured output is the boost pressure. However, special care has to be taken to keep them from reaching dangerously high rotational speeds.[8]

1.4.2 Main Control Loops in Diesel Engines

With emission legislation becoming ever more stringent, exhaust-gas recirculation is needed for NO_x reduction. A closed-loop control system takes care of the EGR path. Even if the objective is EGR control, the closed loop takes the measured air mass-flow as the feedback variable. A Diesel engine produces smoke if the air/fuel ratio falls below a certain value. The air/fuel ratio with the best NO_x reduction under the boundary condition of no increase in smoke generation is mapped over the quantity of fuel injected and the engine speed. Together with the quantity of fuel injected, which is known from the injection table, the reference air mass-flow can be derived. The EGR valve position is determined from the difference between reference and measured air mass-flows. The EGR flow heavily affects the air mass-flow through the states of the intake and outlet receiver. Additional devices, such as throttles, must be used to maintain a pressure difference over the EGR valve.

While engine performance, dynamic behavior and fuel consumption are important criteria for engine manufacturers, all engine system must comply

---

[8] Note that it is not sufficient to limit the boost pressure. Depending on the operating point and on the ambient conditions, e.g. high altitude and the corresponding low ambient pressure, the turbocharger speed may exceed critical limits even while the demanded boost is not attained. Measuring the turbocharger speed, as indicated in Figure 1.9, or calculating it on-line by means of an observer in the boost controller is the only way to reliably resolve this problem.

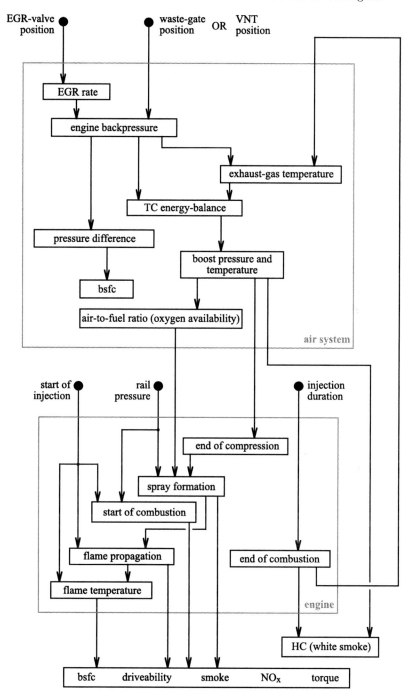

Fig. 1.10. Qualitative input-output relations in a Diesel engine.

with the emission limits. Figure 1.10 summarizes the main input-output relationship and attempts to illustrate the complex network of the underlying physics. In most cases, a single control input affects several outputs.

Three main phenomena are especially important for understanding the key issues of emission control in Diesel engines:

- The thermal efficiency[9] of any combustion process (see Appendix C) depends on the mean combustion temperature ϑ_{com} (determined by the thermodynamic cycle)

$$\eta_{th} = 1 - \frac{\vartheta_{exh}}{\vartheta_{com}} \quad (1.1)$$

  where ϑ_{exh} is the mean exhaust temperature. Obviously, the higher the combustion temperature with respect to the exhaust temperature, the better the thermal efficiency and, therefore, the lower the fuel consumption.

- The rate at which NO is produced can be approximated, according to [97], by

$$\frac{d}{dt}[NO] = \frac{6 \cdot 10^{16}}{\sqrt{\vartheta}} \cdot e^{\frac{-69090}{\vartheta}} \cdot [O_2]^{1/2} \cdot [N_2] \quad (1.2)$$

  where [.] denotes equilibrium concentrations. The strong dependence of the NO formation on the temperature ϑ of the burned gas fraction in the exponential term is evident. High temperatures and oxygen concentrations, therefore, result in high rates of NO formation.

- Diesel particulate matter consists principally of combustion-generated soot absorbing organic compounds. Lubricating oil contributes to the formation of particulate matter. The amount of particulate matter produced during combustion depends on oxygen availability, spray formation and oxidation conditions towards the end of the combustion process.

These facts raise the question of how the electronic control inputs affect the key parameters mentioned:

- *Brake-specific fuel consumption:* With a given injection amount, the main inputs to improve bsfc are start of injection, rail pressure, and boost pressure. An early start of injection results in a fast heat release around top dead center (TDC). Because of the sinusoidal motion of the piston in this area, the volume of the combustion chamber remains almost constant, which results in high gas temperatures and, thus, a good thermal efficiency. Increasing the rail and boost pressures leads to shorter ignition delays and faster burn rates due to faster fuel evaporation and higher in-cylinder temperatures and pressures, respectively.

- *NO formation:* A late start of injection, combined with EGR, yields low in-cylinder temperatures and therefore reduces the NO_x formation. At the same time, relatively high temperatures during the expansion stroke enhance the reduction of NO being formed. Note that these measures are conflicting the ones stated for improved bsfc above.

---

[9] Sometimes also called the Carnot efficiency.

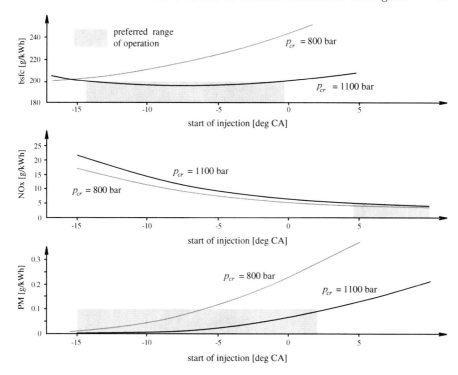

Fig. 1.11. Influence of start of injection on bsfc and on the emission of PM and NO_x, with p_{cr} representing the injection pressure.

- *Particulate matter (PM):* Early start of injection with its fast, hot and complete combustion produces low amounts of particulate matter. Additionally, with an early start of injection, conditions for soot oxidation are good during the long period of the expansion stroke. Due to the good influence on spray formation, high injection pressures also are beneficial for obtaining low amounts of PM. Unfortunately, the high-pressure fuel pump introduces an additional load and thus reduces the engine's overall efficiency.

The tendencies of various input-output relations are summarized in Table 1.1. It shows the difficulty of stating clear control objectives, since nearly every input has a good and a bad effect on the outputs of interest. A well-known method for dealing with the PM-NO_x trade-off is to vary the start of injection from early (e.g., 30 degrees before TDC) to late (e.g., eight degrees after TDC). The optimal injection timing is selected as the one where the trade-off curve crosses the acceptable emission limits defined by the corresponding emission regulations. This approach, however, does not take into account that bsfc should be as low as possible.

Table 1.1. Tendencies in the influence of control inputs on fuel economy (bsfc) and emission quantity.

Control Input	Result
early start of injection	good bsfc low particulate matter high NO_x
high rail pressure	increased NO_x low PM slightly improved bsfc
pilot injection(s)	low noise
smaller VNT area	improved bsfc lower particulate matter higher NO_x
increased EGR	lower NO_x danger of higher PM improved noise equal or slightly increased bsfc

The PM-NO_x trade-off is inherently connected to the principle of the Diesel cycle and creates an obstacle that is difficult to surmount. Figure 1.11 shows the influence of injection on bsfc and on the emission of NO_x and PM. The shaded areas indicate where the emission values are within the regulations and where bsfc is below 200 g/kWh. In this case, even with an injection pressure of 1100 bar, there is no starting point for the injection at which NO_x and PM satisfy current emission limits. Especially in Europe, selective catalytic reduction (SCR) technologies are increasingly used to clean the exhaust gas from NO_x, breaking the PM-NO_x trade-off (see Section 2.9).

For concepts not using SCR, the problem is being tackled in the following order: Sophisticated developments are pursued in the areas of the combustion chamber as well as for the injection, air, and EGR systems. The electronic control system must then guarantee the best use of the given hardware under stationary and transient conditions. Regardless of the selected strategy, model-based controller designs play an important role as an enabling technology.

1.4.3 Future Developments

Diesel engines clearly have a high potential to become much cleaner. On one hand, progress in reducing engine-out emissions will continue. On the other hand, several aftertreatment systems are ready to be introduced on a large scale.

In heavy-duty applications, where fuel economy is a top priority, lean de-NO_x systems using a selective catalytic reduction (SCR) approach are an interesting alternative. Such systems can reduce engine-out NO_x by approximately one order of magnitude. This permits the engine to be calibrated at the high-efficiency/low-PM boundary of the trade-off curve (see Fig. 1.11, early injection angle). The drawback of this approach is, of course, the need for an additional fluid distribution infrastructure (most likely urea).

While such systems are feasible in heavy-duty applications, for passenger cars it is generally felt that a solution with Diesel particulate filters (DPF) is more likely to be successful on a large scale. These filters permit an engine calibration on the high-PM/low-NO_x side of the trade-off. Using this approach, engine-out NO_x emissions are kept within the legislation limits by using high EGR rates but without any further after-treatment systems.

Radically new approaches, such as cold-flame combustion (see e.g., [190]) or homogeneous-charge compression ignition engines (HCCI, see [187], [188], [172] for control-oriented discussion) promise further reductions in engine-out emissions, especially at part-load conditions.

In all of these approaches, feedforward and feedback control systems will play an important role as an enabling technology. Moreover, with ever increasing system complexity, model-based approaches will become even more important.

1.5 Structure of the Text

The main body of this text is organized as follows:

- Chapter 2 introduces mean-value[10] models of the most important phenomena in IC engines.
- Chapter 3 derives discrete-event or crank-angle models for those subsystems that will need such descriptions to be properly controlled.
- Chapter 4 discusses some important control problems by applying a model-based approach for the design of feedforward as well as feedback control systems.

In addition to these three chapters, the three appendices contain the following information:

- Appendix A summarizes, in a concise formulation, most of the control system analysis and synthesis ideas that are required to follow the main text.

---

[10] The term mean-value is used to designate models that do not reflect the engine's reciprocating and hence crank-angle sampled behavior, but which use a continuous-time lumped parameter description. Discrete-event models, on the other hand, explicitly take these effects into account.

- Appendix B illustrates the concepts introduced in the main text by showing the design of a simplified idle-speed controller. This includes some remarks on parameter identification and on model validation using experimental data.
- Appendix C summarizes some control oriented aspects of fuel properties, combustion, and thermodynamic cycle calculation

2

Mean-Value Models

In this chapter, mean-value models *(MVM) of the most important subsystems of SI and Diesel engines are introduced. In this book, the notion of MVM[1] will be used for a specific set of models as defined below. First, a precise definition of the term MVM is given. This family of models is then compared to other models used in engine design and optimization. The main engine sub models are then discussed, namely the air system that determines how much air is inducted into the cylinder; the fuel system that determines how much fuel is inducted into the cylinder; the torque generation system that determines how much torque is produced by the air and fuel in the cylinder as determined by the first two parts; the engine inertial system that determines the engine speed; the engine thermal system that determines the dynamic thermal behavior of the engine; the pollution formation system that models the engine-out emission; and the pollution abatement system that models the behavior of the catalysts, the sensors, and other relevant equipment in the exhaust pipe.*

All these models are control oriented models *(COM), i.e., they model the input-output behavior of the systems with reasonable precision but low computational complexity They include, explicitly, all relevant transient (dynamic) effects. Typically, these COM are represented by systems of nonlinear differential equations. Only physics-based COM will be discussed, i.e., models that are based on physical principles and on a few experiments necessary to identify some key parameters.*

---

[1] The terminology MVM was probably first introduced in [89]. One of the earliest papers proposing MVM for engine systems is [195]. A good overview of the first developments in the area of MVM of SI engine systems can be found in [167]. A more recent source of information on this topic is [44].

2.1 Introduction

Reciprocating engines in passenger cars clearly differ in at least two aspects from continuously operating thermal engines such as gas turbines:

- the combustion process itself is highly transient (Otto or Diesel cycle, with large and rapid temperature and pressure variations); and
- the thermodynamic boundary conditions that govern the combustion process (intake pressure, composition of the air/fuel mixture, etc.) are not constant.

The thermodynamic and kinetic processes in the first class of phenomena are very fast (a few milliseconds for a full Otto or Diesel cycle) and usually are not accessible for control purposes. Moreover, the models necessary to describe these phenomena are rather complex and are not useful for the design of real-time feedback control systems. Exceptions are models used to predict pollutant formation or analogous tasks. Appendix C describes the elementary ideas of engine thermodynamic cycle calculation. (See Sec. 2.5.3 for more details on engine test benches.)

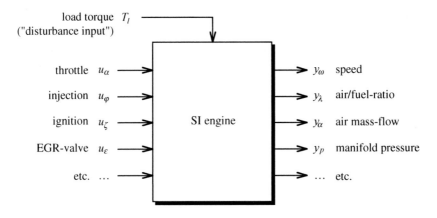

Fig. 2.1. Main system's input/output signals in a COM of an SI engine (similar for Diesel engines).

This text focuses on the second class of phenomena using control-oriented models, and it simplifies the fast combustion characteristics as static effects. The underlying assumption is that, once all important thermodynamic boundary conditions at the start of an Otto or Diesel cycle are fixed, the combustion itself will evolve in an identical way each time the same initial starting conditions are imposed. Clearly, such models are not able to reflect all phenomena (the random combustion pressure variations in SI engines, for example).

As shown in Fig. 2.1, in the COM paradigm, the engine is a "gray box" that has several input (command) signals, one main disturbance signal (the load torque) and several output signals. The inputs are signals, i.e., quantities that can be arbitrarily chosen.[2] Rather than physical quantities, the outputs also are signals that can be used by the controller without the system behavior being affected. The only physical link of the engine to the rest of the power train is the load torque, which in this text is assumed to be known.

The reciprocating behavior of the engine induces another dichotomy in the COM used to describe the engine dynamics:

- *Mean value models* (MVM), i.e., continuous COM, which neglect the discrete cycles of the engine and assume that all processes and effects are spread out over the engine cycle;[3] and
- *Discrete event models* (DEM), i.e., COM that explicitly take into account the reciprocating behavior of the engine.

In MVM, the time t is the independent variable, while in DEM, the crankshaft angle ϕ is the independent variable. Often, DEM are formulated assuming a constant engine speed. In this case, they coincide with classical sampled data systems, for which a rich and (at least for linear systems) complete theoretical background exists. These aspects are treated in detail in Chapter 3.

In MVM, the reciprocating behavior is captured by introducing delays between cylinder-in and cylinder-out effects (see Fig. 2.2). For example, the torque produced by the engine does not respond immediately to an increase in the manifold pressure. The new engine torque will be active Only after the induction-to-power-stroke (IPS) delay [166]

$$\tau_{IPS} \approx \frac{2\pi}{\omega_e} \tag{2.1}$$

has elapsed.[4]

Similarly, any changes of the cylinder-in gas composition, such as air/fuel ratio, EGR ratio, etc. will be perceived at the cylinder exhaust only after the induction-to-exhaust-gas (IEG) delay

$$\tau_{IEG} \approx \frac{3\pi}{\omega_e} \tag{2.2}$$

The proper choice of model class depends upon the problem to be solved. For example, MVM are well suited to relatively slow processes in the engine periphery, constant engine speed DEM are useful for air/fuel ratio feedforward

---

[2] In order to allow full control of the engine, usually these will be electric signals, e.g., the throttle plate will be assumed to be "drive-by-wire."

[3] In MVM, the finite swept volume of the engine can be viewed as being one that is distributed over an infinite number of infinitely small cylinders.

[4] The expression (2.1) is valid for four-stroke engines. Two-stroke engines have half of that IPS delay. As shown in Chapter 3, additional delays are introduced by the electronic control hardware.

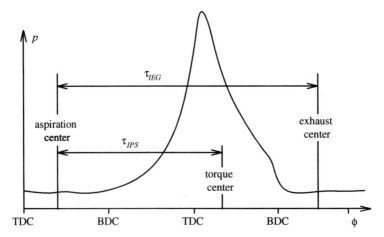

Fig. 2.2. Definition of IPS and IEG delays for MVM using a pressure/crank angle diagram.

control, and crank-angle DEM are needed for misfire detection algorithms based on measurement data of crankshaft speed.

Most MVM are *lumped parameter* models, i.e., system descriptions that have no spatially varying variables and that are represented by ordinary differential equations (ODE). If not only time but location also must be used as an independent variable, *distributed parameter models* result that are described by partial differential equations (PDE). Such models usually are computationally too demanding to be useful for real-time applications[5] such that a spacial discretization is necessary (see e.g., Sect. 2.8).

2.2 Cause and Effect Diagrams

In this section, the internal structure of MVM for SI and CI engines will be analyzed in detail. When modeling any physical system there are two main classes of objects that must be taken into account:

- *reservoirs*, e.g., of thermal or kinetic energy, of mass, or of information (there is an associated *level* variable to each reservoir that depends directly on the reservoir's content); and
- *flows*, e.g., energy, mass, etc. flowing between the reservoirs (typically driven by differences in reservoir levels).

A diagram containing all relevant reservoirs and flows between these reservoirs will be called a *cause and effect diagram* (see, for instance, Fig. 2.5).

---

[5] There are publications which propose PDE-based models for control applications, see for instance [43].

Since such a diagram shows the driving and the driven variables, the cause and effect relations become clearly visible.

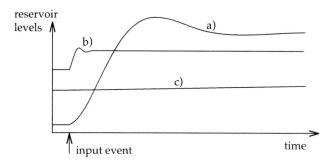

Fig. 2.3. Relevant reservoirs: a) variable of primary interest, b) very fast and c) very slow variables.

A good MVM contains only the *relevant* reservoirs (otherwise "stiff" systems will be obtained). To define more precisely what is relevant, the three signals shown in Fig. 2.3 can be useful. Signal a) is the variable of primary interest (say, the manifold pressure). Signal b) is very fast compared to a) (say, the throttle plate angle dynamics) and must be modeled as a purely static variable which can depend in an algebraic way on the main variable a) and the input signals. Signal c) is very slow compared to a) (say, the temperature of the manifold walls) and must be modeled as a constant (which may be adapted after a longer period). Only in this way a useful COM can be obtained.

Unfortunately, there are no simple and systematic rules of how to decide a priori which reservoirs can be modeled in what way. Here, experience and/or iteration will be necessary, making system modeling partially an "engineering art." Readers not familiar with the basic notions of systems modeling and controller design find some basic information in Appendix A.

2.2.1 Spark-Ignited Engines

Port-Injection SI Engines

A typical *port-injected* SI engine system has the structure shown in Fig. 2.4. In a mean value approach, the reciprocating behavior of the cylinders is replaced by a continuously working volumetric pump that produces exhaust gases and torque. The resulting main engine components are shown in Fig. 2.4. The different phenomena will be explained in detail in the following sections.

However, the main reservoir effects can be identified at the outset:

- gas mass in the intake and exhaust manifold;

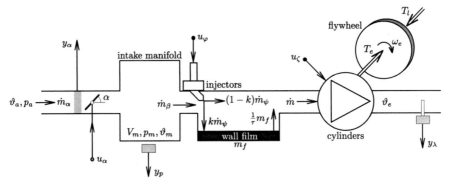

Fig. 2.4. Abstract mean-value SI-engine structure.

- internal energy in the intake and exhaust manifolds;
- fuel mass on the intake manifold walls (wall-wetting effect);
- kinetic energy in the engine's crankshaft and flywheel;
- induction-to-power-stroke delay in the combustion process (essentially an information delay); and
- various delays in the exhaust manifold (including transport phenomena).

Figure 2.5 shows the resulting simplified cause and effect diagram of an SI engine (assuming isothermal conditions in the intake manifold and modeling the exhaust manifold as a pure delay system). In the cause and effect diagram, the reservoirs mentioned appear as blocks with black shading. Between these reservoir blocks, flows are defined by static blocks (gray shading). The levels of the reservoirs define the size of these flows.

Each of these blocks is subdivided into several other parts which will be discussed in the sections indicated in the corresponding square brackets.[6] However, the most important connections are already visible in this representation. Both air and fuel paths affect the combustion through some delaying blocks while the ignition affects the combustion (almost) directly. The main output variables of the combustion process are the engine torque T_e, the exhaust gas temperature ϑ_e, and the air/fuel ratio λ_e.

The following signal definitions have been used in Figs. 2.4 and 2.5:

$\dot{m}_\alpha$ air mass flow entering the intake manifold through the throttle;
$\dot{m}_\beta$ air mass flow entering the cylinder;
p_m pressure in the intake manifold;
$\dot{m}_\psi$ fuel mass flow injected by the injectors;
$\dot{m}_\varphi$ fuel mass flow entering the cylinder;
$\dot{m}$ mixture mass flow entering the cylinder, with $\dot{m} = \dot{m}_\beta + \dot{m}_\varphi$;
T_e engine torque;
ω_e engine speed;

---

[6] The block [x] will be discussed in Sect. 2.x.

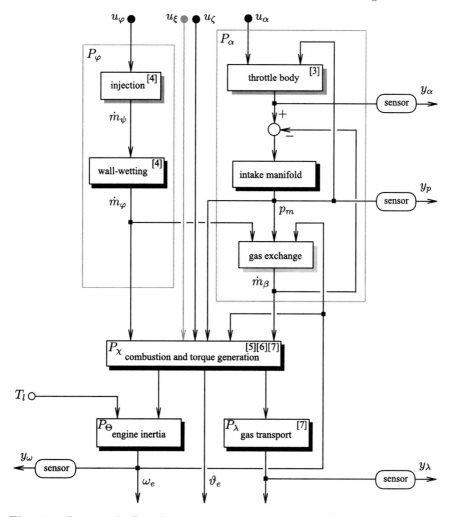

Fig. 2.5. Cause and effect diagram of an SI engine system (numbers in brackets indicate corresponding sections, gray input channel only for GDI engines, see text).

ϑ_e engine exhaust gas temperature; and
λ_e normalized air/fuel ratio.

Direct-Injection SI Engines

Direct-injection SI engines (often abbreviated as GDI engines — for gasoline direct injection) are very similar to port-injected SI engines. The distinctive feature of GDI engines is their ability to operate in two different modes:

- Homogeneous charge mode (typically at high loads or speeds), with injection starting during air intake, and with stoichiometric air/fuel mixtures being burnt.
- Stratified charge mode (at low to medium loads and low to medium speeds), with late injection and lean air/fuel mixtures.

The *static properties* of the GDI engine (gas exchange, torque generation, pollution formation, etc.) deviate substantially from those of a port injected engine as long as the GDI engine is in *stratified charge mode*. These aspects will be discussed in the corresponding sections below.

The main differences from a control engineering point of view are the additional control channel (input signal u_ξ in Fig. 2.5) and the missing wall-wetting block [4] (see [197]). The signal u_ξ controls the injection process in its timing and distribution (multiple pulses are often used in GDI engines) while the signal u_φ indicates the fuel quantity to be injected.

2.2.2 Diesel Engines

As with SI engines, in a mean value approach, CI engines are assumed to work continuously. The resulting schematic engine structure has a form similar to the one shown in Fig. 2.4. The cause and effect diagram of a supercharged direct-injection Diesel engine (no EGR) is shown in Fig. 2.6. Even without considering EGR and cooling of the compressed intake air, its cause and effect diagram is considerably more complex than that of an SI engine.

The main reason for this complexity is the turbocharger, which introduces a substantial coupling between the engine exhaust and the engine intake sides. Moreover, both in the compressor and in the turbine, thermal effects play an important role. However, there are also some parts that are simpler than in SI engines: fuel injection determines both the quantity of fuel injected and the ignition timing, and, since the fuel is injected directly into the cylinder, no additional dynamic effects are to be modeled in the fuel path.[7]

The following *new* signal definitions have been used in Fig. 2.6:

$\dot{m}_c$ air mass flow through the compressor;
$\dot{m}_t$ exhaust mass flow through the turbine;
p_c pressure immediately after the compressor;
p_2 pressure in the intake manifold;
p_3 pressure in the exhaust manifold;
ϑ_c air temperature after the compressor;
ϑ_2 air temperature in the intake manifold;
ϑ_3 exhaust gas temperature in front of the turbine;
ω_{tc} turbocharger rotational speed;
T_t torque produced by the turbine; and

---

[7] For fluid dynamic and aerodynamic simulations, usually a high-bandwidth model of the rail dynamics is necessary, see [127] or [143].

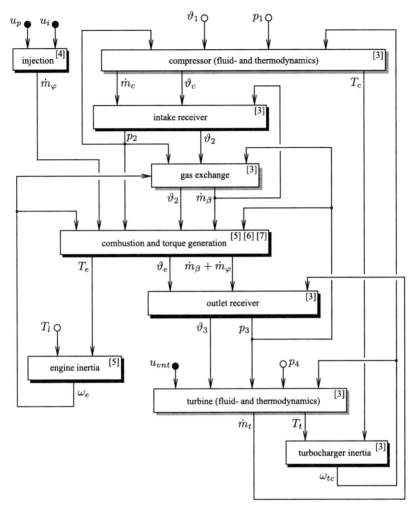

Fig. 2.6. Cause and effect diagram of a Diesel engine (EGR and intercooler not included).

T_c torque absorbed by the compressor.

Compared to an SI engine, there are several additional reservoirs to be modeled in a Diesel engine system. In the intake and exhaust manifolds, for instance, not only masses, but thermal (internal) energy is important. Accordingly, two level variables (pressure and temperature) form the output of these blocks. The turbocharger's rotor, which stores kinetic energy, is an additional reservoir.

If EGR and intercooling are modeled as well, the cause and effect diagram has a similar, but even more complex structure. The most important additional variable is the intake gas composition, i.e., the ratio between fresh air and burnt gases in the intake. If perfect mixing can be assumed, this leads to only one additional reservoir, see Sect. 2.3.4.

2.3 Air System

2.3.1 Receivers

The basic building block in the air intake system and also in the exhaust part is a *receiver*, i.e., a fixed volume for which the thermodynamic states (pressures, temperatures, composition, etc., as shown in Fig. 2.7) are assumed to be the same over the entire volume (lumped parameter system).

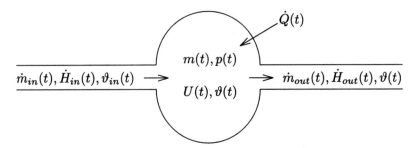

Fig. 2.7. Inputs, states, and outputs of a receiver.

The inputs and outputs are the mass and energy flows,[8] the reservoirs store mass and thermal energy,[9] and the level variables are the pressure and temperature. If one assumes that no heat or mass transfer through the walls and that no substantial changes in potential or kinetic energy in the flow occur, then the following two coupled differential equations describe such a receiver

$$\frac{d}{dt}m(t) = \dot{m}_{in}(t) - \dot{m}_{out}(t) \tag{2.3}$$

$$\frac{d}{dt}U(t) = \dot{H}_{in}(t) - \dot{H}_{out}(t) + \dot{Q}(t) \tag{2.4}$$

Assuming that the fluids can be modeled as ideal gases, the coupling between these two equations is given by the ideal gas law

$$p(t) \cdot V = m(t) \cdot R \cdot \vartheta(t) \tag{2.5}$$

and by the caloric relations

---

[8] Thermodynamically correct enthalpy flows, $\dot{H}(t)$.

[9] Thermodynamically correct internal energy, $U(t)$

$$U(t) = c_v \cdot \vartheta(t) \cdot m(t) = \frac{1}{\kappa-1} \cdot p(t) \cdot V$$

$$\dot{H}_{in}(t) = c_p \cdot \vartheta_{in}(t) \cdot \dot{m}_{in}(t) \tag{2.6}$$

$$\dot{H}_{out}(t) = c_p \cdot \vartheta(t) \cdot \dot{m}_{out}(t)$$

Note that the temperature $\vartheta_{out}(t)$ of the out-flowing gas is assumed to be the same as the temperature $\vartheta(t)$ of the gas in the receiver (lumped parameter approach). The following well-known definitions and connections among the thermodynamic parameters will be used below:

c_p specific heat at constant pressure, units $J/(kg\,K)$;
c_v specific heat at constant volume, units $J/(kg\,K)$;
κ ratio of specific heats, i.e., $\kappa = c_p/c_v$; and
R gas constant (in the formulation of (2.5) this quantity depends on the specific gas), units $J/(kg\,K)$, that satisfies the equation $R = c_p - c_v$.

Substituting (2.5) and (2.6) into (2.3) and (2.4), the following two differential equations for the level variables pressure and temperature (which are the only measurable quantities) are obtained after some simple algebraic manipulations

$$\frac{d}{dt}p(t) = \frac{\kappa \cdot R}{V} \cdot [\dot{m}_{in}(t) \cdot \vartheta_{in}(t) - \dot{m}_{out}(t) \cdot \vartheta(t)]$$

$$\frac{d}{dt}\vartheta = \frac{\vartheta \cdot R}{p \cdot V \cdot c_v} \cdot [c_p \cdot \dot{m}_{in} \cdot \vartheta_{in} - c_p \cdot \dot{m}_{out} \cdot \vartheta - c_v \cdot (\dot{m}_{in} - \dot{m}_{out}) \cdot \vartheta] \tag{2.7}$$

(for reasons of space here, the explicit time dependencies have been omitted in the second line of (2.7)).

The adiabatic formulation (2.7) is one extreme that well approximates the receiver's behavior when the dwell time of the gas in the receiver is small or when the surface-to-volume ratio of the receiver is small. If the inverse is true, then the isothermal assumption is a better approximation, i.e., such a large heat transfer is assumed to take place that the temperatures of the gas in the receiver ϑ and of the inlet gas ϑ_{in} are the same (e.g., equal to the engine compartment temperature). In this case, (2.7) can be simplified to

$$\frac{d}{dt}p(t) = \frac{R \cdot \vartheta(t)}{V} \cdot [\dot{m}_{in}(t) - \dot{m}_{out}(t)]$$

$$\vartheta(t) = \vartheta_{in}(t) \tag{2.8}$$

In Sect. 2.6, some remarks will be made for the intermediate case where a substantial, but incomplete, heat transfer through the walls takes place (e.g., by convection or by radiation). A detailed analysis of the effects of the chosen model on the estimated air mass in the cylinder can be found in [94].

2.3.2 Valve Mass Flows

The flow of fluids between two reservoirs is determined by valves or orifices whose inputs are the pressures upstream and downstream. The difference between these two level variables drives the fluid in a nonlinear way through such restrictions. Of course, since this problem is at the heart of fluid dynamics, a large amount of theoretical and practical knowledge exists on this topic. For the purposes pursued in this text, the two simplest formulations will suffice.

The following assumptions are made:

- no friction[10] in the flow;
- no inertial effects in the flow, i.e., the piping around the valves is small compared to the receivers to which they are attached;
- completely isolated conditions (no additional energy, mass, etc. enters the system); and
- all flow phenomena are zero dimensional, i.e., no spatial effects need be considered.

If, in addition, one can assume that the fluid is *incompressible*,[11] Bernoulli's law can be used to derive a valve equation

$$\dot{m}(t) = c_d \cdot A(t) \cdot \sqrt{2\rho} \cdot \sqrt{p_{in}(t) - p_{out}(t)} \tag{2.9}$$

where the following definitions have been used:

$\dot{m}$ mass flow through the valve;
A open area of the valve;
c_d discharge coefficient;
ρ density of the fluid, assumed to be constant;
p_{in} pressure upstream of the valve; and
p_{out} pressure downstream of the valve.

For *compressible* fluids, the most important and versatile flow control block is the *isothermal orifice*. When modeling this device, the key assumption is that the flow behavior may be separated as follows:

- No losses occur in the accelerating part (pressure decreases) up to the narrowest point. All the potential energy stored in the flow (with pressure as its level variable) is converted isentropically into kinetic energy.
- After the narrowest point, the flow is fully turbulent and all of the kinetic energy gained in the first part is dissipated into thermal energy. Moreover, no pressure recuperation takes place.

The consequences of these key assumptions are that the pressure in the narrowest point of the valve is (approximately) equal to the downstream pressure and that the temperature of the flow before and after the orifice is (approximately) the same (hence the name, see the schematic flow model shown in Fig. 2.8).

Using the thermodynamic relationships for isentropic expansion [147] the following equation for the flow can be obtained

---

[10] Friction can be partially accounted for by discharge coefficients that must be experimentally validated.

[11] For liquid fluids, this is often a good approximation.

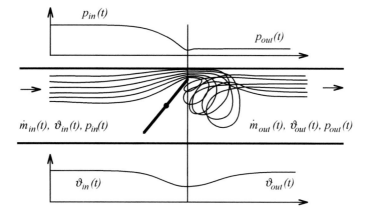

Fig. 2.8. Flow model in an isenthalpic throttle.

$$\dot{m}(t) = c_d \cdot A(t) \cdot \frac{p_{in}(t)}{\sqrt{R \cdot \vartheta_{in}(t)}} \cdot \Psi\left(\frac{p_{in}(t)}{p_{out}(t)}\right) \tag{2.10}$$

where the flow function $\Psi(.)$ is defined by

$$\Psi\left(\frac{p_{in}(t)}{p_{out}(t)}\right) = \begin{cases} \sqrt{\kappa \left[\frac{2}{\kappa+1}\right]^{\frac{\kappa+1}{\kappa-1}}} & \text{for} \quad p_{out} < p_{cr} \\ \left[\frac{p_{out}}{p_{in}}\right]^{\frac{1}{\kappa}} \cdot \sqrt{\frac{2\kappa}{\kappa-1} \cdot \left[1 - \left(\frac{p_{out}}{p_{in}}\right)^{\frac{\kappa-1}{\kappa}}\right]} & \text{for} \quad p_{out} \geq p_{cr} \end{cases} \tag{2.11}$$

and where

$$p_{cr} = \left[\frac{2}{\kappa+1}\right]^{\frac{\kappa}{\kappa-1}} \cdot p_{in} \tag{2.12}$$

is the critical pressure where the flow reaches sonic conditions in the narrowest part.

Equation (2.10) is sometimes rewritten in the following form

$$\frac{\sqrt{\vartheta_{in}}}{p_{in}} \cdot \dot{m} = \frac{c_d \cdot A}{\sqrt{R}} \cdot \Psi\left(\frac{p_{in}}{p_{out}}\right) = \widetilde{\Psi}\left(\frac{p_{in}}{p_{out}}\right) \tag{2.13}$$

For a fixed orifice area A, the relationship described by the function $\widetilde{\Psi}(p_{in}, p_{out})$ can be *measured* for some reference conditions $p_{in,0}$ and $\vartheta_{in,0}$. This approach permits the exact capture of the influence of the discharge coefficient c_d, which, in reality, is not constant. At a later stage, the actual mass flow $\dot{m}$ is computed using the corresponding conditions p_{in} and ϑ_{in} and the following relationship

$$\dot{m} = \frac{p_{in}}{\sqrt{\vartheta_{in}}} \cdot \widetilde{\Psi}\left(\frac{p_{in}}{p_{out}}\right) \tag{2.14}$$

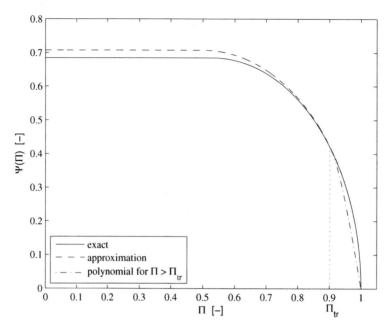

Fig. 2.9. Comparison of (2.11) (black) and (2.15) (dash) for the nonlinear function $\Psi(.)$. Also shown, approximation (2.17) (dash-dot) around $\Pi \approx 1$ for an unrealistic $\Pi_{tr} = 0.9$.

For many working fluids (e.g., intake air, exhaust gas at lower temperatures, etc.) with $\kappa \approx 1.4$ (2.11) can be approximated quite well by

$$
\Psi\left(\frac{p_{in}(t)}{p_{out}(t)}\right) \approx
\begin{cases}
1/\sqrt{2} & \text{for} \quad p_{out} < \frac{1}{2} p_{in} \\[2mm]
\sqrt{\frac{2\,p_{out}}{p_{in}}\left[1 - \frac{p_{out}}{p_{in}}\right]} & \text{for} \quad p_{out} \geq \frac{1}{2} p_{in}
\end{cases}
\tag{2.15}
$$

Figure 2.9 shows the exact curve (2.11) and the approximation (2.15).

As Fig. 2.9 also shows, the nonlinear function $\Psi(.)$ has an infinite gradient (in both formulations) at $p_{out} = p_{in}$. Inserting this relationship into the receiver equation, (2.7) or (2.8), yields a system that is not Lipschitz and which, therefore, will be difficult to integrate for values $p_{out} \approx p_{in}$ [209].

To overcome this problem, a laminar flow condition can be assumed for very small pressure ratios (see [56] for a formulation that is based on incompressible flows). This (physically quite reasonable) assumption yields a differentiable formulation at zero pressure difference and thus eliminates the problems mentioned.

An even more pragmatic approach to avoid that singularity is to assume that there exists a threshold

$$
\Pi_{tr} = \left.\frac{p_{out}}{p_{in}}\right|_{tr} < 1
\tag{2.16}
$$

below which (2.11) or (2.15) are valid. If larger pressure ratios occur, then a
smooth approximation of the form

$$\Psi(\Pi) = a \cdot (\Pi - 1)^3 + b \cdot (\Pi - 1) \tag{2.17}$$

with

$$a = \frac{\Psi'_{tr} \cdot (\Pi_{tr} - 1) - \Psi_{tr}}{2\,(\Pi_{tr} - 1)^3}, \qquad b = \Psi'_{tr} - 3\,a \cdot (\Pi_{tr} - 1)^2 \tag{2.18}$$

is used (Ψ_{tr} is the value of $\Psi(.)$ and Ψ'_{tr} the value of the derivative of $\Psi(.)$
at the threshold Π_{tr}). The special form of (2.17) and (2.18) guarantees a
symmetric behavior around the critical pressure ratio $\Pi = 1$ and a smooth
transition at the departure point Π_{tr}. Figure 2.9 shows an example of this
curve (for reasons of clarity, an unrealistically large threshold $\Pi_{tr} = 0.9$ has
been chosen).

2.3.3 Engine Mass Flows

Regarding the air system, the engine itself can be approximated as a volumetric pump, i.e., a device that enforces a volume flow approximately proportional
to its speed. A typical formulation for such a model is

$$\dot{m}(t) = \rho_{in}(t) \cdot \dot{V}(t) = \rho_{in}(t) \cdot \lambda_l(p_m, \omega_e) \cdot \frac{V_d}{N} \cdot \frac{\omega_e(t)}{2\pi} \tag{2.19}$$

where the new variables are defined as follows:

ρ_{in} density of the gas at the engine's intake (related to the intake pressure and
temperature by the ideal gas law, (2.5));
λ_l *volumetric efficiency*, which describes how far the engine differs from a
perfect volumetric device (see below);
V_d displaced volume; and
N number of revolutions per cycle ($N = 2$ for four-stroke and $N = 1$ for
two-stroke engines, where the formulations presented below will be for
four-stroke engines).

An alternative definition of the volumetric efficiency is obtained by considering the *air* mass flow only

$$\dot{m}_\beta(t) = \rho_m(t) \cdot \dot{V}(t) = \rho_m(t) \cdot \tilde{\lambda}_l(p_m, \omega_e) \cdot \frac{V_d}{N} \cdot \frac{\omega_e(t)}{2\pi} \tag{2.20}$$

where $\dot{m}_\beta$ is the air mass flow aspirated by the engine and ρ_m the intake
manifold density. The experimental identification of (2.20) is simpler than
those of (2.19). The drawback of (2.20) is that it cannot account for varying
fuel mass flows.

The volumetric efficiency determines the engine's ability to aspire a mixture or air and is of special importance at full-load conditions. It is rather

difficult to predict reliably since many effects influence it (internal exhaust gas recirculation, ram effects in the intake runner, resonance in the manifold, cross-coupling between the different cylinders, etc.). Precise models will therefore have to rely on measurements of this quantity (usually as a two-dimensional map over speed and load obtained in steady-state conditions) or on detailed simulations of the fluid dynamic processes in the intake system (see e.g., [43]). Note also that, for gasoline engines, the evaporation of the fuel in the intake port causes a substantial reduction of the temperature of the inflowing mixture. This effect can increase the volumetric efficiency [6].

As a first approximation, a multilinear formulation is sometimes useful, i.e.,

$$\lambda_l(p_m, \omega_e) = \lambda_{lp}(p_m) \cdot \lambda_{l\omega}(\omega_e) \tag{2.21}$$

Assuming perfect gases with constant κ and isentropic processes, the pressure dependent part (which essentially describes the effects caused by the trapped exhaust gas at TDC) for an ideal Otto or Diesel cycle can be approximated as

$$\lambda_{lp}(p_m) = \frac{V_c + V_d}{V_d} - \left(\frac{p_{out}}{p_m}\right)^{1/\kappa} \cdot \frac{V_c}{V_d} \tag{2.22}$$

where p_{out} is the pressure at the engine's exhaust side (assumed to be constant), and V_c is the compression volume, i.e., the volume in the cylinder at TDC.

The advantage of using (2.21) and (2.22) is that, in this case, the experimental validation has to consider changing engine speeds only, such that a reduced number of experiments are sufficient. Other simplified formulations of the volumetric efficiency are in use (see e.g., [206] and [167] for an affine form).

The volumetric efficiency of port and direct injected SI engines is very similar (see [197]). The only obvious difference is the cooling effect of the gasoline evaporation taking place inside the cylinder in the GDI case, which causes a small increase of the volumetric efficiency. The overall behavior, however, is not affected thereby.

Equation (2.19) represents the total mass flow $\dot{m}$ that the engine aspirates into its cylinders. For port-injected engines (e.g., gasoline or especially natural gas fueled SI engines) a certain part of $\dot{m}$ is fuel[12] such that the air mass flow $\dot{m}_\beta$ leaving the intake manifold is smaller than $\dot{m}$. Assuming the fuel to be fully evaporated, the density in (2.19) must be changed in this case, as follows

$$\rho_{in}(t) = \frac{p_m}{R_m \cdot \vartheta_m} \tag{2.23}$$

where the temperature ϑ_m and the gas constant R_m of the mixture are found using the corresponding variables of the fuel (subscript φ) and of the fresh air at the engine inlet valve (subscript β), i.e.,

---

[12] Additional exhaust gas recirculation is discussed in the next section.

$$R_m = \frac{\dot{m}_\beta \cdot R_\beta + \dot{m}_\varphi \cdot R_\varphi}{\dot{m}_\beta + \dot{m}_\varphi} \qquad (2.24)$$

and

$$\vartheta_m = \frac{\dot{m}_\beta \cdot c_{p\alpha} \cdot \vartheta_\beta + \dot{m}_\varphi \cdot c_{p\varphi} \cdot \vartheta_\varphi}{\dot{m}_\beta \cdot c_{p\alpha} + \dot{m}_\varphi \cdot c_{p\varphi}} \qquad (2.25)$$

Of course, these last two equations are based on the implicit assumption that perfect and adiabatic mixing takes place at the intake valve.

Using the mass balance

$$\dot{m} = \dot{m}_\beta + \dot{m}_\varphi \qquad (2.26)$$

and assuming that all of the evaporated fuel is indeed aspirated into the cylinder, the air flowing out of the intake manifold can be computed by solving a quadratic equation which is derived by combining (2.19) with (2.24) through (2.26)

$$\dot{m}_\beta = \frac{-b + \sqrt{b^2 - 4\,a \cdot c}}{2\,a} \qquad (2.27)$$

where the following definitions have been used in (2.27)

$$a = R_\beta \vartheta_\beta c_{p,\alpha}$$
$$b = \left(R_\beta \vartheta_\varphi c_{p,\varphi} + R_\varphi \vartheta_\beta c_{p,\alpha}\right) \dot{m}_\varphi - p_m V_d \lambda_l \tfrac{\omega_e}{2N\pi} c_{p,\alpha} \qquad (2.28)$$
$$c = \left(\dot{m}_\varphi R_\varphi \vartheta_\varphi - p_m V_d \lambda_l \tfrac{\omega_e}{4\pi}\right) c_{p,\varphi} \dot{m}_\varphi$$

Note that for physically meaningful parameters, the expression $b^2 - 4\,a \cdot c$ is positive and that the ambiguity in the solution of the quadratic equation is easily solved.

If the fuel's temperature, specific heat, and gas constant are not known, or if the computations must be limited to a bare minimum, the following approximation of the air mass flow $\dot{m}_\beta$ can be used

$$\dot{m}_\beta = \frac{p_m}{\vartheta_\beta \cdot R_\beta} \cdot \lambda_l(p_m, \omega_e) \cdot \frac{V_d \cdot \omega_e}{N \cdot 2\pi} - \dot{m}_\varphi \qquad (2.29)$$

where the properties of the mixture are approximated by the properties of the air. For gasoline, the error made with such an approximation is only $\approx 5\%$; for other fuels, however, it can be larger.

2.3.4 Exhaust Gas Recirculation

Exhaust gas recirculation (EGR) is either direct or cooled recirculation of the cylinder exhaust gases into the intake manifold. In the following, it is assumed that:

- the exhaust gas is collected immediately beyond the exhaust valve (no dynamic gas mixing or delays in the exhaust part);

- cooled EGR is realized (such that the inlet manifold can be modeled isothermically by (2.8));
- the fresh air and the exhaust gas are ideal gases with identical gas constants R;[13] and
- the mixture is stoichiometric or lean (but not rich).

Since EGR is usually not applied in SI engines at full load (otherwise power would be lost), the last assumption is not restricting. The first and third assumptions are usually satisfied. The second assumption is not always correct, but it is possible to extend the ideas presented below to the adiabatic case (using (2.7) instead of (2.8)).

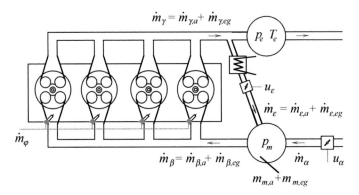

Fig. 2.10. Definitions of variables for the EGR model.

Figure 2.10 illustrates the definitions of the main variables of the EGR model. The index a represents air and the index eg exhaust gas. As usual, mass flows are labeled by $\dot{m}_x$ and masses by m_x where $x = \{a, eg\}$. The index m indicates a variable pertaining to the intake manifold, β to the engine intake, γ to the engine outlet, and ε to the exhaust gas recirculation duct. The three inputs are the fuel mass flow entering the cylinders $\dot{m}_\varphi$, the command signal to the fresh air throttle u_α, and the command signal to the exhaust gas recirculation valve u_ε.

In the following, the air/fuel ratio λ will be used which describes the ratio of air and fuel flow (or mass) into the cylinder in relation to the stoichiometric constant σ_0,[14] i.e.,

---

[13] At ambient temperature, the actual values are $R_{air} = 287.1\ J/kg\,K$ and $R_{eg} = 286.6\ J/kg\,K$, and these values do not change substantially in the relevant temperature range.

[14] For typical RON 95 gasoline, σ_0 is approximately 14.67. This quantity can be calculated for any hydrocarbon fuel having b carbon and a hydrogen atoms (no oxygen) by $\sigma_0 \approx 137 \cdot (b + a/4)/(12 \cdot b + a)$. More details are given in Sect. 2.7.

$$\lambda(t) = \frac{1}{\sigma_0} \cdot \dot{m}_{\beta,a}(t)/\dot{m}_\varphi(t) \tag{2.30}$$

According to the assumptions stated above $\lambda(t) \geq 1$ for all times t. A mass flow balance over the cylinders yields the mass flows

$$\dot{m}_{\gamma,eg}(t) = \dot{m}_{\beta,eg}(t - \tau_{IEG}) + \dot{m}_{\beta,a}(t - \tau_{IEG}) \cdot \frac{1 + 1/\sigma_0}{\lambda(t - \tau_{IEG})} \tag{2.31}$$

and

$$\dot{m}_{\gamma,a}(t) = \dot{m}_{\beta,a}(t - \tau_{IEG}) \cdot \left[1 - \frac{1}{\lambda(t - \tau_{IEG})}\right] \tag{2.32}$$

In these equations the (relatively small) influence of incomplete combustion and pollutant formation has been neglected. Note also the influence of the time delays as discussed in Sect. 2.1 above.

The air/fuel ratio is defined by the control variable u_φ and the fuel path dynamics (see Sect. 2.4.1 below). The total engine-in mass flow $\dot{m}$ is given by (2.19) such that $\dot{m}_{\beta,eg}$ and $\dot{m}_{\beta,a}$ are the only engine-in quantities not yet known. These variables are derived using the assumption of a perfect gas mixing in the inlet manifold, i.e., it is assumed that the EGR ratio of the out-flowing gas corresponds to the EGR ratio inside the intake manifold

$$x_{egr}(t) = \frac{m_{m,eg}(t)}{m_m(t)} = \frac{m_{m,eg}(t)}{m_{m,a}(t) + m_{m,eg}(t)} \tag{2.33}$$

Therefore,

$$\dot{m}_{\beta,eg}(t) = \dot{m}_\beta(t) \cdot x_{egr}(t), \quad \dot{m}_{\beta,a}(t) = \dot{m}_\beta(t) \cdot (1 - x_{egr}(t)) \tag{2.34}$$

The next step is to formulate a first-order mass balance for the intake manifold which comprises two species

$$\frac{d}{dt} m_{m,a}(t) = \dot{m}_\alpha(t) - \dot{m}_{\beta,a}(t) + \dot{m}_{\varepsilon,a}(t)$$
$$\frac{d}{dt} m_{m,eg}(t) = \dot{m}_{\varepsilon,eg}(t) - \dot{m}_{\beta,eg}(t) \tag{2.35}$$

The manifold pressure is found using the ideal gas law (recall that both gas constants are assumed to be equal)

$$p_m(t) \cdot V_m = [m_{m,a}(t) + m_{m,eg}(t)] \cdot R \cdot \vartheta_m(t) \tag{2.36}$$

The total recirculated gas mass flow $\dot{m}_\varepsilon(t) = \dot{m}_{\varepsilon,a} + \dot{m}_{\varepsilon,eg}$ can be found using the exhaust and intake manifold pressures and temperatures (assumed to be known) and the orifice equation (2.10). Assuming again a perfect mixing behavior at the engine outlet, the exhaust gas and air fraction can be found by

$$\dot{m}_{\varepsilon,eg}(t) = \dot{m}_{\gamma,eg}(t) \cdot y_{egr}(t), \quad \dot{m}_{\varepsilon,a}(t) = \dot{m}_{\gamma,a}(t) \cdot y_{egr}(t) \tag{2.37}$$

where y_{egr} is the ratio of the recirculated mass flow to the total exhaust gas mass flow, i.e.,

$$y_{egr}(t) = \frac{\dot{m}_\varepsilon(t)}{\dot{m}_\gamma(t)} \tag{2.38}$$

2.3.5 Supercharger

Thus far, the amount of air in the cylinder has been determined by the "pumping" properties of the engine itself. These *naturally aspirated* engines (both Diesel and Otto) are limited in their power by their volumetric efficiency. Consequently, in most modern engines the maximum torque is closely proportional to the swept volume.

Supercharged[15] engines can produce much more power for a given swept volume by forced air transport into the intake part. Three basically different devices are in use:

- turbochargers, i.e., *fluid-dynamic* devices consisting of a compressor and a turbine on the same shaft (with rotational speeds of more than 200,000 rpm), utilizing some of the exhaust gas exergy to compress the intake air [5];
- pressure wave superchargers, i.e., *gas-dynamic devices* consisting of a barrel whose narrow channels are in direct contact with both engine in and exhaust sides, utilizing the exhaust gas exergy for the boosting process ([6]); and
- mechanical superchargers (preferably with internal compression), which receive the power needed for boosting directly from the engine's crankshaft.

From a control engineering point of view these devices substantially differ only in their static functions (blocks "compressor" and/or "turbine" in Figure 2.6). The interconnections with the receivers and the other storage blocks remain the same. The key assumption is that the dynamic phenomena in the fluid- or gas-dynamic processes are much faster than the rate of change in the thermodynamic boundary conditions (taking place in the receivers to which the superchargers are connected). In this case the fluid- or gas-dynamic processes can be modeled as quasi stationary phenomena ("type b" events in Figure 2.3), and "static maps" can be used to describe the (usually highly nonlinear) behavior of these devices.

Fluid Dynamic Compressors

A fluid dynamic compressor is discussed first (other compressor types are briefly mentioned below). Typically a "compressor map" is used to describe the device, the inputs ("causes") being:

- the pressures at the compressor inlet and outlet ports (starting point 1 in Figure 2.11);
- the compressor rotational speed (curve 2 in Figure 2.11);
- the temperature at the compressor inlet port (needed to compute the corrected mass flow and speed, see below for details).

---

[15] The term supercharging will be used here to describe the forced induction of air or a mixture into an engine by any device.

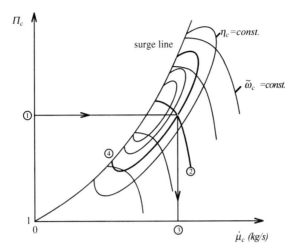

Fig. 2.11. Fluid-dynamic (radial) compressor "map" (normalized mass flows and efficiencies).

Notice that all these variables are "level variables" and are associated with storages (two are mass, one is internal energy, and one is kinetic energy). With this information two outputs ("effects") are calculated, namely the mass flow through the compressor (point 3 in Figure 2.11) and the torque needed to drive the compressor at the actual state.

As an intermediate variable the compressor's isentropic efficiency is also derived from the map (point 4 in Figure 2.11, see below). The following new definitions are thus used in 2.11 (see also Figure 2.6):

- $\Pi_c = p_2/p_1$, the pressure ratio over the compressor;
- $\tilde{\omega}_c = \sqrt{\vartheta_{1,0}/\vartheta_1} \cdot \omega_c$, the corrected compressor speed ($\vartheta_{1,0}$ is the inlet temperature at which the map was originally measured); and
- $\dot{\mu}_c = (p_1/p_{1,0}) \cdot \sqrt{\vartheta_{1,0}/\vartheta_1} \cdot \dot{m}_c$, the corrected compressor mass flow ($p_{1,0}$ is the ambient pressure at which the map was originally measured).

From these relations the compressor mass flow can be computed once the reference and the actual pressures and the temperature are known. To compute the power (or torque) absorbed by the compressor for a given operating point, its (isentropic) efficiency has to be known. Typically this quantity is also plotted in the "maps" as iso-level contours (see point 4 in Figure 2.11). The power absorbed by the compressor is the isentropic power $P_{c,s}$ (i.e., the power that a "perfect" compressor would need to compress the mass flow of gas from its input to its output pressure level), divided by the isentropic efficiency η_c which takes into account all the unavoidable deterioration in the actual technical compressor realization. To be more precise

$$P_c = \frac{P_{c,s}}{\eta_c} = \dot{m}_c \cdot c_p \cdot \vartheta_1 \cdot \left[\Pi_c^{(\kappa-1)/\kappa} - 1\right] \cdot \frac{1}{\eta_c} \qquad (2.39)$$

Since the compressor speed is a *level variable* (i.e., an available output of a storage) the torque absorbed by the compressor can be found easily from equation (2.39) to be

$$T_c = \frac{P_c}{\omega_c} \qquad (2.40)$$

Similarly, the temperature increase due to the compression is formulated as an isentropic (unavoidable) part and an additional increase due to technical imperfections in the compressor realization

$$\vartheta_c = \vartheta_1 + \left[\Pi_c^{(\kappa-1)/\kappa} - 1\right] \cdot \frac{\vartheta_1}{\eta_c} \qquad (2.41)$$

Notice that in equation (2.41) adiabatic conditions have been assumed , i.e., heat losses through the compressor walls and in the compressor outlet will have reduced the gas temperature before entering the intercooler or the intake manifold.

Sometimes it is useful to use "analytic" (closed form) descriptions of the compressor characteristics (e.g., for numerical optimization routines where the gradients have to be supplied to the optimization routines). The following parametrization is a possible choice for the efficiency

$$\eta_c(\Pi_c, \dot{\mu}_c) = \eta_{c,opt} - \chi^T(\Pi_c, \dot{\mu}_c) \cdot Q \cdot \chi(\Pi_c, \dot{\mu}_c) \qquad (2.42)$$

where

$$\chi^T(\Pi_c, \dot{\mu}_c) = [\Pi_c - \Pi_{c,opt}, \ \dot{\mu}_c - \dot{\mu}_{c,opt}] \qquad (2.43)$$

These formulations require six parameters to be fitted to the compressor map at hand.

The mass flow to pressure ratio curves depend on the compressor speed. An implicit formulation (which has the advantage of being easily identifiable with the measured curves in the compressor map) and which is valid for $\dot{\mu}_c > \dot{\mu}_{c,min}$ (see Figure 2.12 below) is

$$\Pi_c(\dot{\mu}_c, \omega_c) = \Pi_{c,max}(\omega_c) - c_{c,c}(\omega_c) \cdot [\dot{\mu}_c - \dot{\mu}_{c,min}(\omega_c)]^2 \qquad (2.44)$$

The "parameters" $\Pi_{c,max}(\omega_c)$, $c_{c,c}(\omega_c)$ and $\dot{\mu}_{c,min}(\omega_c)$ depend only on the compressor speed. Typically, they can be parametrized as low-order polynomials in ω_c. Notice that equation (2.44) has to be resolved for the mass flow $\dot{\mu}_c$ to fit into the cause and effect structure of the compressor model. This is the reason why only second-order polynomials are used in (2.44). Other approaches are possible, see e.g. [20] and below in the Moore-Greitzer model.

Every compressor of course has a limited operating range. As Figure 2.12 shows, there are at least four limiting conditions:

- the mechanical limit, i.e., the maximum rotational speed that is allowed to keep mechanical damage from occurring due to the large centrifugal forces (this is especially critical when vehicles are operated at high altitudes and therefore low air density);
- the choke limit, where the flow in the compressor reaches sonic conditions in its narrowest part;
- the blocking limit, i.e., the behavior of the compressor at zero (or very low) speed where it represents a blocking orifice;
- the surge limit, where fluid-dynamic instabilities destroy the regular flow patterns inside the compressor (see below for more details).

These limitations are very complex to model accurately and are the topic of ongoing research efforts (see e.g. [15] for an overview). For control purposes, however, simple models have to be found which describe the system's behavior even outside the regular operating range[16].

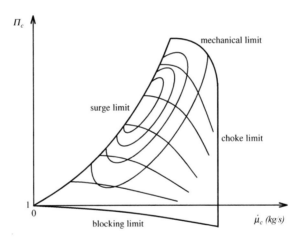

Fig. 2.12. Boundaries of operation of a compressor.

Maximum speed and choked flow usually do not pose any special problems. A simple model for the low-speed regime is to assume that the compressor behaves like a fixed orifice for speeds $\omega_c < \omega_{c,min}$ (see equation (2.10)). In this regime the compressor does not produce any mass flow unless the pressure at its outlet is smaller than the inlet pressure. No changes in the cause and effect structure are necessary, however. Notice also that the minimum speed $\omega_{c,min}$ at which the compressor changes its behavior generally is larger

---

[16] These conditions occur for instance when simulating driving cycles (e.g., engine start-up implies zero compressor speed, and a quick throttle closing often induces a compressor surge).

than zero. A smooth transition between the two regimes (with a hysteresis) is recommended.

Compressor Surge

Surge imposes serious limitations on the operating range of automotive compressors[17]. The behavior of the compressor in the "surge" (or "pumping") mode[18] is very complex. In the following, a simple model will be presented that permits qualitatively realistic simulations with few excursions into the surge regime. However, the model is not able to accurately describe the onset or the effects of compressor surge.

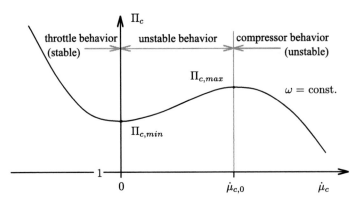

Fig. 2.13. Quasi-static representation of the compressor behavior according to (2.45).

As Figure 2.13 shows for one fixed rotor speed, the relationship between pressure ratio and mass flow is extended from the stable compressor regime into the unstable compressor zone and into the stable throttle regime. The inner unstable zone is never sustained for any longer time intervals[19]. The simplest possible extension of the pressure-ratio to mass-flow characteristic beyond the surge limit is given by the following third-order polynomial:

$$\tilde{\Pi}_c(\dot{\mu}_c) = \Pi_{c,min} + \Pi_{c,1} \cdot \left[\frac{1}{2} + \frac{3}{2} \left(\frac{\dot{\mu}_c}{\dot{\mu}_{c,0}} - \frac{1}{2} \right) - 2 \left(\frac{\dot{\mu}_c}{\dot{\mu}_{c,0}} - \frac{1}{2} \right)^3 \right] \quad (2.45)$$

---

[17] In practice, ancillary devices (e.g. bleed valves) are used to soften these restrictions.

[18] Compressor pumping occurs when the flow in the compressor is stalled, i.e., when the pressure gradient becomes too large to be sustained. Several forms of stalling (starting from local rotating stall to full rotor stall) are known.

[19] In this regime a small negative pressure disturbance leads to smaller mass flow and hence to even smaller pressure after the compressor. This represents a dynamic system with positive feedback, hence this regime is very rapidly crossed once the compressor reaches its surge limit.

(see Figure 2.15 for the definition of the parameters $\Pi_{c,min}$, $\Pi_{c,1}$ and $\dot{\mu}_{c,0}$).

A simplified model of a compressor/manifold/throttle system that is able to reproduce (deep) surge phenomena was proposed in [7] and has been extended in several subsequent publications (system structure shown in Figure 2.14, see [17] for a recent survey). In the simplest case, the model neglects all thermal phenomena and rotating stall and assumes constant compressor speed and incompressible flows.

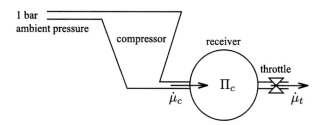

Fig. 2.14. Simplified Moore-Greitzer system compressor/throttle system.

The Moore-Greitzer model assumes that both the pressure in the manifold and the mass flow through the compressor are dynamic variables, i.e., that the system shown in Figure 2.14 is described by a coupled system of differential equations

$$\frac{d}{dt}\dot{\mu}_c(t) = \frac{1}{\tau_c} \cdot \left[\tilde{\Pi}_c(\dot{\mu}_c(t)) - \Pi_c(t) \right]$$

$$\frac{d}{dt}\Pi_c(t) = \frac{1}{\tau_m} \cdot [\dot{\mu}_c(t) - \dot{\mu}_t(\Pi_c(t), A_\alpha(t))]$$

(2.46)

The mass flow $\dot{\mu}_t$, which under the given assumptions represents the turbine mass flow, is approximated using (2.9).

The physical interpretation of these equations is that the compressor has some delays and time lags (inherent dynamics) which limit the rate with which the mass flow $\dot{\mu}_c(t)$ can be changed. These delays are modeled as a first-order lag behavior with a time constant τ_c. A second time constant τ_m characterizes the usual dynamics of the manifold (for a more detailed discussion of these parameters see [17]).

Notice that if the system of (2.46) reaches a stable equilibrium the pressure ratio satisfies the condition $\Pi_c = \tilde{\Pi}_c(\dot{\mu}_c)$, which corresponds to the behavior of the quasi static compressor model introduced above.

In Figure 2.15 three simulations of the Moore-Greitzer system are shown. In one case (black curves) the throttle area A_α has been chosen sufficiently large such that the system is stable. The second case (dashed grey curves), characterized by a smaller throttle area, never attains a stable equilibrium but rapidly reaches a limit-cycle orbit ("mild surge"). By further reducing A_α, the system enters a cycle where even backflow ($\dot{\mu}_c < 0$) occurs (grey curves, "deep surge").

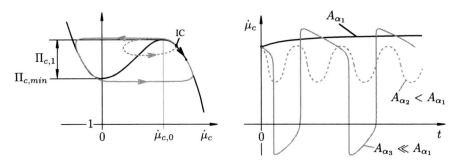

Fig. 2.15. Limit cycle behavior of the Moore-Greitzer system. Left as trajectory in the compressor map, right as function of time. All three cases share the same initial conditions (IC) but have different throttle areas A_α.

The dynamic properties of this orbit are defined by the two time constants τ_m and τ_c, respectively, and the function (2.45). The boundary between stable equilibrium operation and limit cycle behavior is *not* simply found by requiring the throttle to satisfy the condition

$$\dot{\mu}(\Pi_{c,min} + \Pi_{c,1}) = \dot{\mu}_{c,0} \tag{2.47}$$

Mild surge and even "chaotic" behavior can be found for turbine throttle characteristics close to that critical value.

Other Compressor Systems

Other compressors (pressure wave or mechanical) have similar "maps". Especially the input/output relations are identical, such that the formulations presented above can be applied in that case as well.

For example, a mechanical compressor (connected to the crankshaft with a clutch and a 1:2 gear box) will have a map similar to the one shown in Figure 2.16. Again, knowing the pressure ratio (point 1) and the compressor speed (point 2), the volume flow can be found in this map. The efficiency (in the case of Figure 2.16 the total efficiency, i.e., including the mechanical losses) can then be found directly.

Pressure wave superchargers (PWS) represent another class of devices that can be used to boost the in-cylinder pressure. Similar to turbochargers, they utilize the enthalpy of the exhaust gas to drive that process. The main advantage of PWS is that they transform the exhaust gas enthalpy directly to boost pressure by a tuned pressure-wave mechanism. This approach avoids additional time delays and yields high boost pressures even at low engine speeds. For more information on the basic principles of PWS, see [83]; for a modern application of PWS to realize fuel efficient engines, see [82]; and for the modeling and control of PWS, see [214] and [191].

Mechanical and pressure-wave superchargers do not have the "pumping problems" of fluid-dynamic compressors and they can be designed to provide

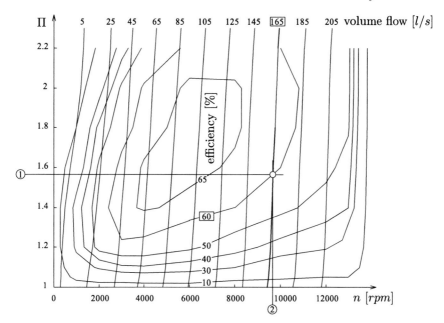

Fig. 2.16. Mechanical supercharger "map" (input volume flow and mechanical efficiencies).

excellent dynamic response times. Problems arise if the compressor is used as a fuel economy device in a "downsizing supercharging" approach (see [25]). In this case the compressor is linked to the engine crankshaft only in high-power phases and the control of the compressor clutch requires a careful design in order to avoid large torque disturbances.

Turbines

Since mechanical superchargers draw their power directly from the engine's crankshaft, the engine exhaust part does not differ from that of a natural aspirated engine. Pressure wave superchargers have no explicit separation between "compressor" and "turbine" function (see [24]). However, in fluid-dynamic turbochargers "turbines" play an important role.

For control purposes, the mass flow behavior of fluid-dynamic turbines can be modeled quite well as orifices (with a variable area in the case of variable nozzle turbines and a blade speed dependent discharge coefficient c_d, see Figure 2.18). Analytic descriptions (similar to equation(2.10)) or "maps" (as shown in Figure 2.17) are commonly used.

In both cases similar considerations hold as in the compressor case, i.e., turbine speed and pressure ratio are the input variables (points 1 and 2 in Figure 2.18) and the (corrected) mass flow is the system's output:

- $\Pi_t = p_3/p_4$, the pressure ratio over the turbine ($p_4 \approx p_1$);
- $\tilde{\omega}_t = \sqrt{\vartheta_{3,0}/\vartheta_3} \cdot \omega_t$, the corrected turbine speed ($\vartheta_{3,0}$ is the inlet temperature at which the map was originally measured);
- $\dot{\mu}_t = p_3/p_{3,0} \cdot \sqrt{\vartheta_{3,0}/\vartheta_3} \cdot \dot{m}_t$, the corrected turbine mass flow ($p_{3,0}$ is the turbine inlet pressure at which the map was originally measured). Notice that for turbines the correction terms play a more important role since both inlet temperature and pressure do vary substantially.

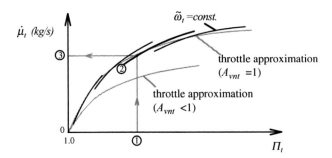

Fig. 2.17. Simplified turbine map — normalized mass flows.

Again, the torque and the turbine-out temperature can only be determined when the turbine's isentropic efficiency is known. Representations for that variable, similar to the compressor case, maps or diagrams as shown in Figure 2.17 are in use. Since turbine efficiency mainly depends on the angle of incidence of the inflowing gas, the "turbine blade speed" ratio is used as the main independent variable

$$\tilde{c}_{us} = \frac{r_t \cdot \omega_t}{c_{us}} \tag{2.48}$$

where

$$c_{us} = \sqrt{2\,c_p \cdot \vartheta_3 \cdot \left[1 - \Pi_t^{(1-\kappa)/\kappa}\right]} \tag{2.49}$$

Once the turbine efficiency is known the power produced by the turbine can be found by

$$P_t = P_{t,s} \cdot \eta_t = \dot{m}_t \cdot c_p \cdot \vartheta_3 \cdot \left[1 - \Pi_t^{(1-\kappa)/\kappa} - 1\right] \cdot \eta_t \tag{2.50}$$

where $P_{t,s}$ is the (maximum) power which would be produced by a perfect (isentropic) turbine. Again, since the turbine speed is a level variable, the torque produced by the turbine can be derived directly from (2.50) to be

$$T_t = P_t/\omega_t \tag{2.51}$$

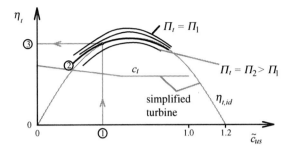

Fig. 2.18. Simplified turbine map — efficiencies and discharge coefficient.

As indicated in Figure 2.18, simplified closed-form descriptions of the turbine's efficiency are quite accurate, i.e., it can be approximated by

$$\eta_t(\tilde{c}_u) = \eta_{t,max} \cdot \left[\frac{2\,\tilde{c}_u}{\tilde{c}_{u,opt}} - \left(\frac{\tilde{c}_u}{\tilde{c}_{u,opt}}\right)^2 \right] \qquad (2.52)$$

In automotive applications typical values are $\eta_{t,max} \approx 0.65\ldots0.75$ and $c_{u,opt} \approx 0.55\ldots0.65$.

For turbines with a variable nozzle area $c_{u,opt}$ depends on the nozzle area, with larger areas yielding larger values for $c_{u,opt}$ (the efficiency η_t is affected only little by nozzle area changes, see for example [2]).

The mechanical behavior of a combined turbocharger group (compressor and turbine on the same shaft) is described by

$$\frac{d}{dt}\omega_{tc}(t) = \frac{1}{\Theta_{tc}} \cdot \left[T_t(t) - T_c(t) - T_f(t) + T_a(t) \right] \qquad (2.53)$$

where Θ_{tc} is the turbocharger's rotational inertia and T_i represents the torques acting on the shaft. The first two torques have been introduced above (equation (2.40) and equation (2.51)). The remaining two torques model additional friction losses and possible external auxiliary torques (e.g. as produced by an electric motor to improve the acceleration behavior, as described in [28]).

Map Fitting

Compressor Mass Flow

Turbocharger data are measured on dedicated test benches, where all the necessary boundary conditions can be set. The results from these measurements is usually recorded in a compressor and a turbine map.

The dashed lines in Fig. 2.19 represent measurement results of a compressor. The surge line is not measured separately. Rather, it is constructed from the very left point of every speed line. Determining surge on a test bench is a difficult task. Some ad-hoc measurements give indications to the onset of surge, i.e., strong fluctuations in pressure or noise produced by the pulsating fluid. More precise results require rather complex approaches [72], [83].

Usually rotational speeds below 10,000 rpm are not measured, because they represent only a small region in the compressor map, and measurement is difficult. The mass flows and the pressure ratios are small, which leads to large relative errors. Furthermore, heat flow from the turbine to the compressor becomes substantial in comparison to the enthalpy flow of the fluids, resulting in substantial deviations of the efficiency calculated. What looks like a small region in the compressor map is very relevant when SI engines are simulated. While idling or under low load conditions the turbocharger speed can fall below 10,000 rpm.

The sum of these statements makes an extrapolation to lower speed regions inevitable. The extrapolation can either be done with a good physical model or with map fitting. Unfortunately, the former does not result in sufficiently accurate models. Several methods for map fitting are presented by [182] and [146]. Good results can be achieved by the inverted approach described in [105]. In that approach, the pressure ratio π_c is expressed as a function of the mass flow μ and the rotational speed ω_{tc}

$$\pi = f(\mu, \omega_{tc}) \tag{2.54}$$

For the fitting two dimensionless variables are defined; namely the head parameter

$$\Psi = \frac{c_p T_a \left(\left(\frac{p_{out}}{p_{in}} \right)^{\frac{\gamma-1}{\gamma}} - 1 \right)}{\frac{1}{2} U_c^2}$$

and the normalized compressor flow rate

$$\Phi = \frac{\overset{*}{m}_c}{\rho_a \frac{\pi}{4} d_c^2 U_c} \quad \text{with} \quad U_c = \frac{\pi}{60} d_c N_{tc}$$

The head parameter is then approximated as follows

$$\Psi = \frac{k_1 + k_2 \Phi}{k_3 - \Phi} \qquad k_i = k_{i1} + k_{i2} \, Ma \tag{2.55}$$

where Ma is the Mach number at the ring orifice of the compressor

$$Ma = \frac{U_c}{\sqrt{\gamma \, R \, T_a}} \tag{2.56}$$

With a least square fitting algorithm, the parameters k_{ij} can be identified with measurement data.

Since (2.55) is invertible, this compressor model can be used in mean value models to describe the compressor:

$$\dot{\mu} = f(\Phi, \omega_{TC})) \quad \text{with} \quad \Phi = \frac{k_3 \, \Psi - k_1}{k_2 + \Psi} \quad , \quad \Psi = f(\omega_{tc}, \Pi_c) \qquad (2.57)$$

The results of this fitting can be seen in Fig. 2.19. The function applied is able to capture the very flat region at low mass flows as well as the very steep characteristic close to the choke line. At very high speeds the measured speed lines no longer peak on the surge line, but rather at higher mass flows. This behavior cannot be captured with this approach, since it has to be invertible. The reference [182] explains this unsteady decrease of the speed lines with gas dynamic effects which falsify the measurement.

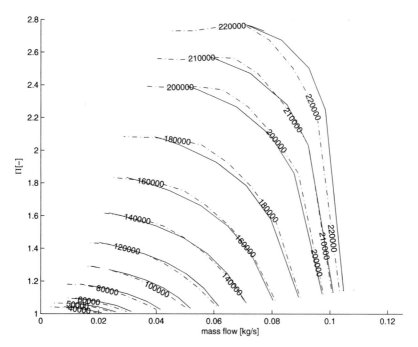

Fig. 2.19. Comparison of the measured and fitted compressor map for a GT14

Fitting the speed lines is the most difficult task in modeling a turbocharger. The three remaining tasks (the fitting of the compressor efficiency, the turbine mass flow, and the turbine efficiency) can be carried out according to the techniques presented in the modeling part.

Compressor Efficiency

The efficiency function can be depicted by (2.42). The parameters to be fitted are $\eta_{c,opt}$, $\dot{\mu}_{c,opt}$, $\Pi_{c,opt}$, and the elements of the symmetric matrix $Q = [q_{11}, q_{12}; q_{12}, q_{22}]$.

The result of the fitting can be improved by rescaling the ordinate:

$$1 + \sqrt{\Pi_c - 1} \Rightarrow \Pi_c \tag{2.58}$$

Turbine Mass Flow

The suggested throttle equation yields a choking behavior above a pressure ratios of 2. Since the measurement data does not reflect this behavior, it can be approximated better with a heuristic function

$$\dot{\mu}_{t,norm} = \frac{p_3}{\sqrt{\vartheta_3}} \cdot c_t \cdot \sqrt{1 - \Pi_t^{k_t}} \tag{2.59}$$

The parameters (c_t, k_t) can be determined with a least square algorithm. A typical value for the latter is $k_t = -2.2$.

Turbine efficiency

The parabolic function from (2.52) can be fitted to measurement data in order to depict the efficiency of a radial turbine. The only two parameters to be fitted are $\eta_{t,max}$ and $\tilde{c}_{u,opt}$.

2.4 Fuel System

2.4.1 Introduction

The fuel path, i.e., the subsystem that provides the cylinder with the necessary fuel for the combustion process, incorporates an additional dynamic subsystem. Both indirect and direct fuel injection systems are in use.[20] Direct injection is the direct transport of fuel into the (main) combustion chamber, which can be the case in Diesel and SI engines, while indirect injection can be realized as port injection (i.e., upstream of the intake valve) in SI engines or as pre-chamber injection in Diesel engines. Modern Diesel engines almost exclusively are fitted with direct-injection fuel systems while most SI engines are equipped with a port-injection configuration.[21]

For direct injection systems, the dynamic effects described below are not present, i.e., the command inputs to the fuel system typically produce "immediate" corresponding fuel quantity changes in the next combustion process.

---

[20] Carburetors are only used in special applications and will therefore not be treated in this text.

[21] Direct-injection SI engines can only reach substantial improvements in fuel economy by running in lean conditions ($\lambda \gg 1$) at low loads. However, in this case the proven and cost-effective TWC technology cannot be used to achieve the required very low tail-pipe emission levels (see Sect. 1.3.2). More complex aftertreatment systems that are able to deal with a lean and non-homogeneous combustion are required in this case.

The aspect of delays due to the engine's reciprocating behavior will be discussed in the next chapter. The higher injection pressures of DI systems induce complex dynamics in the fueling system which, in some situations, can have an influence on the thermodynamic processes inside the cylinder (see [143] or [127]).

In this section, only port-injected SI gasoline engine systems will be discussed. In such systems, the liquid fuel is injected through a solenoid on-off valve in the intake port, which usually is located directly in front of the intake valve. Since the difference of the injection pressure to the pressure in the intake manifold generally is kept constant by a dedicated control loop, the injected fuel mass is (approximately) proportional to the injection times (typically 5-20 ms). In addition, the synchronization of the time based injection process with the crank angle domain and other discrete event constraints (say, in communication with the electronic engine controller) cause delays in the fuel injection transmission path (for more details on this topic see Chapter 3).

2.4.2 Wall-Wetting Dynamics

Phenomenological Models

One of the most important dynamic effects in the fuel path is caused by the *wall-wetting* phenomena. The liquid fuel injected into the intake port only partially enters the cylinder in the next intake stroke. Some of it is stored in fuel puddles of the mass $m_f(t)$ at the intake port walls and at the back face of the intake valve.[22] Of course, fuel also evaporates from these puddles such that a mass balance can be expressed as follows (see [9])

$$\dot{m}_\varphi(t) = (1 - \kappa(\omega_e, p_m, \vartheta_f, \ldots)) \cdot \dot{m}_\psi(t) + \frac{m_f(t)}{\tau(\omega_e, p_m, \vartheta_f, \ldots)} \qquad (2.60)$$

and

$$\frac{d}{dt} m_f(t) = \kappa(\omega_e, p_m, \vartheta_f, \ldots) \cdot \dot{m}_\psi(t) - \frac{m_f(t)}{\tau(\omega_e, p_m, \vartheta_f, \ldots)} \qquad (2.61)$$

where $\dot{m}_\psi$ is the fuel mass injected, $\dot{m}_\varphi$ the fuel mass aspirated into the cylinder, and m_f the mass of fuel stored in the fuel film. The coefficients $\kappa(.)$ and $\tau(.)$ depend on the engine speed and load, as well as on many other variables (mean fuel temperature ϑ_f, etc.). According to [9], this description is often called Aquino model.

---

[22] Of course, gaseous fuels, especially compressed natural gas, have no wall-wetting problems. Nevertheless, dynamic phenomena occur due to substantial back flow of aspirated mixture from the cylinder to the intake manifold. Since these effects are best described in a discrete-event setting, the corresponding models will be introduced in Sect. 3.2.4.

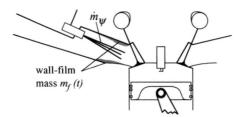

Fig. 2.20. Fuel wall-film model.

The usual approach coping with this variability is to use a pair of *constant* coefficients k and τ only in the neighborhood of the operating point (engine speed and load) for which they were experimentally found. Of course, the values of these coefficients must be updated when the actual operating point of engine moves away from the original operating point. Additionally, a correction factor is used to compensate for the changing engine temperature.

Another approach is to automatically identify the time-varying numerical values of the k and τ parameters (see e.g., [134], [202]), and to apply adaptive control methods (see e.g., [204]).

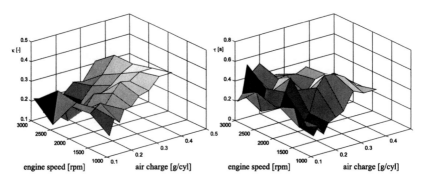

Fig. 2.21. Example of experimentally obtained maps of the coefficients k, τ used in (2.60) and (2.61); fully warmed-up engine (1.8 liter, 4 cylinders, 16V).

Notice that (2.60) and (2.61) inherently satisfy the structural constraint imposed by the mass conservation principle. Other formulations, which assume two (or more) separate wall films (with different time constants τ modeling fast and slow response contributions) or which include the influence of the back flow of exhaust gas, have been proposed as well (see e.g., [203]).

Notice also that the model represented by (2.60) and (2.61) (excluding the non-generic case $\kappa(.) = 1$) is *invertible*, i.e., if the actual wall film mass m_f is known, the fuel mass $\dot{m}_\psi$ which needs to be injected to guarantee a desired fuel mass flow $\dot{m}_{\varphi,des}$ to enter the cylinder can be computed by

$$\dot{m}_\psi(t) = \frac{1}{1 - \kappa(\omega_e, p_m, \vartheta_f, \ldots)} \cdot \left(\dot{m}_{\varphi,des}(t) - \frac{m_f(t)}{\tau(\omega_e, p_m, \vartheta_f, \ldots)} \right) \quad (2.62)$$

This property will be important in the air/fuel ratio control problem (see Chapter 4).

First-Principle Models

Since the experimental identification of the complete set of parameters k and τ is time consuming, approaches based on physical first principles that promise to shorten that time are of great interest. Several publications discuss this problem, ranging from complex simulations [117] to less detailed formulations [85].

In the following, a model proposed in [132], which is based on thermodynamic analogies, is presented. This model will be shown to have the same model structure as the model described by (2.60) and (2.61) while requiring much less experimental effort for the identification of the full set of parameters.

The wall-wetting phenomenon is split into three subproblems:

- the evaporation of droplets while they are airborne;
- the reduction of the number of droplets that collide with the wall; and
- the evaporation of the liquid fuel on the manifold walls.

The change of the mass of fuel on the manifold walls can be expressed as the amount of fuel impinging on the walls minus the amount of fuel evaporating from the walls $\dot{m}_{EVF}$. The first contribution is the difference between the injected mass, $\dot{m}_\psi$, and the mass flow $\dot{m}_{EVD}$ that evaporates from the droplet while airborne. The two evaporation processes are modeled by using well-known thermodynamic analogies [102], i.e., the fuel mass flow resulting from the evaporation process is given by

$$\dot{m}_{EV} = h_m \cdot \rho_f \cdot A_f \cdot \ln(1 + B) \quad (2.63)$$

where h_m is the mass transfer coefficient, ρ_f is the density of the liquid fuel, A_f is the area of the surface on which the evaporation takes place, and B is the *Spalding* number which describes the evaporation properties of the liquid. This last parameter is defined as

$$B = \frac{x_{vf}(\vartheta_f) - x_{vg}}{1 - x_{vf}(\vartheta_f)} \quad (2.64)$$

where $x_{vf}(\vartheta_f)$ is the mass fraction of fuel that has vaporized at the actual temperature of the liquid fuel ϑ_f and x_{vg} is the mass fraction of fuel in the surrounding gas flow. Below, it is assumed that the vaporized fuel is rapidly convected past the liquid fuel zone such that x_{vg} can be assumed to be zero. The function $x_{vf}(\vartheta_f)$ can be determined experimentally. Figure 2.22 shows such an evaporation curve for a standard unleaded RON98 type of gasoline.

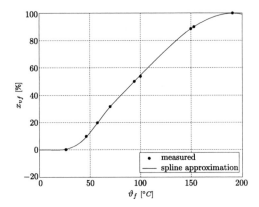

Fig. 2.22. Example of a fuel evaporation curve.

The mass transfer coefficient h_m must be calculated with thermodynamic analogies as well

$$h_m = Sh \cdot \frac{D_{AB}}{d_x} \qquad (2.65)$$

where Sh is the *Sherwood number* (2.68) defined below and D_{AB} is the diffusivity coefficient, which may be taken from tables in literature (see e.g., [102]). Assuming ideal gas behavior, the pressure and temperature dependency of the diffusion coefficient for a binary mixture of gases, such as air and fuel, can be taken into account with the relation

$$D_{AB} \propto \frac{\vartheta_f^{3/2}}{p_m} \qquad (2.66)$$

The characteristic length in the thermodynamic analogy of an evaporation problem is given by d_x where x indicates which specific geometry is relevant ($x = D$ for droplet and $x = F$ for wall film evaporation). The evaporated mass is

$$\dot{m}_{EV} = \frac{\rho_{fvs} \cdot D_{AB} \cdot A_f}{d_x} \cdot Sh \cdot \ln(1 + B) \qquad (2.67)$$

The density ρ_{fvs} refers to the fuel vapor at the surface of the droplet. It is approximated as half of the density of the liquid fuel.

The Reynolds analogy allows for the calculation of the Sherwood number based on the conditions of the air flux, represented by the Reynolds number Re, and on the characteristics of the fluid, represented by the Schmidt number Sc, [102]. To be more specific, the Sherwood number is calculated with the following analogy

$$Sh = c_r \cdot Re^{m_r} \cdot Sc^{n_r} \qquad (2.68)$$

while the Reynolds and the Schmidt numbers are defined by

$$Re = \frac{u_\infty \cdot d}{\nu} \qquad Sc = \frac{\nu}{D_{AB}} \qquad (2.69)$$

The kinematic viscosity ν depends on the temperature of the fluid and can be taken from tables as well. The parameters c_r, m_r, and n_r depend on the specific evaporation problem.

The relevant temperatures for the two evaporation problems cannot be measured. The only temperatures that are accessible in conventional engine management systems are the intake air temperature and the engine coolant temperature. Using the data from these two measurements, the temperatures of the droplets and the fuel film can be estimated using simple models, which will be introduced in Sect. 2.6. Relations for the relevant heat fluxes can be found in [50], [2], [48], [3], and [59].

Droplet Evaporation

For the case of droplet evaporation, [169] provides the following correlations

$$c_r = 2 \cdot 0.276 \qquad m_r = 0.50 \qquad n_r = 0.33$$

For this problem, the droplet diameter d_D is chosen as the characteristic length. For the calculation of the Reynolds number, the velocity of the droplets must be calculated. The Bernoulli approach, introduced in (2.9), is used for that purpose.

The mass balance for one single droplet yields

$$\frac{d}{dt}m_D(t) = \rho_f \frac{d}{dt}V_D(t) = \rho_f \frac{1}{2}\pi\, d_D^2 \frac{d}{dt}d_D(t) = -\dot{m}_{EVD}(t) \tag{2.70}$$

Combining (2.70) and (2.67), and considering $A_{fD} = \pi d_D^2$, the dynamic behavior of the droplet diameter is described by the following first-order differential equation

$$\frac{d}{dt}d_D(t) = -\frac{1}{d_D} \cdot D_{AB} \cdot \ln(1+B) \cdot Sh_D \tag{2.71}$$

Assuming all thermodynamic parameters (diffusivity coefficient, Sherwood number, Spalding Number, etc.) to remain constant, this equation can be solved in closed form

$$d_D(t) = \sqrt{d_0^2 - 2 \cdot D_{AB} \cdot \ln(1+B) \cdot Sh_D \cdot t} \tag{2.72}$$

Using this expression, the mass of evaporated fuel can easily be calculated

$$\dot{m}_{EVD}(t) = \frac{1}{2} \cdot \rho_f \cdot \pi \cdot d_D \cdot D_{AB} \cdot \ln(1+B) \cdot Sh_D \tag{2.73}$$

For the calculation of the total mass resulting from droplet evaporation the number of droplets must be assumed. In [132], an exponential decay for the number of airborne droplets is introduced. It is assumed that all the fuel is injected instantaneously and that all droplets are identical at the beginning.

Then, due to the geometry of the intake runner, some droplets will hit the manifold walls. The time constant of the exponential decay, τ_{DA}, thus depends on the geometry of the engine under investigation and can be identified once and for all. The number of droplets at the beginning of the evaporation process can be calculated using a starting diameter d_0 for the droplets of approximately 100 μm, [135]. The number of airborne droplets $N_{dp}(t)$ can then be calculated as follows.

$$N_{dp}(t) = \frac{m_\psi}{\frac{\pi}{6} \cdot d_0^3 \cdot \rho_f} \cdot e^{-t/\tau_{DA}} \tag{2.74}$$

where m_ψ is the mass of fuel injected into one intake runner during one engine cycle.

To calculate the total mass of evaporated fuel due to droplet evaporation for one injection pulse, the differential equation for the evaporation must be integrated over an interval which starts with the beginning of the injection pulse, t_{is}, and ends with the closing of the intake valve, t_{IVC}). Combining (2.74) with (2.70) yields

$$m_{EVD} = \int_{t_{is}}^{t_{IVC}} N_{dp}(t) \cdot \dot{m}_{EVD} \cdot dt \tag{2.75}$$

Usually, the time required until a droplet is completely evaporated, t_{ev}, is smaller than the length of this interval. Thus, for a fully warmed-up engine it is sufficient to integrate only up to this time.

$$t_{ev} = \frac{d_0^2}{2 \cdot C_D} \tag{2.76}$$

In this expression, all operating-point-dependent factors have been combined into one parameter

$$C_D = D_{AB} \cdot \ln(1 + B) \cdot Sh_D \tag{2.77}$$

Since it was assumed that all droplets have the same diameter at $t = 0$, the time when all droplets have evaporated depends only upon thei starting diameter and not on the amount of fuel injected. Consequently, the mass of fuel that evaporates from all droplets can be expressed by

$$m_{EVD} = (1 - \kappa(\pi_{op})) \cdot m_\psi \tag{2.78}$$

where the parameter $\kappa(\pi_{op})$ defined by

$$1 - \kappa(\pi_{op}) = \frac{3}{d_0^3} \cdot C_D \cdot \int_0^{t_{IVC}-t_{is}} e^{-\frac{t}{\tau_{DA}}} \cdot \sqrt{d_0^2 - 2 \cdot C_D \cdot t} \cdot dt \tag{2.79}$$

summarizes all operating parameters (intake manifold pressure and temperature, engine speed, fuel properties, and engine geometry). For a given set of thermodynamic properties, i.e., temperature and pressure, $\kappa(\pi_{op})$ is constant and thus m_{EVD} is proportional to the amount of fuel injected.

Wall Film Evaporation

The mass $m_\psi - m_{EVD}$ of fuel droplets hitting the manifold wall represents the mass increase for the wall film. This input is therefore dependent on the droplet evaporation model discussed before.

Experiments on a wetted tube, as described in [50] and [201], provide the analogy parameters used in (2.68) for this type of evaporation

$$c_r = 0.023 \quad , \quad m_r = 0.83 \quad , \quad \text{and} \quad n_r = 0.44$$

The fuel film is assumed not to be moving. The evaporation is, therefore, directly dependent upon the convection on the film surface caused by the air mass flow through the duct. Since it is assumed that the wall film forms directly around the intake valve(s), it is the air flow velocity through these valves that is relevant for the convection. For each cylinder, the average air flow velocity $u_{F\infty}$ through the valves can be approximated well by

$$u_{F\infty} = \frac{\dot{m}_\beta}{n_{cyl}} \cdot \frac{1}{\rho_{in} \cdot \frac{\pi}{4} d_F^2} \tag{2.80}$$

where d_F represents the diameter of the intake valve. This diameter is used as the characteristic length in the Reynolds analogy as well.

There are many approaches that describe the geometry of the wall film. One possibility is to represent it by a cylindric ring of fuel around the intake valve with thickness δ_{th} and height h_F as proposed in [132]. The height is the actual state variable of the wall film, whereas the thickness is a parameter which is slowly varying depending on the wall temperature. The thickness can either be identified in advance for a specific engine, or (preferably) identified on-line with an extended Kalman filter.

Assuming $\delta_{th} \ll d_F$, this identification results in a relevant area A_F for the evaporation of the wall film of

$$A_F = \pi \cdot d_F \cdot h_F \tag{2.81}$$

Thus the mass of fuel evaporating from the wall film is

$$\dot{m}_{EVF} = \frac{\rho_{fvs} \cdot A_F}{d_F} \cdot D_{AB} \cdot Sh_F \cdot \ln(1 + B) \tag{2.82}$$

$$= \rho_{fvs} \cdot \pi \cdot h_F \cdot D_{AB} \cdot Sh_F \cdot \ln(1 + B) \tag{2.83}$$

The mass of fuel in the wall film is described by

$$m_f = \rho_f \cdot \pi \cdot d_F \cdot \delta_{th} \cdot h_F \tag{2.84}$$

The change in the mass of the wall film is then calculated as the difference between the mass of fuel of the injection which has not evaporated in the

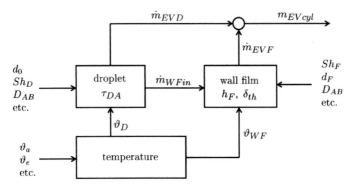

Fig. 2.23. Interconnections between droplet model and wall film model

droplet phase and the mass of fuel evaporated from the wall film. The interconnections between droplet model and wall film model are illustrated in
Fig. 2.23.

Notice that the fuel injection process takes place at discrete points in time
and that the droplet evaporation processes start after that event. This is an
inherently discrete-event formulation and, as such, not compatible with the
mean-value paradigm used in this chapter. Nevertheless, once the corresponding fuel mass (2.78) has been computed, a continuous process is assumed,
neglecting the exact location of the injection on the time axis. By that approximation, a mean-value formulation is obtained.

For instance, the mass balance for the fuel in the wall film can be formulated as follows

$$\frac{d}{dt} m_f(t) = \kappa(\pi_{op}) \cdot \dot{m}_\psi(t) - \rho_{fvs} \cdot \pi \cdot h_F \cdot D_{AB} \cdot Sh_F \cdot \ln(1 + B)$$

$$= \kappa(\pi_{op}) \cdot \dot{m}_\psi(t) - \dot{m}_{EVF}(t) \tag{2.85}$$

This equation can be transformed to the Aquino structure defined by (2.60)
and (2.61). Again, utilizing the assumption $\rho_{fvs} = \rho_f/2$, the time constant of
the evaporation is found to be

$$\frac{1}{\tau(\pi_{op})} = \frac{D_{AB} \cdot Sh_F \cdot \ln(1 + B)}{2 \cdot d_F \cdot \delta_{th}} \tag{2.86}$$

Figures 2.24 and 2.25 show the results obtained with the approach presented here for two very different engine temperatures. Table 2.1 lists all
relevant parameters of the two experiments and of the simulations shown in
these figures. Case 1 represents a fully warmed up engine in an operating point
with low engine speed and load. For the measurements of case 2, the engine's
thermostat has been removed. This results in a large coolant flow through the
engine, which prevents the engine from warming up.

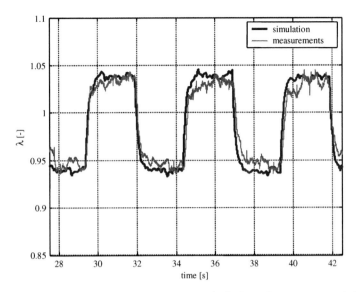

Fig. 2.24. Air/fuel ratio for warmed engine (85°C), fuel injection modulation.

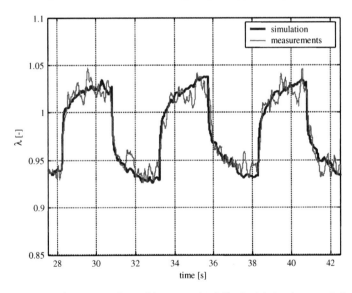

Fig. 2.25. Air/fuel ratio for cold engine (35°C), fuel injection modulation.

Table 2.1. Parameters for the wall film model.

Parameter	Case 1	Case 2
ω_e [rad/s]	157	157
m_β [g/cyl]	0.2	0.2
m_ψ [g/cyl]	$0.0136 \pm 5\%$	$0.0138 \pm 5\%$
p_m [bar]	0.578	0.586
ϑ_e [$^\circ K$]	335	303
ϑ_f [$^\circ K$]	430	390
ρ_f [kg/m^3]	747	747
Droplets		
c_r [-]	0.552	0.552
m_r [-]	0.5	0.5
n_r [-]	0.33	0.33
D_{AB} [m^2/s]	$2.2e^{-5}$	$1.8e^{-5}$
B [-]	0.35	0.0067
Sh_D [-]	1.30	1.32
τ_{DA} [s]	0.012	0.012
t_{ev} [s]	0.00029	0.0157
Film		
c_r [-]	0.023	0.023
m_r [-]	0.83	0.83
n_r [-]	0.44	0.44
D_{AB} [m^2/s]	$2.8e^{-5}$	$2.6e^{-5}$
B [-]	8.5	1.5
Sh_F [-]	1.95	2.10
d_F [m]	0.052	0.052
δ_{th} [m]	0.00085	0.00137
Aquino Parameters		
k [-]	0.81	0.53
τ [s]	0.36	1.42

Direct Injection Engines

Obviously, the wall-wetting effect is not present in direct injection engines. The mass of fuel entering the cylinder $\dot{m}_\varphi$ for this class of engines is an immediate control input. Moreover, not only can the mass of fuel be assigned for each cycle individually, but the timing of the different fuel injections pulses can also be used to control the engine's behavior (torque, pollutant formation, etc., see below).

The injection pressure (fuel rail pressure) is another control variable that influences the mixture formation inside the cylinder and, hence, the engine behavior. Simple models of the fuel rail pressure dynamics are easy to derive (lumped parameter gasoline reservoir, gasoline as incompressible fluid, etc.). Precise fuel line dynamics (including compressibility effects and fluid dynamic oscillation phenomena) have also been developed (see for instance [24]). Such models, however, are not applicable to real-time control objectives.

2.4.3 Gas Mixing and Transport Delays

In the last sections, the air and fuel paths up to the cylinder entrance have been introduced. The next dynamic block will deal with the air/fuel ratio formation and propagation to the oxygen sensor. Several dynamic effects must be considered in that part (see Fig. 2.26 for some of the definitions used below) that will be modeled in this section using an MVM approach. Obviously, such a formulation is not able to correctly model single-cylinder effects. These aspects will be discussed in Sect. 3.2.6.

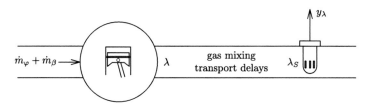

Fig. 2.26. Definitions of variables in the air/fuel ratio path.

The discrete engine operation introduces the induction to exhaust gas delay τ_{IEG} as defined in (2.2) such that, neglecting all other phenomena, immediately after the cylinder the air/fuel ratio should be

$$\widetilde{\lambda}(t) = \frac{\dot{m}_\beta(t - \tau_{IEG})}{\dot{m}_\varphi(t - \tau_{IEG})} \cdot \frac{1}{\sigma_0} \tag{2.87}$$

However, additional dynamics are introduced by the engine's gas exchange behavior. As shown above (see for instance (2.22)), a substantial part of the burnt gases remains inside the cylinder, especially at low loads. This gas fraction carries the air/fuel ratio of the present cycle on to the next one. A possible formulation[23] of an MVM for this phenomenon is

$$\lambda(t) = \frac{\widetilde{\lambda}(t) \cdot m(t) + \lambda(t - \tau_{IEG}) \cdot m_{eg}(t)}{m(t) + m_{eg}(t)} \tag{2.88}$$

where $m(t)$ is the mass of fresh charge that enters the cylinder and $m_{eg}(t)$ is the mass of burnt gas that remains in the cylinder. Note that (2.88) is a recursive formulation. The fresh charge is obtained by (Euler) integration over one segment, i.e.,

---

[23] In Sect. 3.2.5 a more detailed analysis of this problem will be made using a discrete-event approach. The formulation (2.88) will be shown to be a good approximation of the EGR dynamics.

$$m(t) = \dot{m}(t - \tau_{IEG}) \cdot \frac{2\,\pi \cdot N}{n_{cyl} \cdot \omega_e} \qquad (2.89)$$

where the engine speed is assumed to be constant over this short period. An estimation of $m_{eg}(t)$ is possible when the cylinder pressure p_z and the burnt gas temperature ϑ_z at the exhaust valve closing are known

$$m_{eg} = V_c \cdot \frac{p_z}{R_{eg} \cdot \vartheta_z} \qquad (2.90)$$

(see Sect. 2.6 below for details on ϑ_z). After exiting the cylinder, the exhaust gases must reach the oxygen sensor, a device with its own dynamics that must be considered in an air/fuel ratio loop model. The exhaust gas transport can be approximated by a combination of delay and gas mixing, i.e., using a frequency domain formulation

$$\lambda_s(s) = e^{-\tau_1 \cdot s} \cdot \frac{1}{\tau_2 \cdot s + 1} \cdot \lambda(s) \qquad (2.91)$$

The time constants τ_1 and τ_2 depend upon the engine system's geometry (volume and length of exhaust pipes) and upon the actual operating point (especially exhaust gas mass flow and temperature), as will be discussed in Sect. 2.6.

The air/fuel ratio sensor is rather fast, provided it is actively heated and built using state-of-the-art planar technology. A first-order lag is usually a good approximation of its dynamics [71]

$$x_\lambda(s) = \frac{1}{\tau_s \cdot s + 1} \cdot \lambda_s(s) \qquad (2.92)$$

where the time constant τ_s is in the order of $10 - 20\ ms$ (depending on the type of protection cap used). The output voltage y_λ depends upon the fictitious intermediate signal x_λ in a nonlinear but static way. Both switching and continuous sensors are in use.

2.5 Mechanical System

2.5.1 Torque Generation

The primary objective of an engine is to produce mechanical power. Its speed is a level variable, i.e., it is not assignable arbitrarily. The torque, however, can be changed at will, provided that the amount of mixture in the cylinder and/or its composition, for instance, can be changed arbitrarily. The mean-value engine torque is a nonlinear function of many variables, such as fuel mass in cylinder, air/fuel ratio, engine speed, ignition or injection timing, EGR rate, etc.

$$T_e = f(\dot{m}_\varphi, \lambda, \omega_e, \zeta, x_{egr}, \dots) \qquad (2.93)$$

Detailed thermodynamic simulations are necessary to correctly predict the engine torque. For control purposes, however, such simulations are too time-consuming. Thus, alternative approaches have been investigated. An obvious solution might be to grid the space of all possible operating conditions, to measure or compute the engine torque in each vertex, and to store that information for later on-line use in appropriate maps. For reasons of computing time, this solution is not practicable, however.[24] Another, more promising approach is to use physical insight to separate the various influencing variables and to divide the modeling task into several low-dimensional problems.

Before entering into a detailed discussion of this approach, two definitions are introduced. Since engine torque mainly depends on engine size, i.e., its displaced volume V_d, it is advantageous to use a normalized formulation introducing the *brake mean effective pressure*

$$p_{me} = \frac{T_e \cdot 4\,\pi}{V_d} \tag{2.94}$$

and the *fuel mean effective pressure*

$$p_{m\varphi} = \frac{H_l \cdot m_\varphi}{V_d} \tag{2.95}$$

where the mass of fuel, with lower heating value H_l, burnt per combustion cycle m_φ in (2.95) is related to the mean-value fuel mass flow by

$$\dot{m}_\varphi(t) = m_\varphi(t) \cdot \frac{\omega_e(t)}{4\,\pi} \tag{2.96}$$

The brake mean effective pressure p_{me} is the pressure that has to act on the piston during one full expansion stroke to produce the same amount of work as the real engine does in two engine revolutions (assuming a four-stroke engine). The fuel mean effective pressure $p_{m\varphi}$ is that brake mean effective pressure that an engine with an efficiency of 1 would produce with the fuel mass m_φ burnt per engine cycle (with perfect conversion of the fuel's thermal energy into mechanical energy).

Obviously, the engine's efficiency can then be written as

$$\eta_e = \frac{p_{me}}{p_{m\varphi}} = \frac{T_e \cdot 4\,\pi}{m_\varphi \cdot H_l} \tag{2.97}$$

With this definition, (2.94) can be rewritten as

$$p_{me} = \eta_e(m_\varphi, \omega_e, \lambda, \zeta, x_{egr}, \ldots) \cdot p_{m\varphi} \tag{2.98}$$

---

[24] Taking into consideration only the five independent variables mentioned in (2.93), using for each variable say, 20 support points, and assuming a measurement duration of one minute per vertex would require six years of consecutive experimentation.

and an approximation of the following form

$$\eta_e \approx \eta_{e0}(m_\varphi, \omega_e) \cdot \eta_\lambda(\lambda) \cdot \eta_\zeta(\zeta) \cdot \eta_{egr}(x_{egr}) \cdot \ldots \qquad (2.99)$$

of the efficiency can be used. The first factor $\eta_{e0}(m_\varphi, \omega_e)$ is the engine's nominal efficiency map, while the remaining factors represent the changes occurring when the respective inputs deviate from their nominal values.

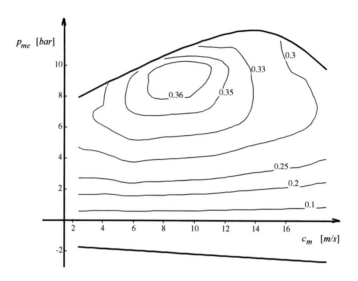

Fig. 2.27. Engine map of a modern naturally aspirated 2-liter SI engine.

For existing engines, the nominal engine efficiency $\eta_{e0}(m_\varphi, \omega_e)$ is often available in the form of a measured engine map (see Fig. 2.27 for a typical example). However, in control applications this representation has several drawbacks:

- At low loads, efficiency is not precisely measurable since unavoidable small errors in measured torque produce relatively large errors in estimated brake power.
- The efficiency is displayed as a function of mean piston speed c_m[25] (which does not pose any problems) and mean effective pressure p_{me}, which, according to Fig. 2.5, is not a level variable, but the *output* of the combustion block.

Other forms are therefore necessary to display the engine's nominal torque production properties. Figure 2.28 shows several possibilities, starting with the usual engine map in the top left corner. The form shown in the bottom right corner is especially useful for mean-value engine models since it interfaces directly with the main inputs (see Fig. 2.5).

---

[25] Mean piston speed is defined by $c_m = S \cdot \omega_e / \pi$, where S is the engine's stroke.

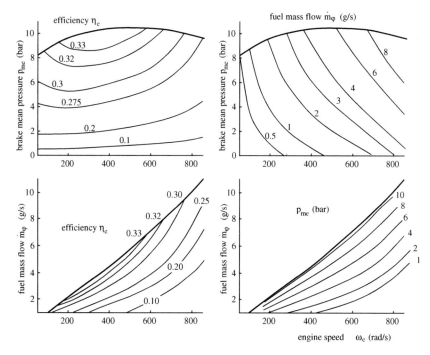

Fig. 2.28. Various alternative representations of an engine map.

Willans Approximation

A useful simplification of the engine's torque and efficiency characteristics is often feasible. In comprehensive measurements and process simulations, it has been observed [77] that the following simple relationship between mean fuel pressure and mean effective pressure approximates the real engine behavior astonishingly well:

$$p_{me} = e(m_\varphi, \omega_e, \lambda, \zeta, x_{egr}, \ldots) \cdot p_{m\varphi} - p_{me0}(\omega_e, \vartheta_e, \ldots) \qquad (2.100)$$

The efficiency coefficient $\hat{e}(.)$ represents the thermodynamic properties of the engine (i.e., the deviations from a perfect conversion from chemical energy in the fuel to mechanical work), whereas the second part $p_{me0}(.)$ incorporates the friction and gas-exchange losses. Quite often the simplification

$$\frac{\partial e(m_\varphi, \omega_e, \lambda, \zeta, x_{egr}, \ldots)}{\partial m_\varphi} = 0 \qquad (2.101)$$

is possible (especially when excluding full-load conditions). If, in addition, the equation

$$\frac{\partial p_{me0}(\omega_e, \vartheta_e, \ldots)}{\partial m_\varphi} = 0 \qquad (2.102)$$

is satisfied, (2.100) becomes affine in the injected fuel mass (and therefore in $p_{m\varphi}$, see equation (2.95)) as visualized in Fig. 2.29.

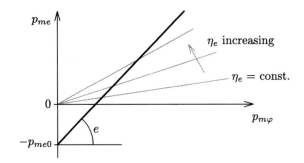

Fig. 2.29. Affine Willans model of the engine efficiency.

Aside from its simplicity, (2.100) has another important advantage: It clearly distinguishes between the contributions of the internal (thermodynamic) and external effects (friction, gas exchange, etc.) on engine efficiency. This separation simplifies the modeling of the individual influences as introduced in (2.99).

The next section will provide certain approximations for the case in which varying gas exchange work (i.e., (2.102) does not hold) has to be taken into account.

Gas-Exchange and Friction Losses

The total external losses are made up by the friction in the engine and the pumping work required during the gas exchange process:

$$p_{me0}(\omega_e, \dot{m}_\beta, \vartheta_e, \ldots) = p_{me0f}(\omega_e, \vartheta_e, \ldots) + p_{me0g}(r_l) \tag{2.103}$$

where $p_{me0g}(r_l)$ is the cycle-averaged pressure difference between the intake and exhaust port of the engine. Its argument is the *relative load*, given by the actual torque divided by the maximum possible torque for this engine speed:

$$r_l(\omega_e) = \frac{T_e}{T_{e,max}(\omega_e)} \tag{2.104}$$

Since in an SI engine the load is controlled by (and proportional to) the air mass, the relative load can be defined by using the actual and the maximum air charge in the cylinder. For a Diesel engine, which is controlled by the amount of injected fuel, this quantity is used.

$$r_l(\omega_e)|_{SI} = \frac{m_\beta}{m_{\beta,max}(\omega_e)} \tag{2.105}$$

$$r_l(\omega_e)|_{Diesel} = \frac{m_\varphi}{m_{\varphi,max}(\omega_e)} \tag{2.106}$$

When the engine is motored (with no fuel injection and completely closed throttle valve for SI engines), the following values are reasonable estimations

$$\text{Otto}: p_{me0g}(0) \approx 0.9 \; bar, \quad \text{Diesel}: p_{me0g}(0) \approx 0.1 \ldots 0.2 \; bar \quad (2.107)$$

In Diesel engines, $p_{me0g}(r_l)$ is primarily a function of the turbocharger characteristics, i.e., the total gas exchange losses[26] are the pressure rise from the intake to the exhaust side of the engine plus the $0.1 \ldots 0.2$ bar of $p_{me0g}(0)$, which model flow-friction losses (intake and exhaust valves, intercooler, etc.).

In naturally aspirated SI engines, the exhaust pressure varies less such that an approximation of the gas-exchange losses can be derived by *assuming* a constant volumetric efficiency of the engine and thus, according to (2.29), a linear dependency between the intake pressure and the relative load:

$$p_{me0g}(r_l) = p_{me0g}(0) \cdot [1 - 0.8 \ldots 0.9 \cdot r_l] \quad (2.108)$$

Inserting the definitions of the relative load for an SI engine (2.105) and of the fuel mean-effective pressure (2.95), the pumping work can be writte as

$$p_{me0g}(\lambda, \omega_e, p_{m\varphi}) = p_{me0g}(0) \cdot \left[1 - 0.8 \ldots 0.9 \cdot \frac{\lambda \cdot \sigma_0 \cdot V_d / H_l}{m_{\beta,max}(\omega_e)} \cdot p_{m\varphi}\right] \quad (2.109)$$

The second term is linear in $p_{m\varphi}$ and can therefore be added to $\hat{e}$ in (2.100), yielding a modified efficiency coefficient

$$e = \hat{e} + 0.8 \ldots 0.9 \cdot p_{me0g}(0) \cdot \frac{\lambda \cdot \sigma_0 \cdot V_d / H_l}{m_{\beta,max}(\omega_e)} \quad (2.110)$$

which is effective up to the maximum torque in naturally aspirated (NA) engine operation. The total efficiency coefficient is given by the thermodynamic efficiency $\hat{e}$, increased by the reduction of the pumping work for higher loads.

If supercharged engine operation is considered, i.e., when the value of $p_{m\varphi}$ is higher than the corresponding value of the NA full load value, gas exchange losses rise again. If a large turbocharger and/or a "waste-gate open" strategy is used, the value of the back-pressure p_3 will stay close to the one of p_2 (see [62] for details) for a moderate engine-out volumetric exhaust gas flow. Thus, $e = \hat{e}$ holds up to values of p_2 where engine knock prevents an efficient stoichiometric combustion. In order to prevent knock, usually various countermeasures are implemented, e.g. fuel enrichment (lowering the overall in-cylinder temperatures by increasing the total mass in the cylinder) or increasing the ignition delay (lower peak pressure and temperatures during combustion). These considerations as well as the efficiency improvement due to higher loads (as described above) are represented by the "extended Willans model" shown in Fig. 2.30.

---

[26] Of course, for well-dimensioned turbocharged engine systems, in certain operating points, the pressure difference can be positive, i.e., positive gas-exchange loops can occur.

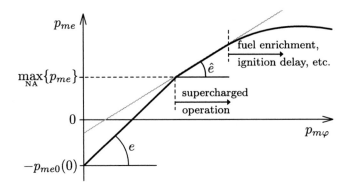

Fig. 2.30. Extended Willans model for supercharged SI engines.

A more detailed investigation of the low-load and motoring region can be found in [158]. Note that for both Diesel and SI engines, the intake and exhaust pressure levels become readily available if models such as those depicted in Figures 2.5 and 2.6 are implemented, respectively. In that case, the relation

$$p_{me0g} = p_3 - p_2 \tag{2.111}$$

can be used directly. The approximation presented in Fig. 2.30 is only used in less detailed simulations, for instance in quasi-static estimations of the fuel consumption [80].

Engine friction, as represented by the $p_{me0f}(\omega_e, \vartheta_e, \dots)$ term in (2.103), is discussed next. Several models have been proposed for this part (see e.g., [161]). The friction model developed in [193] is used in this text. It has the following form

$$p_{me0f}(\omega_e, \vartheta_e, \dots) = k_1(\vartheta_e) \cdot \left(k_2 + k_3 \cdot S^2 \cdot \omega_e^2 \right) \cdot \Pi_{e,max} \cdot \sqrt{\frac{k_4}{B}} \tag{2.112}$$

Typical values for the parameters are given in Table 2.2. The variable $\Pi_{e,max}$ is the maximum boost ratio for which the engine is designed to operate at low

Table 2.2. Parameters of the ETH friction-model.

	SI	Diesel
$k_1(\vartheta_\infty)$	$1.44 \cdot 10^5$ (Pa)	$1.44 \cdot 10^5$ (Pa)
k_2	0.46 $(-)$	0.50 $(-)$
k_3	$9.1 \cdot 10^{-4}$ (s^2/m^2)	$1.1 \cdot 10^{-3} (s^2/m^2)$
k_4	0.075 (m)	0.075 (m)

speeds,[27] B is the engine's bore diameter and S its stroke. The temperature behavior of k_1 is assumed to vary as shown in Fig. 2.31, where the engine temperature ϑ_e is an average value that describes the thermal state of the engine (see Sect. 2.6).

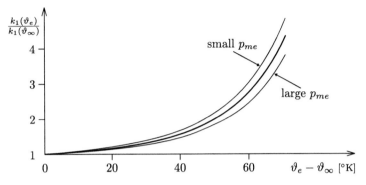

Fig. 2.31. Temperature dependency of mechanical friction. Mean engine temperature ϑ_e, with ϑ_∞ being the temperature in fully warmed conditions.

Thermodynamic Efficiencies

According to the approach in (2.99), the internal (thermodynamic) efficiency is separated into its components as follows

$$e(\omega_e, \lambda, \zeta, x_{egr}, \varepsilon \ldots) = e_\omega(\omega_e) \cdot e_\lambda(\lambda) \cdot e_\zeta(\zeta) \cdot e(\varepsilon) \cdot e_{egr}(x_{egr}) \cdot \ldots \quad (2.113)$$

where the first factor $e_\omega(\omega_e)$ is centered at the optimum point of the other variables $\lambda, \zeta, x_{egr}, \ldots$

Engine Speed

Typically, the factor $e_\omega(\omega_e)$ has a parabolic form, as shown in Fig. 2.32. At very low speeds, the relatively large heat losses through the wall reduce engine efficiency, while at very high speeds, the combustion times become unfavorably large compared to the available interval in the expansion stroke. Note that $e_\omega(\omega_e)$ has a magnitude substantially smaller than 1 since it incorporates the basic thermodynamic efficiency mechanisms, while the other factors by design are around 1.

---

[27] Higher boost ratios require larger crankshaft bearings, etc., thus leading to increased friction.

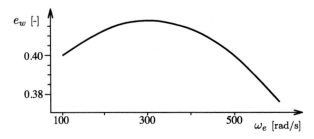

Fig. 2.32. Engine speed influence on efficiency for a SI engine (for a Diesel engine, similar form but maximum value around 0.46).

Air/Fuel Ratio

The factor $e_\lambda(\lambda)$ models the influence of the changing air/fuel ratio on the thermodynamic efficiency. This influence is, of course, very different for SI and CI engines.

In SI engines, rich to lean mixtures are possible. For rich mixtures, incomplete combustion and water/gas shift reactions[28] substantially reduce the thermodynamic efficiency. For mixtures more than slightly lean, sufficient oxygen is available for complete combustion such that efficiency is not affected by changing values of λ. For intermediatevalues, $\lambda_1 \leq \lambda \leq \lambda_2$, a smooth transition is observed. For $\lambda > \lambda_{max}$, finally, misfires start to interfere such that these operating points are not modeled.

This behavior can be expressed quantitatively as

$$e_\lambda(\lambda) = \begin{cases} \gamma_1 \cdot \lambda - \gamma_0 & \text{for } \lambda_{min} < \lambda < \lambda_1 \\ e_{\lambda,1} + (1 - e_{\lambda,1}) \cdot \sin\left(\frac{\lambda - \lambda_1}{1 - \lambda_1}\right) & \text{for } \lambda_1 < \lambda < \lambda_2 \\ 1 & \text{for } \lambda_2 < \lambda < \lambda_{max} \end{cases} \qquad (2.114)$$

where $e_{\lambda,1} = \gamma_1 \cdot \lambda_1 - \gamma_0$. Moreover, for continuity reasons, the following equation must be satisfied

$$\lambda_2 = \lambda_1 + \frac{\pi}{2} \cdot (1 - \lambda_1) \qquad (2.115)$$

Figure 2.33 shows the form of $e_\lambda(\lambda)$ for realistic parameter values.

In Diesel engines, which are always operated in lean conditions, the air/fuel ratio has little influence on efficiency, with the exception of the extreme levels. Both at very rich ($\lambda < 1.4$) and very lean conditions, the efficiency slightly drops. However, for most control purposes these extreme operating regions can be neglected.

---

[28] The conversion of fuel (hydrocarbon) into H_2 and CO consumes some of the fuel's enthalpy, such that it is not available for the thermodynamic processes.

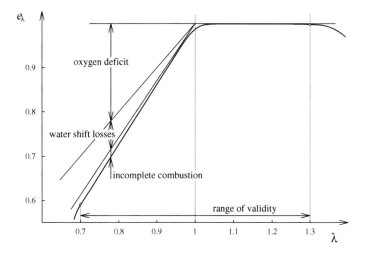

Fig. 2.33. Air/fuel ratio influence on engine efficiency. Parameters $\lambda_{min} = 0.7$, $\lambda_1 = 0.95$, $\lambda_2 = 1.0285$, $\lambda_{max} = 1.3$, $\gamma_0 = 0.373$, $\gamma_1 = 1.373$.

Ignition/Injection Timing

The timing of the ignition in SI and of the injection in CI engines has a similar influence on the production of torque:

- There is an optimal (maximum break torque, MBT) ignition angle and start of injection angle, denoted as $\zeta_0(\omega_e, p_{me})$;
- This MBT angle varies strongly with engine operating conditions (especially speed and load);
- For ignition or injection angles other than MBT, the torque drops almost parabolically; and
- The form of this reduction, in first a approximation, is independent of the operating point and of the engine type [70]. An example of that reduction for an SI engine is shown in Fig. 2.34.

The factor $e_\zeta(\zeta)$ in (2.113) can thus be written as

$$e_\zeta(\zeta) = 1 - k_\zeta \cdot (\zeta - \zeta_0(\omega_e, p_{me}))^2 \qquad (2.116)$$

Note that in both engine types, MBT settings might not be attainable due to knocking phenomena or pollutant emission restrictions (see Sect. 2.7).

Compression Ratio

The factor $e_\varepsilon(\varepsilon)$ models the influence of the compression ratio. It is important to take into account the change of compression ratio, for instance when downsizing and supercharging an engine for fuel consumption reduction. Due to the higher cylinder pressures and cylinder temperatures the engine starts

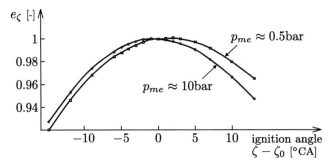

Fig. 2.34. Influence of ignition angle deviation from MBT settings on engine torque at constant engine speed for almost idle to full-load conditions.

to show knocking already at part load because this load corresponds to the full-load of the identical engine without supercharger. Therefore the compression ratio has to be reduced to avoid damage to pistons and other important engine parts.

A straightforward approximation for e_ε is given by

$$e_\varepsilon = \frac{1 - \frac{1}{\varepsilon^{k_\varepsilon}}}{1 - \frac{1}{\varepsilon_0^{k_\varepsilon}}} \tag{2.117}$$

where k_ε can be estimated by the use of engine cycle calculations.

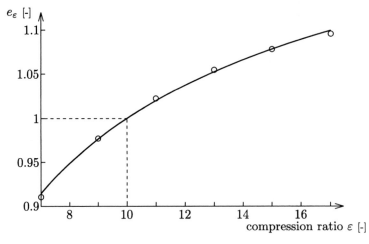

Fig. 2.35. Influence of compression ratio ε on efficiency e_ε. The solid line shows the approximation of (2.117), the circles represent values derived by an engine-cycle calculation.

The data depicted in Fig. 2.35 is calculated for an engine speed of 3000 rpm. Taking a reference compression ratio $\varepsilon_0 = 10$, a value $k_\varepsilon = 0.55$ can be found.

Burnt Gas Fraction

The factor $e_{egr}(x_{egr})$ models the influence of the changing burnt gas fraction on the thermodynamic efficiency. This influence, of course, is very much different for SI and Diesel engines. In both engine types, burnt gases are always present (internal EGR). Their effect on the thermodynamic efficiency is already included in the formulations above.

Additional (external or internal) EGR reduces the formation of nitrogen oxide, but negatively affects the thermodynamic efficiency of SI engines. In fact, increased EGR rates lead to slower burning speeds and, therefore, to reduced factors $e_{egr}(x_{egr})$. The qualitative effect of this rule is shown in Fig. 2.36. Note that the reduced burning speed has a more negative influence at high engine speeds, where relatively less time for charge burning is available. Therefore, a speed-dependent formulation of the EGR term of the following form is recommended

$$e_{egr}(x_{egr}) = 1 - k_{egr,1} \cdot (1 + k_{egr,2} \cdot \omega_e) \cdot x_{egr}^2 \qquad (2.118)$$

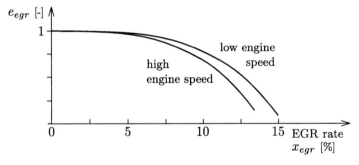

Fig. 2.36. Influence of EGR rate on SI engine efficiency.

In Diesel engines the EGR rate has — with the exception of very large ratios — almost no effect on efficiency. The Diesel-specific diffusion-controlled burning mechanism and the always abundant oxygen render this type of combustion much more tolerant to high EGR, which is one of the reasons why EGR in modern Diesel engines is a standard feature. However, as stated in the Introduction, higher EGR rates lead to increased particulate matter formation, such that this trade-off is often an important limiting factor in Diesel engines.

Direct-Injection SI Engines

All the assertions made above are valid, in principle, for direct-injection gasoline engines operating in lean and stratified mode alike (in homogeneous mode there are almost no differences at all). The differences that can be observed concern the specific shape of the functions relating torque and input parameters, but not the structural properties. Of course, for GDI engines, more degrees of freedom are available.

Injection timing ξ, together with ignition timing ζ, substantially influences engine torque in the stratified mode. This can be captured, for instance, by adding an additional term in (2.113)

$$e(\omega_e, \lambda, \zeta, x_{egr}, \xi, \ldots) = e_\omega(\omega_e) \cdot e_\lambda(\lambda) \cdot e_\zeta(\zeta) \cdot e_{egr}(x_{egr}) \cdot e_\xi(\xi) \cdot \ldots \quad (2.119)$$

or by modifying the ignition factor (2.116) to include also the injection timing effects

$$e_\zeta(\zeta, \xi) = 1 - k_\zeta(\xi) \cdot (\zeta - \zeta_0(\omega_e, p_{me}, \xi))^2 \quad (2.120)$$

2.5.2 Engine Speed

System Dynamics

In a mean value setting, modeling the engine speed behavior is a straightforward process. In fact, the engine inertia is assumed to be constant and all engine friction phenomena are already included in the torque model described by (2.100).

The only relevant reservoir is the engine flywheel. It stores kinetic energy, and the differential equation for the corresponding level variable $\omega_e(t)$ is

$$\Theta_e \cdot \frac{d}{dt}\omega_e(t) = T_e(t) - T_l(t) \quad (2.121)$$

The load torque T_l is assumed to be known. In a more general setting, this variable would depend on engine speed, drive train properties, and external loads like road slope or wind disturbances. However, as mentioned in the Introduction, these aspects are not analyzed in this text.

Sometimes a crank-angle-based formulation of (2.121) is preferred, which is obtained using the transformation $t \rightarrow \phi$, where ϕ is the engine's crank angle

$$\Theta_e \cdot \frac{d}{d\phi}\omega_e(\phi) = \frac{1}{\omega_e(\phi)} \cdot [T_e(\phi) - T_l(\phi)] \quad (2.122)$$

Note that for torques that depend linearly on engine speed, (2.121) is linear, while (2.122) — describing the same system — is nonlinear. On the other hand, many engine phenomena (e.g., air intake) are easier to describe with the crank angle as the independent variable, since the influence of the engine speed on the dynamic phenomena is then substantially reduced. More details on these aspects will be given in Chapter 3.

Note also that, for diagnostic purposes (e.g., in on-board diagnosis systems for misfire detection), much more detailed models are in use. In these models, the reciprocating structure (both of the combustion and of the crank mechanism) is correctly modeled such that the engine speed variations during one engine cycle are predicted (see the original papers [174] and [164], or [113] and [37] for more recent publications).

Constant Speed Approximation

In discrete-event models, the mean engine speed is often assumed to be almost constant during the next few (say 10) cycles such that the discrete-*event* model can be approximated by a discrete-*time* model in which the crank angle and the time evolve in parallel.[29] For this assumption to be realistic, the speed dynamics described by (2.121) must be much slower than the other dynamic effects of interest. For naturally aspirated SI engines, one of the most important of these other variables is the torque, which, according to (2.100), depends mainly on manifold pressure. All other variables have much faster time constants or are linked to the manifold pressure (air/fuel ratio, EGR ratio, etc.). In the following, only SI engines will be analyzed (see [207] for a similar problem formulation arising in Diesel engines).

As a first step, the slowest part of the manifold pressure dynamics has to be identified, for which a constant engine speed $\omega_e(t) = \omega_{e0}$ is assumed temporarily. The mean-value intake manifold dynamics for sonic conditions and isothermal[30] gas behavior in the manifold can be derived from (2.8), (2.10), (2.15) and (2.19) to be

$$\frac{d}{dt}p(t) = \frac{R \cdot \vartheta_m}{V_m} \cdot \left(A(t) \frac{p_a}{\sqrt{R \cdot \vartheta_m}} \cdot \frac{1}{\sqrt{2}} - \frac{p(t)}{R \cdot \vartheta_m} \cdot \frac{V_d \cdot \omega_{e0}}{4\pi} \right) \quad (2.123)$$

(for the sake of simplicity, discharge coefficient and volumetric efficiency have been set to 1), or for subsonic conditions

$$\frac{d}{dt}p(t) = \frac{R\vartheta_m}{V_m} \left(A(t) \frac{p_a}{\sqrt{R\vartheta_m}} \sqrt{\frac{2p(t)}{p_a}[1 - \frac{p(t)}{p_a}]} - \frac{p(t)}{R\vartheta_m} \frac{V_d \omega_{e0}}{4\pi} \right) \quad (2.124)$$

For constant engine speeds, $\omega_e(t) = \omega_{e0}$, (2.123) is linear, with the time constant

$$\tau_l = \frac{4\pi \cdot V_m}{\omega_{e0} \cdot V_d} \quad (2.125)$$

Equation (2.124) is nonlinear. It is linearized around the equilibrium operating point $\{A_0, p_0\}$

---

[29] Analyzing an engine system as a truly discrete-event system requires sophisticated mathematical tools, see, for instance, [20] or [21].

[30] The same conclusions are drawn for the adiabatic formulation.

$$A_0 = \frac{V_d \cdot \omega_e \cdot p_0}{\sqrt{R \cdot \vartheta_m} \cdot 4\pi \cdot \sqrt{2\,p_0 \cdot [p_a - p_0]}} \tag{2.126}$$

Linearizing (2.124) around this operating point yields the following time constant

$$\tau_{nl} = \frac{4\pi \cdot V_m}{\omega_{e0} \cdot V_d} \cdot \frac{2\,(p_a - p_0)}{p_a} \tag{2.127}$$

If these time constants are much faster than the speed dynamics for all operating points possible, then the postulated simplification of the discrete-event system to a discrete-time system is indeed valid. As Fig. 2.37 shows, the critical case (the largest time constant) occurs at very low loads or manifold pressures.

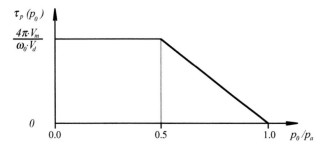

Fig. 2.37. Time constants of linearized manifold system for different nominal points.

Of course, it could be argued that the linear system behavior is not relevant since large pressure swings do occur often. However, as Fig. 2.38 shows, in the nonlinear case (dark curves in the middle plot), the manifold pressure converges even faster to its steady-state value than in the linear case (gray curve). In fact, the switch between the sonic and the subsonic conditions accelerates the convergence of the manifold pressure because inlet air mass flow is reduced more than in the linear case where downstream pressure is not influencing throttle air mass flow (see (2.123) and (2.124)).

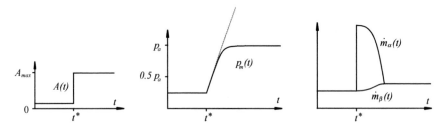

Fig. 2.38. Large-signal system behavior of the manifold pressure dynamics.

In summary, the worst case occurs at:

- low loads and small deviations in manifold pressure from equilibrium (slowest manifold pressure dynamics); and
- open clutch, i.e., zero load torque, since in that case only the engine inertia is active and the vehicle's power train inertia and mass do not delay the speed changes (fastest engine speed dynamics).

If in those situations the engine speed dynamics are much slower than the torque dynamics, the simplification of the engine's discrete-event behavior as a discrete-time system is indeed justified. This scenario will be analyzed in the rest of this section.

The starting point consists of the combined and simplified manifold pressure and speed dynamics

$$\Theta_e \cdot \frac{d}{dt}\omega_e(t) = \frac{e \cdot H_l}{4\pi} \cdot \frac{1}{1+\sigma_0} \cdot \frac{V_d}{R \cdot \vartheta_m} \cdot p_m(t) - \frac{V_d}{4\pi} \cdot p_{me0} \qquad (2.128)$$

and

$$\frac{d}{dt}p_m(t) = \frac{R \cdot \vartheta_m}{V_m} \cdot \left(A(t) \cdot \frac{p_a}{\sqrt{R \cdot \vartheta_m}} \cdot \frac{1}{\sqrt{2}} - \frac{p_m(t)}{R \cdot \vartheta_m} \cdot \frac{V_d}{2} \cdot \frac{\omega_e(t)}{2\pi} \right) \qquad (2.129)$$

To simplify the subsequent computations, the Willans coefficients (e, p_{me0}) are assumed to be constant, the volumetric efficiency and the air/fuel ratio are set to 1, the IPS delay is neglected,[31] and the gas properties of mixture and air are assumed to be identical.

To simplify the notation, the following abbreviations are introduced

$$k_1 = \sqrt{R \cdot \vartheta_m/2} \cdot p_a/V_m$$

$$k_2 = V_d/(4\pi \cdot V_m)$$

$$k_3 = e \cdot V_d \cdot H_l/(4\pi \cdot (1+\sigma_0) \cdot R \cdot \vartheta_m \cdot \Theta_e)$$

$$k_4 = V_d \cdot p_{me0}/(4\pi \cdot \Theta_e)$$

$$(2.130)$$

Based on these definitions, the combined system can be written compactly as

$$\frac{d}{dt}p_m(t) = k_1 \cdot A(t) - k_2 \cdot \omega_e(t) \cdot p_m(t)$$

$$\frac{d}{dt}\omega_e(t) = k_3 \cdot p_m(t) - k_4$$

$$(2.131)$$

If, for the moment, the assumption holds that the speed dynamics (the second equation) are much slower than the torque dynamics (the validity of this assumption will be verified later), then a separation[32] of the two equations is possible as follows:

---

[31] Even at idle speed the IPS delay is smaller than the manifold time constant. However, as discussed in detail in Appendix B, the delays present in the system have a substantial influence on the system behavior.

[32] In system theory, this approach is known as singular perturbation, [209].

- Case a) Assuming a constant engine speed $w_e(t) = w_{e0}$ and a dynamic behavior of the manifold pressure.
- Case b) Assuming an algebraic relationship between engine speed and manifold pressure, i.e., the derivative of $p_m(t)$ is set to zero, and a corresponding dynamic behavior of the engine speed is derived.

In case a), the manifold pressure satisfies the linear first-order ODE

$$\frac{d}{dt}\tilde{p}_m(t) = k_1 \cdot A(t) - k_2 \cdot w_{e0} \cdot \tilde{p}_m(t) \tag{2.132}$$

and in case b), the engine speed satisfies the nonlinear first-order ODE

$$\frac{d}{dt}\tilde{w}_e(t) = \frac{k_1 \cdot k_3}{k_2} \cdot \frac{A(t)}{\tilde{w}_e(t)} - k_4 \tag{2.133}$$

where the tilde is used to denote the corresponding variables in the fully separated limit case. Using the two equilibrium values

$$p_{m,\infty} = \frac{k_4}{k_3}, \qquad A_\infty = \frac{k_2 \cdot k_4}{k_1 \cdot k_3} \cdot w_{e0} \tag{2.134}$$

to linearize (2.133) — the system (2.132) is always linear — yields the two separated time constants

$$\tilde{\tau}_p = \frac{1}{k_2 \cdot w_{e0}}, \qquad \tilde{\tau}_w = \frac{w_{e0}}{k_4} \tag{2.135}$$

Inserting the definitions (2.130), the ratio of these two time constants is then found to be

$$\chi = \frac{\tilde{\tau}_p}{\tilde{\tau}_w} = \frac{p_{me0} \cdot V_m}{\Theta_e \cdot w_{e0}^2} \tag{2.136}$$

Inserting typical numerical values[33] for the critical case, i.e., at idle speed, values of χ in the order of 0.1 are obtained for typical naturally aspirated SI engines. In other words, in the *worst case*, the manifold dynamics are faster than the speed dynamics; in all other cases (higher engine speeds or loads, large transients, etc.), the ratio χ is *much* smaller. This justifies in most situations the time scale separation introduced above and, consequently, the assumption that in most situations engines can be well approximated as discrete-time dynamic systems.

Note that the non-dimensional parameter χ is the ratio of two types of energies. The numerator is proportional to a fictitious pneumatic energy stored in the manifold, while the numerator is proportional to the kinetic energy of the engine's flywheel. Therefore, the simplification of the engine as a discrete-time system is based on the fact that, in most situations, much more energy is stored in the engine's flywheel than in its intake manifold.

---

[33] For a 50 kW SI engine the following values are reasonable estimates $p_{me0} \approx 10^5$ Pa, $V_m \approx 2 \cdot 10^{-3}$ m^3, $\Theta_e \approx 0.2$ kg/m^2 and $w_{e0} \approx 100$ rad/s.

If the parameters of the engine system are such that the value of χ is not substantially smaller than 1, the system (2.131) has to be analyzed without adopting the simplifying assumption used in deriving (2.132) and (2.133). Linearizing equation (2.131) at the equilibrium $\{p_0, A_0(\omega_0)\}$ defined by

$$p_0 = \frac{k_4}{k_3}, \quad A_0 = \frac{k_2 \cdot k_4}{k_1 \cdot k_3} \cdot \omega_0 \tag{2.137}$$

yields a second-order linear system whose state-space description (A.34) includes the following matrix A

$$A = \begin{bmatrix} -k_2 \cdot \omega_0 & -k_2 \cdot k_4/k_3 \\ k_3 & 0 \end{bmatrix} \tag{2.138}$$

The two eigenvalues of this matrix are

$$\lambda_{1,2} = \frac{k_2 \cdot \omega_0}{2} \cdot \left(-1 \pm \sqrt{1 - \frac{4 \cdot k_4}{k_2 \cdot \omega_0^2}} \right) \tag{2.139}$$

For a given engine, i.e., for fixed values of k_2 and k_4, and for increasing nominal speeds ω_0 these two eigenvalues will converge to the negative inverse of the time constants defined in (2.135)

$$\lambda_{1,2} \rightarrow \{ -1/\tilde{\tau}_p, \ -1/\tilde{\tau}_w \} \tag{2.140}$$

Note that the eigenvalues defined in (2.139) are real only if the following condition is satisfied

$$\chi < \frac{1}{4} \tag{2.141}$$

If the parameters $\{p_{me0}, V_m, \Theta_e, \omega_{e0}\}$ are such that χ is larger than 0.25, oscillations in the engine speed and manifold pressure must be expected.[34] Note that in this case the constant speed assumption is not well satisfied. In such situations, approaches similar to the one proposed in [20] are preferable.

2.5.3 Rotational Vibration Dampers

Almost all engine systems today are first tested on a dynamometer before being installed in a vehicle. This is a much more convenient and also economical way to test and optimize the functionality of an engine-system. A standard dynamometer setup can be represented by the simplified model shown in Fig. 2.39. It comprises two rotational inertias connected by a shaft. The left-hand inertia thereby represents the engine, including its flywheel. The right-hand one incorporates the brake itself, usually including the torque measuring device. The spring stiffness c and the damping coefficient d define the characteristics of the shaft linking the two inertias.

---

[34] Due to the several simplifications adopted in the derivation of the equations (2.131), the onset of an oscillatory behavior can be observed for values of χ that are smaller than 0.25.

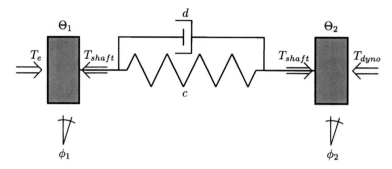

Fig. 2.39. Simplified test-bench setup. Symbols defined in the text.

Being a reciprocating machine, a combustion engine produces torque in a pulsating manner. During the power stroke of each cylinder, approximately four times the average torque (divided by the number of cylinders) is produced, whereas during the remaining three strokes even a negative torque (caused by compression friction, gas exchange, and other losses) is transferred to the crankshaft. These torque pulsations lead to variations of the rotational speed of the crankshaft. Its amplitude strongly varies with the number of cylinders and firing order and is most significant in single-cylinder four-stroke engines.

Increasing the rotational inertia of the flywheel reduces the speed variations of the crankshaft, but the torque pulsations remain. In order to reduce these pulsations and therefore minimize the speed oscillations in the test-bench brake, the connection to the brake is designed as an elastic shaft. However, this elasticity induces a resonance frequency to the system, which during operation has to be avoided to prevent damage to the engine, shaft, and test-bench.

In reality, all the connecting parts in a test-bench setup are elastic. Therefore, the total elasticity of the connection between engine and brake is always larger than expected, resulting in a lower resonance frequency than calculated. Thus, taking only the known elasticity of the shaft into account, an upper bound for the resonance frequency can be estimated. If the rotational speed of the shaft remains at least twice as high as the resonance frequency, its operation can be expected to be safe.

If a perfectly rigid link between engine and brake could be established, the torque oscillations would remain unchanged, but the speed oscillations would almost vanish due to the significantly larger total inertia.[35] In reality, the always present but rather stiff elasticities in such a configuration would result in a high resonance frequency, most probably near or equal to the operating region of the system. Running the system in this resonance frequency would destroy the weakest (usually the most elastic) component after having caused

---

[35] In a well-dimensioned dynamic test-bench setup, the engine inertia is several times smaller than that of the test-bench brake.

severe damage to other devices. Thus, trying to build a rigid system is not recommendable.

The simplified test-bench setup illustrated in Fig. 2.39 is described by the following differential equations

$$\Theta_1 \ddot{\phi}_1 = T_e - T_{shaft}$$
$$= T_e - c(\phi_1 - \phi_2) - d(\dot{\phi}_1 - \dot{\phi}_2) \qquad (2.142)$$
$$\Theta_2 \ddot{\phi}_2 = c(\phi_1 - \phi_2) + d(\dot{\phi}_1 - \dot{\phi}_2) - T_{dyno} \qquad (2.143)$$

Introducing the new variable $\Delta\phi = \phi_1 - \phi_2$ and combining the two equations yields

$$\Delta\ddot{\phi} = \frac{T_e}{\Theta_1} - c\left(\frac{1}{\Theta_1} + \frac{1}{\Theta_2}\right)\Delta\phi - d\left(\frac{1}{\Theta_1} + \frac{1}{\Theta_2}\right)\Delta\dot{\phi} + \frac{T_{dyno}}{\Theta_2} \qquad (2.144)$$

Since the torque acting on the shaft is made up by the two contributions from the engine T_e and the test-bench brake T_{dyno}, the two transfer functions

$$G_{T_e \to T_{shaft}}(s) = \left.\frac{T_{shaft}(s)}{T_e(s)}\right|_{T_{dyno}(s) \equiv 0}$$

$$G_{T_{dyno} \to T_{shaft}}(s) = \left.\frac{T_{shaft}(s)}{T_{dyno}(s)}\right|_{T_e(s) \equiv 0}$$

are of interest. Based on (2.142), the first of these transfer functions is

$$\frac{T_{shaft}(s)}{T_e(s)} = 1 - \frac{\Theta_1 s^2 \phi_1(s)}{T_e(s)} = \frac{T_e(s) - \Theta_1 s^2 \phi_1(s)}{T_e(s)}$$
$$= \frac{c\Delta\phi(s) + d\,s\,\Delta\phi(s)}{T_e(s)} \qquad (2.145)$$

Solving (2.144) for T_e, inserting the result into (2.145) and using the abbreviation

$$\frac{1}{\Theta_{repr}} = \frac{1}{\Theta_1} + \frac{1}{\Theta_2} \qquad (2.146)$$

yields

$$\frac{T_{shaft}(s)}{T_e(s)} = \frac{c\Delta\phi(s) + d \cdot s \cdot \Delta\phi(s)}{\Theta_1\left[s^2 \cdot \Delta\phi(s) + \frac{c}{\Theta_{repr}} \cdot \Delta\phi(s) + \frac{d}{\Theta_{repr}} \cdot s \cdot \Delta\phi(s) - \frac{T_{dyno}(s)}{\Theta_2}\right]}$$

This finally yields the transfer function

$$G_{T_e \to T_{shaft}}(s) = \frac{1}{\Theta_1} \cdot \frac{ds + c}{s^2 + \frac{d}{\Theta_{repr}}s + \frac{c}{\Theta_{repr}}} \qquad (2.147)$$

Similarly, the transfer function for T_{dyno} can be obtained. The influence of T_{dyno} on T_{shaft} can usually be neglected for the calculation of the oscillations since even highly dynamic test-bench brakes, with a torque control bandwidth in the range of $100\frac{rad}{s}$, are not able to act on an engine-cycle timescale. Therefore, the dynamometer torque is assumed to be equal to the mean of the engine torque.

The resonance frequency ω_0 of the system (2.147) lies at

$$\omega_0 = \sqrt{\frac{c}{\Theta_{repr}}} = \sqrt{c\left(\frac{1}{\Theta_1} + \frac{1}{\Theta_2}\right)} \tag{2.148}$$

and the damping ratio δ can be calculated as

$$\delta = \frac{d}{2 \cdot \Theta_{repr} \cdot \omega_0} = \frac{d}{2\sqrt{c \cdot \Theta_{repr}}} \tag{2.149}$$

Choosing the damping for the connecting shaft is a difficult task. Figure 2.40 shows the influence of the damping ratio on the frequency-dependent magnitude $\left|G_{T_e \to T_{shaft}}(j\omega)\right|$. The resonance frequency is dominated by the

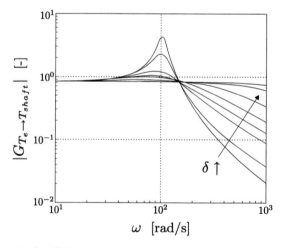

Fig. 2.40. Magnitude of the transfer function from engine torque to shaft torque, $\delta = 0.1, 0.2, 0.5, \frac{\sqrt{2}}{2}, 1, 2, 5, 10$. $\Theta_1 = 0.1 kgm^2$, $\Theta_2 = 0.5 kgm^2$, $c = 833\frac{Nm}{rad}$ (realistic values for a small engine test-bench).

stiffness c of the connecting shaft and the *smaller* of the two inertias (usually the one of the engine), see (2.148). Increasing the damping does not significantly change the resonance frequency but reduces the decoupling of the inertias from the shaft. This lessens the amplitude of the oscillations in the resonance region since they are absorbed by the inertias, which change their speed accordingly. For higher frequencies, however, the variations in the engine torque are transferred to the shaft to a larger extent when higher levels of damping are present.

Since strong damping is difficult to realize and the additional amplitude reduction at high frequencies is desirable, usually a low resonance frequency and a rather low levels of damping are chosen.

The torque excitation introduced by the combustion engine has a characteristic frequency of[36]

$$\omega_{exc} = \frac{\omega_e \cdot n_{cyl}}{N} \tag{2.150}$$

During one engine cycle, n_{cyl} pulses are generated. One cycle corresponds to one revolution for a two-stroke engine ($N = 1$) and two revolutions for a four-stroke engine ($N = 2$). A more detailed description of the torque variations follows in Sect. 3.3.4.

Note that for a two-cylinder engine running at twice the resonance frequency, the failing of one cylinder (misfires, no spark, loss of compression, etc.) for more than a few consecutive cycles is fatal since the system operating at its resonance frequency, almost instantly destroys the connecting shaft.

Dimensioning the shaft requires consideration of the *peak* torque rather than the average one, which is usually specified by the manufacturer. For example, the average torque produced by a $250ccm$ one-cylinder engine running at $p_{me} = 10bar$ is roughly $20Nm$. However, the maximum torque transferred to the shaft is approximately four times larger, namely $80Nm$.

In general, for choosing the damping ratio as well as the size of the shaft, safety margins have to be introduced and close interactions with the manufacturer of the shafts is mandatory. Nevertheless, the above considerations offer a starting point in the setup process for a well-dimensioned engine test-bench.

2.6 Thermal Systems

2.6.1 Introduction

Clearly, the thermal behavior inside the cylinder is of paramount importance for the engine's performance. In steady-state, these effects are already taken into consideration in the engine maps discussed above. The thermodynamic boundary conditions, however, evolve much more slowly such that, from an engine system point of view, these effects must be modeled as dynamic processes [10]. Only with such a model a precise control of the engine's thermal behavior becomes possible (see Chapter 4 and [137], [123]).

Specifically, in this section, the following effects are discussed in more detail:

---

[36] Assuming an even firing order and no simultaneous firings.

- enthalpy present in the engine's exhaust gases;
- thermal behavior of the exhaust manifold;
- simplified engine thermal behavior (one reservoir); and
- detailed engine thermal behavior (several reservoirs and delays).

2.6.2 Engine Exhaust Gas Enthalpy

The engine's exhaust gases carry a substantial part of the fuel's energy. This enthalpy is used in turbocharged engines for boosting and in catalytic converters to maintain the necessary temperature levels. Future engine systems will take advantage of this enthalpy by utilizing heat exchangers to partially recover some of it for various purposes [118].

The temperature of the exhaust gas is primarily a function of the air/fuel ratio and of the operating point's indicated efficiency. Diesel engine exhaust gas is, therefore, usually much cooler (less than $900\ K$) than that of stoichiometrically operated SI engines (up to $1300\ K$). Since several effects (blowdown at exhaust valve opening, heat transfer to the engine block, etc.) influence the exhaust gas temperature, it is quite difficult to accurately predict this variable. Therefore, a combination of measurements (made with standard control settings, say MBT ignition in SI engines) and correcting functions (that model the influence of deviations in control settings) are often used. A different approach would be to carry out a complete thermodynamic cycle simulation which, of course, would be computationally much more demanding.

Returning to the description of the nominal engine-out temperature, the form of the measured map for an SI engine resembles that shown in Fig. 2.41. At very high power levels, cooling is achieved by running the engine in rich conditions, as can be seen in Fig. 2.41 in the upper right-hand corner (close to wide-open throttle). Neglecting these deviations, a qualitative approach would be to assume the engine-out temperature to be an affine function of the engine's mechanical power output.

More precise modeling can be achieved by analyzing the engine-out enthalpy. Knowing the temperature and the mass flow of the exhaust gas, the engine-out enthalphy can be computed since in a first approximation, despite the relatively high temperatures, the exhaust gases can be assumed to be a perfect gas. The enthalpy flow is thus well approximated by

$$\dot{H}_{eg}(t) = c_p \cdot \dot{m}_e(t) \cdot \vartheta_{eg}(t) \qquad (2.151)$$

with a constant specific heat c_p.

In order to formulate an engine-out enthalpy model, it is useful to compute the specific per-cycle enthalpy difference over the engine, divided by the displacement volume V_d, $\frac{\Delta h_{eg}(n, p_{me})}{V_d}$, in each operating point. Figure 2.42 shows this variable for the same engine as that of Fig. 2.41. In this example, $\Delta h_{eg}(n, p_{me})$ can be approximated accurately (the maximum errors in the stoichiometric part are less than 10%) by an affine approach, i.e.,

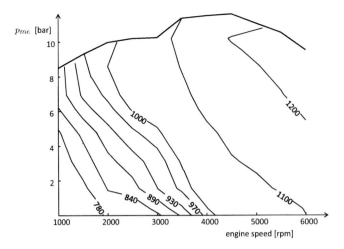

Fig. 2.41. Exhaust gas temperatures ϑ_{eg} (degrees K) for a 3-liter SI engine.

$$\Delta h_{eg}(t) = h_{eg,\omega} \cdot \omega_e(t) + h_{eg,p} \cdot p_{me}(t) + h_{eg,0} \qquad (2.152)$$

where only three constants ($h_{eg,\omega}$, $h_{eg,p}$, $h_{eg,0}$) must be fitted to the engine
to be modeled. This simplification is applicable to other engine classes (see
for instance Fig. 2.47). It reflects the observation that the heat transfer to the
cylinder wall is essentially a function of engine load [97], [86].

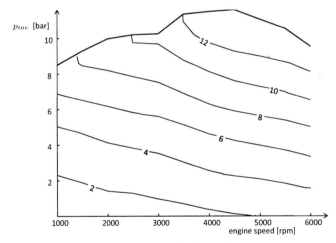

Fig. 2.42. Exhaust gas specific enthalpy $\frac{\Delta h_{eg}(n,p_{me})}{V_d}$ in [bar] for a 3-liter SI engine.

As mentioned above, deviations from the nominal behavior caused by non-
nominal control settings can often be modeled by algebraic correction functions. An example of such an approach is the modeling of the influence of

the ignition angle on the exhaust gas temperature. The nominal engine-out temperature is described by a map $\vartheta_{eg,0}(\omega_e, p_m)$ similar to that shown in Fig. 2.41. The temperature increase $\Delta\vartheta_{eg}(\zeta)$ is related to (2.116). In fact, using the first law of thermodynamics and assuming that the engine can be described using the Willans simplification, the following approximation can be derived

$$\Delta\vartheta_{eg}(\zeta) \approx k_e \cdot \frac{H_m}{c_p} \cdot e_\omega(\omega_e) \cdot k_\zeta \cdot [\zeta - \zeta_0(\omega_e, p_{me})]^2 \qquad (2.153)$$

Here, H_m is the mixture's lower heating value,[37] c_p is the specific heat capacity of the exhaust gas, $k_e \approx 0.5$ is a factor that quantifies how much of the available heat flux is diverted to the exhaust gas,[38] and ζ_0 and k_ζ are parameters that have been defined in (2.116).

2.6.3 Thermal Model of the Exhaust Manifold

The temperatures of the gases in the exhaust manifold can be computed using an approach similar to the one described by the two coupled differential equations in (2.3) and (2.4). The system to be analyzed is shown in Fig. 2.43. The in- and out-flowing exhaust gas masses are given by the engine (see (2.19)) and the downstream elements (modeled for instance as throttles (2.10)), respectively. The only change needed is the inclusion of an additional term in the enthalpy losses, as follows

$$\dot{H}_{out}(t) = \dot{m}_o(t) \cdot c_p \cdot \vartheta_o(t) + \dot{Q}_{in}(t) \qquad (2.154)$$

The heat flow term $\dot{Q}_{in}(t)$ models the heat transfer from the exhaust gases to the exhaust manifold walls. It is assumed that this manifold has a mass m_w, a homogenous temperature ϑ_w, and a constant specific heat c_w, such that a simple enthalpy balance yields the differential equation for the mean wall temperature

$$m_w \cdot c_w \cdot \frac{d}{dt}\vartheta_w(t) = \dot{Q}_{in}(t) - \dot{Q}_{out}(t) \qquad (2.155)$$

The heat flows $\dot{Q}_{in}(t)$ to and $\dot{Q}_{out}(t)$ from the manifold walls contain convection and radiation terms. A detailed modeling of all of these heat transfer terms is beyond the scope of this text. Interested readers can find more information in [23] or in [165]. In the following, just the most basic ideas will be introduced briefly such that the subsequent modeling examples can be followed.

---

[37] The heating value of the mixture is equal to $H_m = H_l/(\lambda \sigma_0 + 1)$ where H_l is the fuel's lower heating value and σ_0 the fuel's stoichiometric constant. For a gasoline/air mixture at $\lambda = 1$ the heating value of the mixture H_m is equal to 2.7 MJ/kg.

[38] The remaining heat flux is carried away by the cooling water.

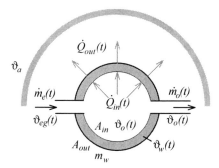

Fig. 2.43. Lumped parameter model of the thermal behavior of a receiver.

The radiation terms are proportional to the difference in the fourth power of the temperatures. Assuming the ambient of the exhaust manifold to be at a temperature ϑ_a that is much smaller than the exhaust manifold wall temperature ϑ_w, this heat flux can be approximated by

$$\dot{Q}_{out,rad}(t) \approx k_{out} \cdot A_{out} \cdot \vartheta_w^4(t) \tag{2.156}$$

The radiation from the ambient to the exhaust manifold and the radiation from the exhaust gas to the manifold walls and vice versa are neglected in this approximation.

The convection terms are linear in the temperature difference, i.e.,

$$\dot{Q}_{in,con}(t) = \alpha_{in} \cdot A_{in} \cdot (\vartheta_o(t) - \vartheta_w(t))$$
$$\dot{Q}_{out,con}(t) = \alpha_{out} \cdot A_{out} \cdot (\vartheta_w(t) - \vartheta_a(t)) \tag{2.157}$$

where the heat transfer coefficients depend upon the flow conditions. A simple approximation of the heat transfer coefficients valid for the conditions encountered in engine systems is given by the following expression

$$\alpha_{in,out} = \begin{cases} 28.6 + 4 \cdot v_g & \text{for } v_g < 5 \ m/s \\ 21 \cdot v_g^{0.52} & \text{for } v_g \geq 5 \ m/s \end{cases} \tag{2.158}$$

where v_g is a representative flow velocity at the interfaces between the exhaust gas and the inner manifold walls and the outer manifold walls and the ambient, respectively. This approximation has been derived from published approaches (see e.g. [101], [133]). The numerical values of the parameters have been adapted to the standard SI engine exhaust manifold conditions.

2.6.4 Simplified Thermal Model

In this subsection, a simple model of the engine's thermal behavior is introduced first, which can be applied, for instance, in combination with the curve shown in Fig. 2.31 to model temperature-dependent engine friction. The starting point is to treat the engine (engine block, cooling water, oil, etc.) in a simplified approach as one single thermal lumped-parameter element. An enthalpy balance over the engine then yields

$$m_e \cdot c_e \cdot \frac{d}{dt} \vartheta_e(t) = \dot{H}_w(t) - \alpha \cdot A \cdot [\vartheta_e(t) - \vartheta_a] \qquad (2.159)$$

where m_e and c_e are the engine's mass and average specific heat capacity, $\dot{H}_w(t)$ represents the heat transfer to the engine block, α the (convective) heat transfer coefficient, A the active heat-exchange area and ϑ_a the ambient air temperature.

The variable $\dot{H}_w(t)$ can be derived using a first-law argument. In fact, once the total fuel enthalpy entering the system (2.60), the engine's actual efficiency $\eta_e(t)$ (using one of the several approaches introduced in Sect. 2.5.1), and the enthalpy exiting the system with the exhaust gases (2.151) are known, the missing information is derived as follows

$$\dot{H}_w(t) = (1 - \eta_e(t)) \cdot H_l \cdot \dot{m}_\varphi(t) - \dot{H}_{eg}(t) \qquad (2.160)$$

Admittedly this model is quite crude due to the simplified heat transfer to the environment, neglected temperature influence on wall heat transfer, etc.. Nevertheless, it yields a useful first estimate of the main thermal effects. For instance, it can be used to estimate the engine's thermal time constant

$$\tau_e = \frac{m_e \cdot c_e}{\alpha \cdot A} \qquad (2.161)$$

which is in the order of several thousand seconds.

An active cooling circuit containing radiator, water pump, and by-pass valve can be included quite easily in this formulation by modeling the radiating surface to be a function $A(u)$ of a control signal u that determines how much of the engine cooling liquid flows through the radiator.

2.6.5 Detailed Thermal Model

Introduction

Based upon the work described in [45], a more detailed model of the engine's thermal behavior including the external heat radiation equipment is developed in this section. Several other publications analyze this problem as well, such as [10] or [189]. In this section, both the external circuit bypass valve and the water pump speed can be used as control quantities, while the engine-in and engine-out water temperature are measured variables, as shown in Fig. 2.44. That system may be subdivided into two main blocks for analysis, as Fig. 2.45 proposes. The inner structure of the block, termed *internal cooling circuit*, is detailed further in Fig. 2.46. The system is approximated using a

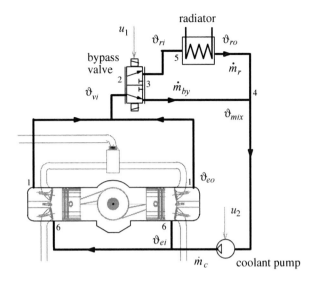

Fig. 2.44. Engine cooling system layout.

lumped parameter approach, i.e., the various temperature distributions are assumed to be represented by a finite number of characteristic temperatures.

The following variables are been used in Figs. 2.44 and 2.45:

u_1 The ratio of the mass flow $\dot{m}_{by}$ that bypasses the radiator and the total coolant mass flow $\dot{m}_c$.

$\dot{m}_r$ The mass flow through the radiator, i.e., $\dot{m}_r = \dot{m}_c - \dot{m}_{by}$.

u_2 The water pump speed, assumed to be proportional to the coolant flow $\dot{m}_c$.

z_1 The heat flux $\dot{Q}_{g,w}$ from the combustion chamber to the cylinder walls (see Fig. 2.46).

z_2 The heat flux to the engine block that is caused by engine friction $\dot{Q}_{if}$ minus the heat flux $\dot{Q}_{eb,a}$, which models the heat losses of the engine block to its environment.

z_3 The heat flux from the radiator to the environment that depends on vehicle speed, ambient temperature, etc.

$\vartheta_{...}$ Several coolant temperatures.

Using these definitions, the mass flows are static multilinear functions of the inputs, i.e.,

$$\dot{m}_r(t) = [1 - u_1(t)] \cdot u_2(t)$$

$$\dot{m}_{by}(t) = u_1(t) \cdot u_2(t) \, . \tag{2.162}$$

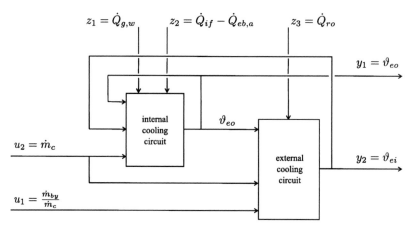

Fig. 2.45. Inputs, outputs, and main blocks of the thermal engine model.

Modeling of the Internal Component

The model as illustrated in Fig. 2.46 comprises three state variables, namely:

ϑ_w the cylinder wall temperature;
ϑ_c the coolant temperature; and
ϑ_{eb} the engine block temperature (including the engine oil).

For each state, an enthalpy balance can be formulated[39] i.e. for the wall

$$\frac{d}{dt}\vartheta_w(t) = \frac{1}{c_w \cdot m_w} \cdot [z_1(t) - \dot{Q}_{w,c}(t)] \qquad (2.163)$$

for the coolant[40]

$$\frac{d}{dt}\vartheta_{eo}(t) = \frac{1}{c_c m_c} \left[c_c \dot{m}_c \{\vartheta_{ei}(t) - \vartheta_{eo}(t)\} + \dot{Q}_{w,c}(t) - \dot{Q}_{c,eb}(t) \right] \qquad (2.164)$$

and, assuming $\vartheta_{eb} \approx \vartheta_{oil}$, for the engine block

$$\frac{d}{dt}\vartheta_{eb}(t) = \frac{1}{c_{eb} \cdot m_{eb} + c_{oil} \cdot m_{oil}} \cdot \left[z_2(t) + \dot{Q}_{c,eb}(t) \right] \qquad (2.165)$$

The specific heat capacities and masses used in (2.163) to (2.165) are attributed to the following sub blocks:

$c_w \cdot m_w$: the cylinder walls;
$c_c \cdot m_c$: the cooling water inside the engine;
$c_{eb} \cdot m_{eb}$: the engine block; and
$c_{oil} \cdot m_{oil}$: the engine oil.

---

[39] There are no mass storage effects in this case, i.e., all fluids are assumed to be incompressible.

[40] As usual, the coolant is assumed to be perfectly mixed, such that the output temperature coincides with the internal temperature.

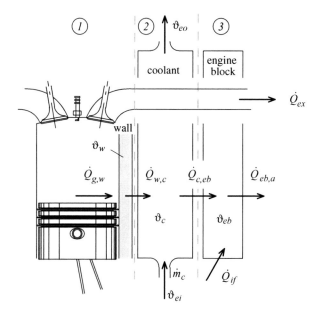

Fig. 2.46. Main heat flows inside the engine.

These parameters must be identified experimentally in order for the engine to be analyzed.

The heat flux $\dot{Q}_{g,w}$ from the combustion chamber to the cylinder wall is a function of several engine parameters, the most important being load p_{me}, speed ω_e, and temperature of the cylinder wall ϑ_w. Since this heat flux is very difficult to measure, it has to be estimated using a thermodynamic process simulation. For the engine described in [82], the results shown in Figs. 2.47 and 2.48 were obtained, as published in [45]. These calculations can be carried out beforehand and the results can be stored in appropriate maps for later use in system simulation and controller design.

A further simplification is indicated in Fig. 2.48. This diagram shows that the cylinder wall temperature influences the heat transfer from the cylinder to the wall almost linearly. Thus, this influence can be included easily as a correcting factor to the nominal case shown in Fig. 2.47.

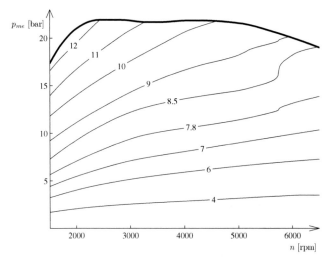

Fig. 2.47. Heat flux from the cylinder to the wall (as mean pressure [bar]) at nominal wall temperature of the engine, as described in [45].

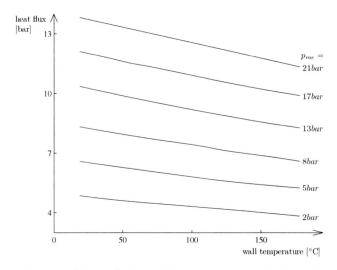

Fig. 2.48. Influence of the cylinder wall temperature on the heat flux from the cylinder to the wall (as mean pressure [bar]), for different loads. Engine speed is $2000rpm$.

As usual, the two internal heat flux values $\dot{Q}_{w,c}$ and $\dot{Q}_{c,eb}$ are modeled using the following formulation

$$\dot{Q}_{w,c}(t) = \alpha_c \cdot A_c \cdot \left[\vartheta_w(t) - \frac{\vartheta_{eo}(t) + \vartheta_{ei}(t)}{2}\right] \qquad (2.166)$$

and

$$\dot{Q}_{c,eb}(t) = \alpha_{eb} \cdot A_{eb} \cdot \left[\frac{\vartheta_{eo}(t) + \vartheta_{ei}(t)}{2} - \vartheta_{eb}(t)\right] \qquad (2.167)$$

where the heat transfer coefficients α_c and α_{eb} can be estimated using (2.158) or experimental data (for other approaches see [45]). The contact areas, A_c between the cylinders and the cooling water, and A_{eb} between the cooling water and the engine block, respectively, can be inferred from the geometrical data of the engine.

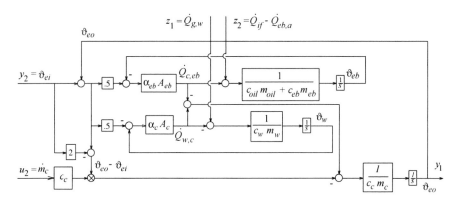

Fig. 2.49. Detailed structure of the engine thermal model.

The heat fluxes that are due to engine friction and to losses to the ambient are the last two unknown variables. The friction term can be derived using (2.112), as follows

$$\dot{Q}_{if}(t) = p_{me0f}(t) \cdot V_d \cdot \frac{\omega_e(t)}{4\,\pi} \qquad (2.168)$$

In (2.168), it is assumed that all of the generated friction heat enters the engine block.

The heat losses to the environment can be modeled using the approach (2.157)

$$\dot{Q}_{eb,a}(t) = A_{eb} \cdot \alpha_{eb,a} \cdot \{\vartheta_{eb}(t) - \vartheta_a(t)\} \qquad (2.169)$$

The heat transfer coefficient $\alpha_{eb,a}$ can be estimated using (2.158) or experimental data. This concludes the modeling of the engine part. The resulting block diagram is shown in Fig. 2.49.

Model of the External Components

The model of the external components (radiator, piping, etc.) has a simpler structure and its dynamics are mainly determined by *time delays*. All connecting pipes are assumed to be adiabatic. Only the radiator exchanges heat with the environment, leading to

$$\frac{d}{dt}\vartheta_{ro}(t) = \frac{1}{c_c \cdot m_r} \cdot [\dot{Q}_{ri}(t) - \dot{Q}_{ro}(t)] \tag{2.170}$$

The heat flux to the radiator is given by

$$\dot{Q}_{ri}(t) = c_c \cdot \dot{m}_r(t) \cdot [\vartheta_{ri}(t) - \vartheta_{ro}(t)] \tag{2.171}$$

where the mass flow $\dot{m}_r$ is a function of the inputs, see (2.162). The heat flux from the radiator to the ambient strongly depends upon the vehicle speed v_v and the ambient temperature ϑ_a

$$\dot{Q}_{ro}(t) = \alpha_r(v_v(t)) \cdot A_r \cdot \left[\frac{1}{2}(\vartheta_{ri}(t) + \vartheta_{ro}(t)) - \vartheta_a(t)\right] \tag{2.172}$$

The heat transfer coefficient can be modeled using (2.158) or more detailed approaches, such as those described in [45] or [22].

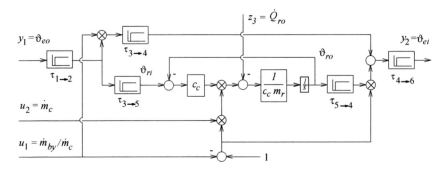

Fig. 2.50. Detailed structure of the model of the external cooling circuit.

From a control engineering point of view, the main dynamic effects are caused by the finite flow velocities in the system's pipes. The flow is assumed to be a perfect *plug flow*, i.e., the time delay $\tau_{i\to j}$ between the junctions i and j is defined by the implicit equation

$$\int_{t-\tau_{i\to j}}^{t} v_{i\to j}(\sigma)\, d\sigma = l_{i\to j} \tag{2.173}$$

where $v_{i\to j}$ is the flow velocity (assuming constant densities it is proportional to the mass flow) and $l_{i\to j}$ is the length of the piping between the junctions i and j as shown in Fig. 2.44.

Assuming that the velocities (control inputs) change slowly compared to the magnitude of the time delays, the variable $\tau_{i \to j}$ can be approximated by

$$\tau_{i \to j} \approx \frac{l_{i \to j}}{v_{i \to j}} \qquad (2.174)$$

Perfect mixing is assumed at junction 4; therefore, the temperature of the water flowing out of that junction can be found using the following relation

$$\vartheta_{mix}(t) = \frac{1}{\dot{m}_c(t)} \cdot [\dot{m}_r(t) \cdot \vartheta_{ro}(t - \tau_{5 \to 4}) + \dot{m}_{by}(t) \cdot \vartheta_{eo}(t - \tau_{1 \to 4})] \qquad (2.175)$$

The resulting detailed block diagram, which corresponds to the *external cooling circuit* block in Fig. 2.45, is shown in Fig. 2.50. (The junction labels used in this illustration were introduced in Fig. 2.44.)

Figure 2.51 shows a comparison of a measured and a simulated step response. As that illustration shows, the model presented above is able to describe the dynamic effects that are relevant for the design and optimization of an engine-in and engine-out temperature controller.

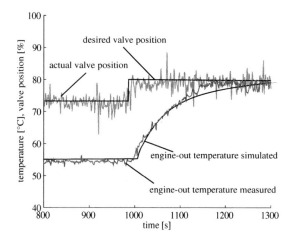

Fig. 2.51. System and model responses to a step change in by-pass valve position.

2.7 Pollutant Formation

2.7.1 Introduction

The formation of pollutants in internal combustion engines is rather difficult to predict theoretically or by numerical simulation. This is primarily due to the fact that these phenomena are governed by the detailed spatial and temporal distribution of the mixture composition, temperature, and pressure inside the combustion chamber.

Control-oriented models, therefore, often rely on the results of experiments which are summarized in appropriate maps. Once these maps are available, slowly varying state variables (engine speed, manifold pressure, EGR rate, etc.) and accessible control variables (injection pressure, ignition timing, etc.) are used to derive the corresponding engine-out pollution values.

The specific form of such maps is not unique. Starting with two-dimensional graphic representations, other forms like polynomial regressions, neural nets, etc. have been proposed in the literature, see, for instance, [11], [14], [219].

All these approaches rely on the fundamental assumption that the pollution formation process depends, in a deterministic way, on the control inputs and on the thermodynamic boundary conditions mentioned above. Of course, this determinism is not entirely true in real situations, such that only average estimations of engine-out pollutant concentration levels are possible. Moreover, aging and other, not easily modeled effects, cause large deviations in pollutant formation such that, in general, prediction errors are quite large. Control-oriented engine-out pollution models are, therefore, often used as a means to predict the *relative* impact of a specific controller structure or algorithm on pollutant emission.

In this section, the most important pollutant formation mechanisms are discussed using *qualitative* arguments. Both SI and Diesel engines are analyzed starting with a discussion of the air/fuel ratio needed for a stoichiometric combustion. As an example, the last part of this section shows a *quantitative* COM of the NO_x formation in homogeneous-charge SI engines.

2.7.2 Stoichiometric Combustion

While the actual reactions are much more complicated and involve a large number of intermediate species, the combustion of hydrocarbon $H_a C_b$ occurs according to the following overall chemical reaction formula

$$1 \text{ mol } H_a\,C_b + c \text{ mol air} \rightarrow d \text{ mol } H_2O + e \text{ mol } CO_2 + f \text{ mol } N_2 \quad (2.176)$$

Assuming the ambient air to consist of 79% N_2 and 21% O_2, the following relations can be derived for a stoichiometric combustion

$$d = \frac{a}{2}, \quad e = b, \quad f = 3.76\,(b + \frac{a}{4}), \quad c = 4.76\,(b + \frac{a}{4}) \quad (2.177)$$

Accordingly, the stoichiometric mass of air needed to burn 1 kg of a hydrocarbon fuel $H_a\,C_b$ is given by the expression

$$m_{air,stoich} = \frac{137\,(1+y/4)}{12+y}\ kg \qquad (2.178)$$

where $y = a/b$ is the fuel's hydrogen-to-carbon ratio.

Figure 2.52 shows the form of function (2.178) and the specific values for three important fuels (the limit values are 11.4 kg of air for pure carbon and 34.3 kg of air for pure hydrogen).

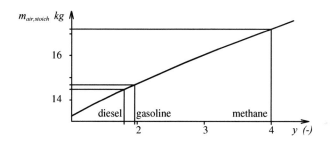

Fig. 2.52. Stoichiometric mass of air required to burn 1 kg of hydrocarbon fuel $H_a\,C_b$, with $y = a/b$ being the fuel's hydrogen-to-carbon ratio.

The complete combustion assumed in (2.176) is not attained even in perfect thermodynamic equilibrium conditions. In reality, traces (a few hundred to several tens of thousands of ppm) of pollutant species are always emitted. The main pollutant species are:

• products of an incomplete oxidation: carbon monoxide CO (with a concentration of typically one order of magnitude larger than those of any other pollutant in the exhaust gases) and hydrocarbon (either unburnt fuel or intermediate species) which — depending on the fuel — can occur as a simple or complex molecular species including cancer-causing components; and

• products of an unwanted oxidation: mostly nitric oxide, NO, but also some NO_2.

In diffusion-flame combustion systems (Diesel and stratified charge GDI engines) a substantial number of particulates is generated as well (sized from a few tens of nanometers to several hundred microns in size). In order to better understand the structural relations in pollutant formation, the basic mechanisms are briefly reviewed in section 2.7.4. For a more detailed discussion, see [97].

2.7.3 Non-Stoichiometric Combustion

For a stoichiometric air-to-fuel ratio the equilibrium of the reaction (2.176) is approximately reached after a catalytic converter. However, for non-stoichiometric combustion other chemical species have to be considered, i.e., CO, NO, NO_2, N_2O, H_2 and C_xH_y. For these species the demand for oxygen and its availability are listed in Table 2.3.

Table 2.3. Oxygen demand and availability of species present in the gas path of an engine.

species	oxygen demand of contained C and H	oxygen availability
CO	2	1
CO_2	2	2
H_2	1	0
C_xH_y	$2x + \frac{y}{2}$	0
H_2O	1	1
O_2	0	2
NO	0	1
NO_2	0	2
N_2O	0	1

Balancing all oxygen demands and availabilities of the components in the exhaust gas, the following equation for the air-to-fuel ratio λ is derived:

$$\lambda \cdot \sum O_{demanded} = \sum O_{available} \tag{2.179}$$

with the sums being derived directly from the oxygen demands and availabilities in Table 2.3:

$$\sum O_{demanded} = 2[\text{CO}] + 2[\text{CO}_2] + [\text{H}_2] + \dots$$
$$\sum \left(2x + \frac{y}{2}\right)[\text{C}_x\text{H}_y] + [\text{H}_2\text{O}] \tag{2.180}$$

$$\sum O_{available} = [\text{CO}] + 2[\text{CO}_2] + [\text{H}_2\text{O}] + 2[\text{O}_2] + \dots$$
$$[\text{NO}] + 2[\text{NO}_2] + [\text{N}_2\text{O}] \tag{2.181}$$

The molecule C_xH_y describes any unburnt hydrocarbon in the gas mixture. Note that the list of species considered is not exhaustive and has to be extended if other components, e.g. NH_3, arise in significant concentrations. Reorganizing (2.179) yields λ as a function of measured exhaust gas concentrations:

$$\lambda = \frac{\sum O_{available}}{\sum O_{demanded}} \qquad (2.182)$$

In [31] a slightly simplified version of this equation is introduced, which is later refined in [32].[41]

Since (2.182) is derived from simple chemical equations, it is valid throughout the full gas path of an engine, i.e., it holds for the unburnt mixture as well as for the exhaust gas upstream or downstream of the catalytic converter. In the first case (unburnt mixture), the fuel is assumed to be fully vaporized and the gas concentrations have to be calculated accordingly.

If the fuel itself contains oxygen, as is the case for ethanol, the equations have to be modified. By using the fuel's oxygen-to-carbon ratio $\zeta_{O/C}$ and assuming that all carbon comes from the fuel,[42] (2.182) becomes

$$\lambda = \frac{\sum O_{available} - [C] \cdot \zeta_{O/C}}{\sum O_{demanded} - [C] \cdot \zeta_{O/C}} \qquad (2.183)$$

Note that the oxygen in the fuel is also accounted for in the sum of available oxygen. To illustrate the calculation, two examples are discussed next.

Example 1: Determine the air-to-fuel ratio for the following (fictitious) chemical reaction in every step:

$$\underbrace{CH_4 + 3O_2}_{A} \rightarrow \underbrace{CO + H_2O + H_2 + 2O_2}_{B} \rightarrow \underbrace{CO_2 + 2H_2O + O_2}_{C} \qquad (2.184)$$

The oxygen-to-carbon ratio $\zeta_{O/C}$ of methane CH_4 is zero (it does not contain any oxygen). Calculating λ for all three steps A, B, and C by applying equation (2.182) therefore yields

$$\lambda = \underbrace{\frac{3 \cdot 2}{1 \cdot \left(2 + \frac{4}{2}\right)}}_{A} = \underbrace{\frac{1 \cdot 1 + 1 \cdot 1 + 2 \cdot 2}{1 \cdot 2 + 1 \cdot 1 + 1 \cdot 1}}_{B} = \underbrace{\frac{1 \cdot 2 + 2 \cdot 1 + 1 \cdot 2}{1 \cdot 2 + 2 \cdot 1}}_{C} = \frac{6}{4} = 1.5$$

This result of course agrees with the fact that completely oxidizing one mole of methane requires two moles of oxygen, but there are three available (and thus $\lambda = \frac{3}{2} = 1.5$).

Example 2: Determine the air-to-fuel ratio for the following reaction, this time for a fuel containing oxygen:

$$CH_3OH + \frac{5}{2}O_2 \rightarrow CO_2 + 2H_2O + O_2 \qquad (2.185)$$

---

[41] This publication extends the concept presented here to sulfur-containing fuels and lubrication oil being burned with the fuel. Furthermore, it considers particulate matter as well as additional species in the exhaust gas.

[42] The mass fraction of CO_2 in air amounts to approximately 0.06% and can therefore be neglected without introducing any significant error.

Methanol CH_3OH has an oxygen-to-carbon ratio of 1. Therefore, (2.183) has to be evaluated for both sides of the reaction:

$$\lambda = \frac{1 \cdot 1 + \frac{5}{2} \cdot 2 - 1 \cdot 1}{1 \cdot \left(2 + \frac{4}{2}\right) - 1 \cdot 1} = \frac{1 \cdot 2 + 2 \cdot 1 + 1 \cdot 2 - 1 \cdot 1}{1 \cdot 2 + 2 \cdot 1 - 1 \cdot 1} = \frac{5}{3}$$

Again, this complies with the consideration that the complete oxidation of one mole of methanol uses $\frac{3}{2}$ moles of oxygen but $\frac{5}{2}$ are at hand. Note that the exhaust gas, although having the same composition as in Example 1, shows a different value for λ because of the oxygen-containing fuel.

2.7.4 Pollutant Formation in SI Engines

Figure 2.53 shows typical pollutant concentrations in the exhaust gas of a port-injected SI engine as a function of the normalized air/fuel ratio λ. *Qualitative explanations for the shape of the curves shown in this figure are given below.*

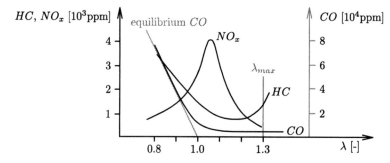

Fig. 2.53. Typical engine-out pollutant emission of a port-injected SI engine, with λ_{max} being the lean combustion limit where misfires start (notice the different scales for CO and the other species).

Hydrocarbon and Carbon Monoxide

In rich conditions, the lack of oxygen produces very high HC and CO concentrations. When the engine is operated under extremely lean conditions, misfires start to occur which, of course, cause increased HC emission. The misfires also reduce in-cylinder and exhaust temperatures, which lead to less post-combustion oxidation. Of course, the amount of CO does not increase due to misfires (when the mixture fails to ignite, no CO can be formed).

The reason that at intermediate air/fuel ratios the HC and CO concentrations do not reach the (almost) zero equilibrium levels can be explained as follows:

- Due to the quick cooling of the cylinder's charge during the expansion stroke, the necessary time required to reach chemical equilibrium is not available (high CO and HC concentrations are "frozen" shortly after their formation during combustion).
- During the compression stroke, the mixture is forced into several crevices present in the cylinder, and part of the mixture is absorbed into the oil film covering the cylinder walls (Fig. 2.54). As soon as the pressure decreases during the expansion stroke (and after combustion has terminated), these HC molecules are released into the cylinder. Since cylinder charge temperatures have already fallen substantially, a non-negligible portion of these HC molecules escapes post-flame oxidation.
- When the flame front reaches the proximity of the cylinder walls (a few $0.1\ mm$ distance), large heat losses to the wall quench the flame before all the mixture has been oxidized.

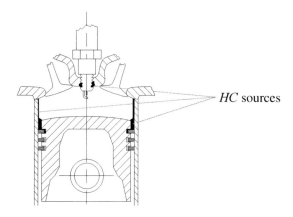

HC sources

Fig. 2.54. Crevices and other sources of engine HC emissions.

Since a substantial part of the HC and CO is always oxidized in the post-flame phase inside the cylinder, and even during blow-down in the exhaust manifold, all effects that influence the conditions for this phenomenon (available oxygen, temperature, time, etc.) also affect the engine-out emission of HC and CO.

The pollutant formation in direct-injection SI engines in homogeneous mode is essentially the same as in port-injection engines. In stratified mode, however, substantial differences appear (for instance, the late injection may eliminate the crevice losses). Figure 2.55 shows that this does not necessarily lead to reduced pollutant levels. In this figure, HC emission are plotted as a function of ignition timing. While the port-injected engine has decreasing HC levels due to later ignition (the thermodynamic efficiency decreases, hence the burnt gases are hotter, which improves post-combustion oxidation), the

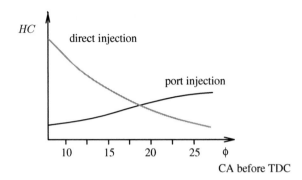

Fig. 2.55. Hydrocarbon emission of a port-injection and a direct-injection stratified-charge SI engine. The first engine operates at $\lambda = 1$ and at MBT ignition of 25 degrees BTDC. The second engine operates at $\lambda = 2$, see [197].

direct-injection engine displays the inverse behavior. This shows that in direct-injection engines the ignition point may not be shifted without an appropriate change in injection parameters.

Note that in modern engines "blow-by" gases, i.e., that part of the mixture that enters the crank shaft compartment due to leakages of the piston rings, are collected and fed back to the combustion process.

Nitrogen Oxide

Engine-out NO_x gases (mostly NO, but also some NO_2) are formed during the peak-temperature phases, i.e., when the burnt gas in the cylinder reaches temperatures of more than 2000 K. The three variables that have the greatest influence on that process are *oxygen availability, cylinder pressure* and, most importantly, *burnt gas temperature.*

All of these parameters define the *equilibrium* concentration of NO shown in Fig. 2.56.[43] Clearly, the gas temperature has a dominant influence on the NO concentration. The influence of the oxygen availability also seems strong, but Fig. 2.56 is somewhat misleading in that aspect. In fact, for a *constant* amount of fuel, a larger air/fuel ratio implies cooler burnt-gas temperatures and, hence, smaller NO concentrations. Finally, the influence of the gas pressure is noticeable, but clearly less important than the contributions of the previous two variables.

The NO concentrations shown in Fig. 2.56 are reached under equilibrium conditions, i.e., after an infinitely long period. However, the reciprocating behavior of the engine limits the time available for these reactions to take place. Therefore, the *rate of formation* of NO has to be taken into account. The equilibrium conditions are reached only if the associated characteristic time constants τ_r of these reaction kinetics are much smaller than the time

---
[43] The data and figures shown in this section have been taken from [29]

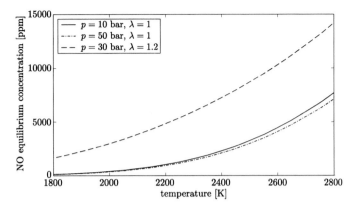

Fig. 2.56. Equilibrium NO concentration as a function of burnt-gas temperature (main parameter), oxygen availability and cylinder pressure.

available during the expansion stroke (for a precise definition of τ_r see [97]). As shown in Fig. 2.57, this is only the case for temperatures much greater 2000 K, irrespective of the actual air/fuel ratios, cylinder pressures and engine speeds.

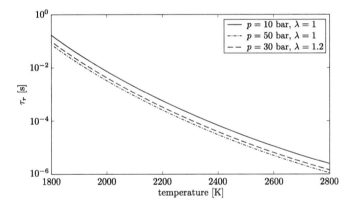

Fig. 2.57. Characteristic time constants of the NO formation rate in burnt gases.

This dependency of the reaction rate on the gas temperature is the main cause of the large NO concentrations found in the exhaust gas. In fact, due to the short interval available in the expansion phase, the NO rapidly formed in the high-temperature phases around TDC are not reduced to the equilibrium values that are expected at the cooler temperatures prevailing before the exhaust valve opens. As Fig. 2.57 shows, all reactions are essentially "frozen" below 2000 K.

An exact modeling of these phenomena requires a thermodynamic process simulation based on the energy conservation laws. The heat release due to the combustion, the piston work, and the heat transfer through the cylinder walls has to be considered together with the NO formation laws in a crank-angle based formulation [97]. The results of such a model yield the profiles of the transient temperature and NO concentration shown in Fig. 2.58. Unfortunately, the numerical solution of the system of ODE that describes that process requires substantial computational power such that real-time applications are not yet possible with standard ECUs.[44]

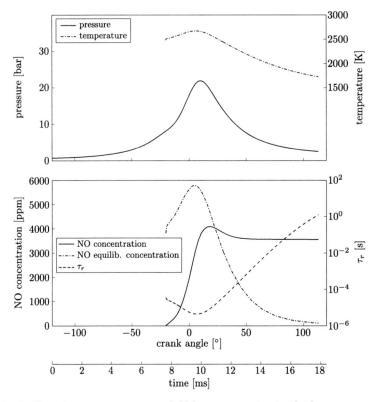

Fig. 2.58. Transient temperature and NO concentration in the burnt-gas zone. SI engine speed and load at $n = 2000$ rpm and $p_m = 0.5$ bar, respectively.

The basic NO formation mechanisms explained above indicate that the engine-out NO emissions depend on almost all engine-system variables such as engine speed and load, air/fuel ratio, spark timing, etc. Since reaction rates depend exponentially on the temperature of the burnt gas, small changes in the engine control variables that affect this signal have dramatic effects on the engine-out levels of NO. The temperature of the burnt gases depends on many

---

[44] In [4], a simplified approach is presented that permits the explicit solution of the ODE being avoided.

parameters. The highest temperatures are obtained for close-to-stoichiometric mixtures, but since oxygen availability is also important for the formation of NO, maximum values of NO concentrations are usually obtained in slightly lean conditions (see Fig. 2.53).

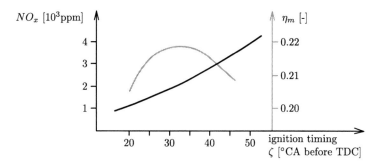

Fig. 2.59. Example of the engine-out NO emission and of the engine efficiency as a function of ignition timing (port-injected SI engine, medium load and speed).

Another important parameter is ignition timing. Shifting ignition timing toward early ignition typically improves engine efficiency. Unfortunately, this comes at the price of an increased NO formation rate, as shown in Fig. 2.59. Again, this is easy to see from the previous discussion. In fact, earlier ignition causes higher peak pressures and, hence, higher burnt-gas temperatures.

Finally, exhaust gas recirculation (EGR) has an important impact on the formation of NO as well. External EGR increases the mass of the cylinder charge and therefore acts as "thermal inertia." This additional mass reduces peak temperatures (especially if the external EGR has been cooled) and, hence, the formation of NO. External EGR is limited, however. In port-injected SI engines, this limit is approximately 15%; otherwise misfiring and unacceptably high levels of HC emission occur.

Summary

In summary, at least the following engine and control parameters affect the pollution formation of port-injected SI engines:

- Air/fuel ratio — most important parameter, strongly affects all pollutants.
- Engine speed — determines available reaction times and, hence, the formation and reduction of all pollutants.
- Engine load — determines peak and post-combustion oxidation temperatures and, hence, all pollutant species.
- External EGR rate — affects peak temperatures and, hence, especially NO_x formation.

- Ignition timing — determines peak temperatures and, hence, especially NO_x, but also has an impact on post-combustion temperatures and, hence, on HC.
- Engine temperature — influences the post-flame oxidation of CO, but also NO_x formation (this temperature is not the oil or the cooling water temperature, but a fictitious characteristic mean temperature of cylinder wall and head).

In direct-injection engines operated in stratified charge mode, several other influencing parameters must be included, the most important being injection timing (start and end of the multiple-injection events). Note that in stratified mode the engines are operated at very lean air/fuel ratios — well beyond the ignition limits of homogeneous charge SI engines ($\lambda = 2$ and higher are possible). This, of course, leads to lower engine-out NO_x levels (the cooling effect dominates the higher availability of oxygen).

HC levels, however, can be much larger (by a factor of 2-3). Particulate matter, which in homogeneous-charge SI engines is not an issue, is a problem in stratified-charge operation as well. Both effects are caused by the unstable flame structure at the very lean borders of the spray.

So far, warmed engines which are operated under stationary conditions have been assumed. Unfortunately, during cold start, engine-out emission can be much higher. This is especially the case for HC, which, due to the not-yet warmed cylinder and exhaust ports, are not oxidized after formation.

Additionally, deterioration of engine-out pollution levels is observed during transients. Without precise compensation of the dynamic variables involved (wall film mass, temperatures, speeds, ignition timing, etc.) unfavorable operating conditions can be imposed on the engine with the corresponding increase in pollution levels. These problems will be discussed in Chapter 3.

2.7.5 Pollutant Formation in Diesel Engines

Diesel engines have relatively low engine-out emissions. In particular, hydrocarbon and carbon monoxide can usually be neglected. The main pollutant species in the exhaust gas are nitrogen oxide NO_x and particulate matter (PM), see Fig. 2.60. As in SI engines, the air/fuel ratio is one primary factor that affects the formation of pollutants. Since Diesel engines are load-controlled by variations of the air/fuel ratio, this parameter plays an even more important role here than for homogeneous-charge SI engines. Other important parameters are injection timing and pressure, as well as the EGR rate.

Figure 2.60 shows the typical relation between air/fuel ratio and engine-out pollutants. As Diesel engines are always operated under lean conditions,[45] CO

---

[45] For the regeneration of lean NOx-trap catalytic converters, a stoichiometric-to-rich operation has been proposed, realized, for instance, by a late additional injection in common-rail systems.

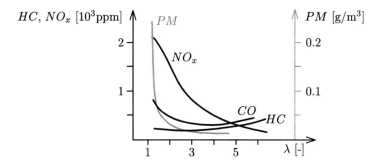

Fig. 2.60. Engine-out emission of Nitrogen oxide NO_x, hydrocarbon HC, and particulate matter PM of a direct-injection Diesel engine as a function of air/fuel ratio.

and HC emission are relatively low. At very high air/fuel ratios, i.e., at very low loads, combustion is no longer complete (low temperatures, too lean local air/fuel mixture composition, etc.) and part of the fuel enters the cylinder too late to be burned. At very high loads, regions with insufficient oxygen supply exist, which leads to the increased emission of CO.

At a first glance, Diesel engine NO_x emission levels are surprisingly high. In fact, the extrapolation of the NO_x levels of a homogeneous-charge SI engine to $\lambda \approx 2$ (see Fig. 2.53) predicts levels of less than 100 ppm. As Fig. 2.60 shows, the measured values are one order of magnitude higher than that.

This apparent contradiction can be resolved by the following argument. As Fig. 2.61 shows, the diffusion combustion process in a direct-injection Diesel engine, by its nature, is inhomogeneous. Global air/fuel ratios are not representative of the local air/fuel ratio at which the actual combustion process takes place. Locally, very rich to pure air zones can be found. As a consequence, during inhomogeneous combustion, there are always some regions where combustion takes place at conditions favorable for NO_x formation. Once this species has been formed, the charge cools down so rapidly that the reverse reaction becomes "frozen."

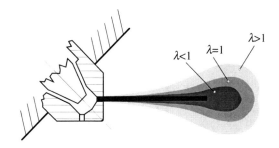

Fig. 2.61. Schematic representation of local air/fuel ratio distribution in the combustion chamber of a Diesel engine.

The inhomogeneity of the air/fuel mixture is also the primary cause of the formation of particulate matter. When the outer layer of a fuel droplet begins to burn, its core becomes so hot that cracking processes start. The shorter hydrocarbon chains produced by this cracking are not easily inflammable such that some escape combustion and appear at the exhaust port as soot particles. Typically, some hydrocarbon compounds are absorbed in these particles. The prediction of the actual PM emission is very difficult because a large amount is first formed and later burnt during the combustion process, see [181]. The actual engine-out emission, therefore, is the difference between two large numbers, such that small changes in these numbers can have substantial effects on the net difference.

2.7.6 Control-Oriented NO Model

In this section, an example of a COM of the NO emissions of an SI engine is described in some detail. The emphasis is not on the well-known basic NO formation laws that have been briefly described above, but on models useful for real-time applications where the computational burden must be limited to the bare minimum.

In this example, it is assumed that only four influencing parameters must be included in the COM, $viz.$, the manifold pressure p_m, the engine speed ω_e, the air/fuel ratio λ, and the ignition angle ζ. Since these four variables are mutually independent, the choice of 20 support points for each variable results in $20^4 = 1.6 \cdot 10^5$ measurements necessary to find the complete four-dimensional "map" that describes the NO engine-out emissions. Assuming one minute per measurement would lead to 110 consecutive days of engine testing being required to obtain this data. Clearly, this straightforward approach is not feasible in practice.[46]

Instead, an approximation of the following form can be used to estimate the concentrations of the engine-out NO emissions

$$[NO](\omega_e, p_m, \lambda, \zeta) \approx [NO]_0(\omega_e, p_m, \lambda_0, \zeta_0) \cdot \qquad (2.186)$$

$$(1 + \Delta_\lambda(\omega_e, p_m, \lambda - \lambda_0) + \Delta_\zeta(\omega_e, p_m, \zeta - \zeta_0))$$

where the *base map* $[NO]_0(\omega_e, p_m, \lambda_0, \zeta_0)$ is measured using the nominal values λ_0 and ζ_0. The two correcting functions $\Delta_\lambda(\omega_e, p_m, \lambda - \lambda_0)$ and $\Delta_\zeta(\omega_e, p_m, \zeta - \zeta_0)$ are determined using process simulations as described above.[47] The advantage of measuring such a base map, an example of which

---

[46] In reality, more parameters such as engine temperature, EGR ratio, etc., must be included, making such an approach even less likely to succeed.

[47] Only 400 measurements are needed in this case. If, instead of a process simulation, actual measurements are made, the number of experiments is $2 \cdot 20^3 = 1.6 \cdot 10^4$, i.e., one order of magnitude less than the full mapping approach described above.

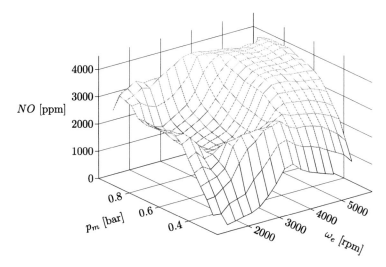

Fig. 2.62. Measured base map of engine-out NO emissions (3.2-liter V6 SI engine).

is shown in Fig. 2.62, is that all subsequent computations can be validated at nominal operating points.

The structure of $[NO]_\lambda(\omega_e, p_m, \lambda - \lambda_0)$ and $[NO]_\zeta(\omega_e, p_m, \zeta - \zeta_0)$ can be derived using the underlying first laws of physics. For the air/fuel ratio, a parabolic function with its maximum at slightly lean conditions follows from the equilibrium values shown in Fig. 2.56 and the comments made in the respective text (see Fig. 2.63).

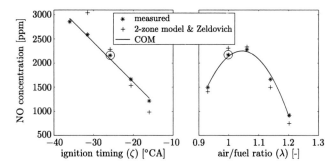

Fig. 2.63. Example of the engine-out NO emission for variations in ignition timing (nominal value $\zeta_0 = -25°$) and air/fuel ratio (nominal value $\lambda_0 = 1$).

Placing the ignition at an early angle causes the temperature to increase during the combustion and, hence, increases the level of engine-out NO. Interestingly, as Fig. 2.63 shows, this increase is approximately linear rather than exponential, as would be expected from the equilibrium values of Fig. 2.56

considering *peak* temperatures only. This apparent contradiction can be re-solved with the illustration shown in Fig. 2.57. In fact, at sufficiently high temperatures the reaction time constants are so small that the NO levels in the burnt gases are close to their equilibrium value even during the first part of the expansion phase. However, in this phase the temperature of the burnt gases decreases and the equilibrium levels of NO follow that trend. Once the limit of approximately 2000 K is reached, the reactions become so slow that almost no further NO reduction takes place. This results in the engine-out NO levels observed during the experiments.

These observations form the basis for the formulation of the structure of the correcting functions in (2.186)

$$\Delta_\lambda(\omega_e, p_m, \lambda - \lambda_0) = k_{\lambda,2} \cdot (\lambda - \lambda_0)^2 + k_{\lambda,1} \cdot (\lambda - \lambda_0) \qquad (2.187)$$
$$\Delta_\zeta(\omega_e, p_m, \zeta - \zeta_0) = k_{\zeta,1} \cdot (\zeta - \zeta_0) \qquad (2.188)$$

The coefficients $k_{\lambda,i}, k_{\zeta,i}$, which depend on the engine operating point, are chosen such that they minimize the error between the data obtained with the thermodynamic process simulation or measurements and the prediction of the COM (2.186). A least-squares approach proves to be useful for this minimization process, as shown in Appendix A.

Table 2.4 shows an example of the NO concentrations resulting from variations of the air/fuel ratio and spark advance . The symbol λ_- designates a richer mixture than λ_0, while λ_+ indicates a leaner mixture. Similarly, the symbol $\theta_{SA,-}$ indicates an earlier spark angle than $\theta_{SA,0}$, while $\theta_{SA,+}$ denotes a later spark angle. The concentrations in each field of Table 2.4 are shown in *ppm*. The first value in each field is the actual measured value. The second value is the NO concentration obtained using the detailed, but computationally demanding two-zone process simulation. Finally, the third value is calculated using the COM proposed in this section.

Since the measured value in the center field (2162 *ppm*) is used to calibrate both the detailed and the control-oriented model, the agreement shown in that field is to be expected. The values shown in the center row of the other eight fields are calculated using the two-zone model, without changing any of its calibration parameters. The measurement data shown in the fields (1,2), (2,1), (2,3), and (3,2) are used to calibrate the additional parameters of the COM. Finally, the fields (1,1), (1,3), (3,1) and (3,3) contain the predictions obtained without changing the parameters of the COM model.

As Table 2.4 shows, neither the detailed model nor the COM is able to precisely predict the measured NO concentrations. In most cases, however, the simplified model yields the correct trend. As mentioned in the introduction to this section, this is the best that can be expected from COM of the pollutant formation.

Note that (2.186) quantifies the *concentration* of the pollutants in the exhaust gas, i.e., the number of NO molecules relative to the total number of parts in the exhaust (usually *mol/mol*). However, the legislations of most

Table 2.4. Comparison of measured and modeled NO concentrations for varying air/fuel ratio and spark advance.

λ	$\phi_{SA} = -34°$			$\phi_{SA} = -26°$			$\phi_{SA} = -18°$		
$\downarrow$	meas.	calc.	COM	meas.	calc.	COM	meas.	calc.	COM
0.9	1378	1396	1677	1495	1409	1591	884	620	584
1.0	2857	3836	3018	2162	2285	2162	1217	985	1362
1.1	2773	2948	2629	1661	1500	1830	1044	872	1487

NO concentrations are in [ppm], the operating-point is given by the engine speed of $1910 rpm$ and the intake-manifold pressure of $0.5 bar$. The used abbreviations are listed below:

meas. = measured values;

calc. = calculated by 2-zone model and Zeldovich NO-formation
mechanism; and

COM = results of the control-oriented model described in (2.186).

countries limit the total *mass* of each pollutant species emitted during the corresponding test cycles.

To obtain this total mass the total engine mass flow, (2.19) has to be combined with the NO_x concentration. For small NO_x concentrations, the following approximation can be used to compute the total engine-out pollutant mass flow

$$\dot{m}_{NO_x}(t) = [NO_x](t) \cdot \frac{M_{NO_x}}{M_{eg}} \cdot \dot{m}(t) \qquad (2.189)$$

where $\dot{m}(t)$ is the total engine mass flow (2.19), $[NO_x](t)$ the concentration of NO_x in the exhaust gas, and M_{eg} and $M_{NO_x^*}$ designate the molar mass of the exhaust gas and of the NO_x gas. Since more than 90% of the NO_x is NO, a value of $M_{NO_x} \approx 30 \ kg/kMol$ is often used. For stoichiometric gasoline mixtures, a value of $M_{eg} \approx 29 \ kg/kMol$ is a reasonable estimate. Integrating the variable (2.189) during the entire test cycle yields the total mass of NO_x emitted.

Note that certification measurements must follow precise procedures that often utilize dilution tunnels and fill bags to collect all tail pipe emissions. These procedures are described in detail in [8].

2.8 Pollutant Abatement Systems

2.8.1 Introduction

Since 1980, the permitted levels for the concentrations of carbon monoxide CO, hydrocarbons HC, and NO_x (NO, NO_2) have been reduced significantly. A continuation of this trend is to be expected. Table 2.5 shows the development of these limits in Europe and the similar standards for California.

Similar legislations are imposed in Japan and other countries. In contrast to the European regulations, which limit the total hydrocarbon emissions, the US regulations consider the non-methane hydrocarbons only. This choice reflects the fact that the methane emissions by passenger cars are very low compared to the agricultural emissions. However, the US standards regulate formaldehyde (HCHO) levels, which are not explicitly limited in the European Union.

Common to all of these emission limits is the fact that they may not be exceeded during a predefined driving cycle, comprising a warm-up phase, transients and idle periods. Also, the boundary conditions (ambient temperature and pressure, etc.) are prescribed accurately.

As discussed in the previous section, the oncoming, ever more stringent emission regulations are difficult to meet. Only homogeneous charge SI engines operated with a stoichiometric air-to-fuel ratio λ can attain easily the required emission levels by using three-way catalytic converters. All other engines (Diesel, direct-injection SI engines, etc.) must be combined with more complex exhaust gas aftertreatment systems such as lean NO_x traps, selective catalytic reduction systems, and filters for particulates.

In this section, three-way catalytic converters and selective catalytic reduction converters are analyzed from a control engineering point of view. Readers interested in lean-NO_x traps are referred to [126], [210], [111] [199], and [115] and details on particulate filters may be found in [121], [122] and [116].

2.8.2 Three-Way Catalytic Converters, Basic Principles

The most common pollution abatement system for SI port-injection engines is the three-way catalytic converter (TWC). It derives its name from its ability to simultaneously reduce NO_x and oxidize CO and HC. State-of-the-art systems are capable of removing more than 98% of the pollutants. This can only be achieved by operating the engine within very narrow air-to-fuel ratio limits, as will be shown in the following subsections.

An ideal TWC enhances the following three elementary chemical reactions

$$NO + CO \rightarrow \tfrac{1}{2}N_2 + CO_2 \tag{2.190}$$

$$CO + \tfrac{1}{2}O_2 \rightarrow CO_2 \tag{2.191}$$

$$H_aC_b + (b + \frac{a}{4})O_2 \rightarrow bCO_2 + \frac{a}{2}H_2O \tag{2.192}$$

In reality, the system of reactions is far more complex [106]. Several dozen reactions are involved, and other species, apart from those mentioned, are produced (mostly during intermediate steps). For example, the concentration of N_2O is often increased by the reactions taking place inside the TWC.

Generally, the reactants are first adsorbed on the catalytic surface. Catalytic substances, such as platinum (Pt), rhodium (Rh) or palladium (Pd), weaken the bonds of the adsorbed species and thereby allow for the formation

Table 2.5. European (top) and Californian (bottom) emission standards.

Tier		Euro I	Euro II	Euro III	Euro IV	Euro V	Euro VI
Date		1992	1996	2000	2005	2009	2014
SI Engines							
CO	[g/km]	2.72	2.2	2.3	1.0	1.0	1.0
HC	[g/km]			0.2	0.1	0.1	0.1
NO_x	[g/km]			0.15	0.08	0.06	0.06
$HC + NO_x$	[g/km]	0.97	0.5				
PM	[g/km]					$0.005^{(i)}$	$0.005^{(i)}$
Diesel Engines							
CO	[g/km]	3.16	1.0	0.64	0.5	0.5	0.5
HC	[g/km]						
NO_x	[g/km]			0.5	0.25	0.18	0.08
$HC + NO_x$	[g/km]	1.13	0.7	0.56	0.3	0.23	0.17
PM	[g/km]	0.18	0.08	0.05	0.025	0.005	0.005
Heavy-Duty Diesel Engines							
CO	[g/kWh]	4.5	4.0	2.1	1.5	1.5	1.5
HC	[g/kWh]	1.1	1.1	0.66	0.46	0.46	0.13
NO_x	[g/kWh]	8.0	7.0	5.0	3.5	2.0	0.4
PM	[g/kWh]	0.36	0.15	0.1	0.02	0.02	0.01
Smoke	$[\text{m}^{-1}]$			0.8	0.5	0.5	

Emission classes[(ii)]		LEV	ULEV	SULEV	ZEV
Passenger Cars					
CO	[g/mile]	4.2	2.1	1.0	0
$NMOG$	[g/mile]	0.09	0.055	0.01	0
$HCOH$	[g/mile]	0.018	0.011	0.004	0
NO_x	[g/mile]	0.07	0.07	0.02	0
PM	[g/mile]	0.01	0.01	0.01	0

[(i)] applicable only to vehicles using direct injection
[(ii)] LEV II introduced since 2004

of the desired products, according to the law of minimizing the (chemical) potential energy. The products are then desorbed and released to the gas phase.

Reactions (2.190) to (2.192) show that optimal conversion rates can only be achieved if exactly as much oxygen is available as is required for the oxidation of CO and HC. It would be favorable if (2.190) was dominant. In fact, in the presence of excess oxygen ($\lambda > 1$), reaction (2.191) always occurs in parallel to the others and therefore, under lean conditions, usually inhibits the reduction of NO. During a shortage of oxygen ($\lambda < 1$), all NO is converted,

but the removal of HC and CO is incomplete. This implies that only operation in a narrow band around $\lambda = 1$ yields a satisfactory performance (i.e.,
long-term efficiency) of the TWC. Figure 2.64 illustrates this by showing the
gas concentrations of various species in the feedgas and exit of a TWC as a
function of λ. (The upstream gas concentrations in Fig. 2.64 correspond to
those shown in Fig. 2.53.)

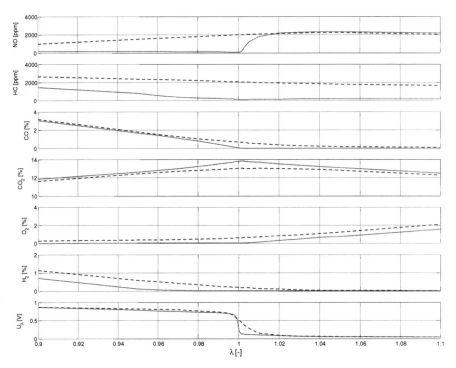

Fig. 2.64. Concentrations of NO, HC, CO, CO_2, O_2, H_2, and the switch-type λ-
sensor voltage; upstream (dashed) and downstream (solid) of an aged three-way
catalytic converter.

Especially during transients, when the wall-wetting dynamics and other
disturbances render a complete compensation of the varying air mass flow
difficult, keeping λ within this band is not practicable. Therefore, the catalyst
must be able to cope with temporary excursions to the lean or to the rich side.
This is achieved by incorporating a reservoir for oxygen, which lies beneath
the catalytically active surface and which consists of special materials, e.g.,
cerium. Excess oxygen can then temporarily be stored in these materials as
well as on the catalytic surface. The oxygen storage within cerium can be
described by the following chemical reaction

$$Ce_2O_3 + O^* \rightleftharpoons 2\,CeO_2 \tag{2.193}$$

The adsorbed oxygen O^* combines with Ce_2O_3 to form a new grid structure. Note that the adsorbed oxygen may also stem from oxygen-containing species other than O_2, such as NO, H_2O, or CO_2. The influence of H_2O and CO_2 acting as oxidizing species on the dynamic behavior of the oxygen storage mechanism is discussed in Section 2.8.3. Of course, the catalyst can only compensate for lean excursions if its oxygen storage is not full already, i.e., if not all ceria has been oxidized to $2\,CeO_2$ yet. The same holds for temporary deviations towards the rich region in the case of a completely empty (or more precisely, "reduced") oxygen storage. To keep the catalyst in a state where both lean and rich excursions can be buffered a feedback controller is needed. This is where the dynamics of the oxygen storage become relevant. While a high storage capacity is favorable for increased conversion rates, the opposite is true considering the robustness of the catalyst controller due to the increased lag of the system.

Significant aging phenomena can be observed during the lifetime of a catalyst, caused mainly by sintering of the noble metal due to high temperatures and so-called catalyst poisoning. [48]

In addition to the oxidation of CO and HC and the reduction of NO, three-way catalysts promote the water-gas shift reaction:

$$CO + H_2O \rightleftharpoons CO_2 + H_2 \tag{2.194}$$

Given the conditions present in a TWC, the chemical equilibrium lies on the right-hand side, which is favorable for the reduction of CO. However, since λ sensors exhibit a significant cross-sensitivity to hydrogen, this effect has to be taken into account when interpreting the downstream λ signal. In fact, for a rich mixture, the air-to-fuel ratio sensor may be used as a hydrogen sensor. Further details on this topic may be found in [17]. Since the catalytically active surface decreases with aging, new three-way catalysts promote the water-gas shift reaction significantly more than aged ones. This results in a varying H_2/CO ratio during the lifetime of a TWC.

2.8.3 Modeling Three-Way Catalytic Converters

Models of TWCs usually serve to predict the conversion rates of the main species and/or the composition of the output gas. They are used to gain a better understanding of the transient behavior of TWCs and to investigate external influences such as "catalyst poisoning" or aging. Further applications are the optimization of the catalyst system and the development of improved controllers.

---

[48] The expression "catalyst poisoning" is used to describe the reduction of catalytic efficiency by unwanted species occupying active surface sites and thereby preventing the desired species from adsorbing on these sites. Poisoning can occur irreversibly with substances such as lead (Pb), but also reversibly, for example, with sulfur dioxide (SO_2).

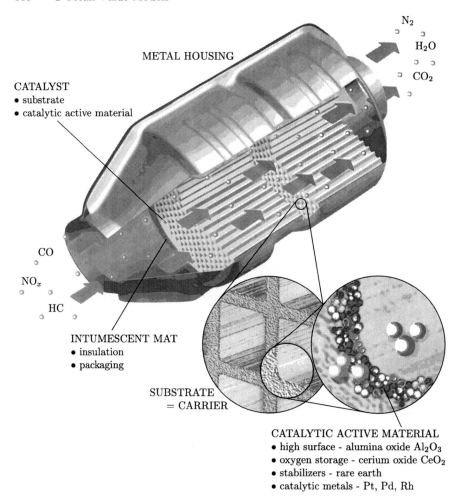

Fig. 2.65. Schematic view of a TWC's structure, reprinted with the permission of Umicore AG.

Predicting the conversion rates is very important for the on-board diagnosis (OBD) of the TWC. An OBD system must be able to warn the driver if the catalytic converter is no longer able to sufficiently remove the pollutants in the exhaust gas. In order to reach the ULEV or Euro IV standards, conversion rates of more than 96% are necessary. Accordingly, the results of the OBD algorithms must be very accurate, which is a challenging objective when considering the few sensor signals available in series production cars. Accurate catalyst models, which do not impose a large computational burden on the ECU, can help achieve these objectives.

Another very important characteristic of a TWC is its warm-up behavior. A cold TWC is almost inactive. TWCs start working efficiently only after

having reached the light-off temperature.[49] The catalyst is heated by the hot
exhaust gases and the reaction enthalpy produced by the exothermic chemical
reactions in the catalytic converter itself. When the light-off temperature is
reached while the TWC is still sufficiently active, the chemical reactions provide enough heat to further warm the TWC, which in turn boosts the catalytic
activity [87]. This chain reaction leads to a significant temperature increase after light-off. In order to efficiently reduce overall emission levels, the warm-up
time must be minimized. Legislation will become more stringent, since cold-
start temperatures will be lowered from $+20°C$ to $-7°C$. Models of TWCs
therefore also have to contain an accurate thermal model.

Detailed models of TWCs are complex. State-of-the-art models incorporate
up to 40 reactions with approximately 10 gas species and between 10 and 15
adsorbed species [106, 39, 99, 33]. These models are very stiff, since the time
constants of most of the chemical reactions are much smaller than those of
the heat and mass flux phenomena.

Control-oriented TWC models usually are much simpler. Often, the dynamics of the temperature are neglected, and most of the reactions are lumped
into the two or three most relevant ones. These simplifications make the models applicable for the use in currently available control systems. Deciding which
are the relevant processes and dynamics and which may be omitted is a deli-
cate task and is discussed, for instance, in [176, 120] and [18].

When developing a control-oriented TWC model, the main objective is
to predict the oxygen storage level because it cannot be measured directly.
Thus, it has to be estimated by an observer, usually supported by a λ sensor
downstream of the TWC [27]. The goal of all TWC control systems is to keep
the TWC in a state from which excursions to the lean and to the rich side
can be handled equally well. It is generally believed that this is achieved by
keeping the oxygen storage level at half the available capacity. However, this
is a rather simplified modeling approach since in reality, not only oxygen, but
also a number of other components are stored.

For these reasons, feedback control and OBD systems necessitate more sophisticated (and thus complex) models. In fact, a complete feedback control
system for the oxygen level would be realizable only if a sensor directly measuring this variable was available. Unfortunately, the only sensor currently
available is a downstream λ sensor (usually switch-type). This sensor detects
and measures the remaining oxygen concentration in the exhaust gas if there
is any. Moreover, when the engine is operated under rich conditions, this sensor is very sensitive to a number of other species, mainly hydrogen, but also
CO and HC [28], [15]. Therefore, in order to allow a correct interpretation
of the output of this sensor, these disturbances of the sensor output must
be modeled. Additionally, the H_2/CO ratio correlates with the cerium/noble

---

[49] The light-off temperature of a TWC is defined as the temperature at which a
conversion efficiency of 50% is attained for HC.

metal surface and therefore changes with the aging of the TWC. This has to be taken into account as well.

Physical Reactor Model

In the following, the modeling of a TWC is described, following a so-called one-dimensional one-channel modeling approach. The TWC consists of a carrier (usually ceramic or metallic substrate) and coating material. First, the washcoat (aluminum oxide) is applied to enlarge the surface of the TWC, followed by the additional oxygen storage material (cerium oxide). Finally, stabilizers and noble metals are placed on top. The better the distribution of the various components, the higher the activity of the TWC.

A catalytic converter consists of many channels that are much longer than their diameter (see Fig. 2.65). Therefore, a lumped-parameter modeling approach is not possible. Instead, a formulation using partial differential equations must be used to describe the physical phenomena within the channels. Fortunately, the gas in the channels may be assumed to be perfectly mixed in the direction *perpendicular* to the flow such that only partial derivatives in the axial direction need to be considered. Often a uniform flow and temperature distribution over all channels is assumed, which reduces the model to the calculation of one channel.

The effects encountered at the channel wall are illustrated in Fig. 2.67. Using this approach, the TWC model consists of three elements:

- Gas phase: Only convective mass and heat transfer in the axial direction are considered.
- Pores/Washcoat: No convection takes place in the direction of the gas flow, but mass transport to and from the gas phase and to and from the solid phase by desorption and adsorption are modeled.
- Solid phase: Heat transfer occurs through heat conduction in the solid phase, heat exchange with the gas phase, and heat production by the chemical reactions between the adsorbed species.

Mass balances for each species can be formulated for the gas phase and the pores while heat balances are formulated for the gas phase and the solid phase.

The concentrations of the components that react in the catalytic converter are much smaller than those of nitrogen, carbon dioxide, and water in the exhaust gas. Therefore, a constant mass flow $\dot{m}$ through the catalytic converter may be assumed. If no additional mass flow into or out of the control volume occurs, the following mass balance for species i may be derived from the schematic diagram shown in Fig. 2.66

$$\underbrace{\varepsilon \cdot A \cdot \partial x \cdot \rho_g}_{\text{mass in control vol.}} \cdot \frac{\partial \xi_{i,g}}{\partial t} = \dot{m} \cdot (\xi_{i,g}(x,t) - \xi_{i,g}(x + \partial x, t)) \qquad (2.195)$$

where ε denotes the volume fraction of the catalyst filled with exhaust gas (which has density ρ_g), A is the cross-section area of the individual channels, and $\xi_{i,g}$ the mass fraction of species i in the exhaust gas.

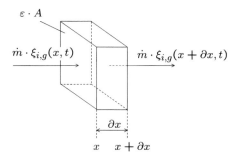

Fig. 2.66. Mass flow through an infinitesimal volume in the channel of a catalytic converter.

Taking the limit case of an infinitesimally small cell size ($\partial x \to 0$), (2.195) yields a partial differential equation (PDE) of the form

$$\varepsilon \cdot A \cdot \rho_g \cdot \frac{\partial \xi_{i,g}}{\partial t} = -\dot{m} \cdot \frac{\partial \xi_{i,g}}{\partial x} \qquad (2.196)$$

for each species i. All mass conservation equations are deduced from this PDE, and the energy conservation equations are derived in a similar way. Its numerical solution is well documented, see e.g. [153]. However, to obtain state space models suitable for control-system design, spatial discretization is necessary (to obtain a finite number of states).

Mass Balances

Figure 2.67 shows the mass and energy flows in the catalytic converter. Flow "1" is the exhaust-gas mass-flow through the channel. Its dynamic is described by combining the basic mass conservation (2.196) with the additional mass flow between the pores of the washcoat and the gas phase as indicated by flow "2", and, it defines the concentration of species i in the gas phase:

$$\underbrace{\varepsilon \cdot A \cdot \rho_g \cdot \frac{\partial \xi_{i,g}}{\partial t} = -\dot{m} \cdot \frac{\partial \xi_{i,g}}{\partial x}}_{\text{cf. (2.196), flow "1"}} \underbrace{-k_{i,g} \cdot A \cdot a_V \cdot \rho_g \cdot (\xi_{i,g} - \xi_{i,pores})}_{\text{to/from pores, flow "2"}} \qquad (2.197)$$

where $k_{i,g}$ is the mass transfer coefficient between gas phase and the pores of the washcoat, a_V is the corresponding relative (scaled by the catalytic-converter volume) cross-sectional area, and $\xi_{i,pores}$ is the concentration of species i in the pores. Note that the diffusion in the axial direction is beeing neglected and the partial pressures of the components are assumed to be constant.[50]

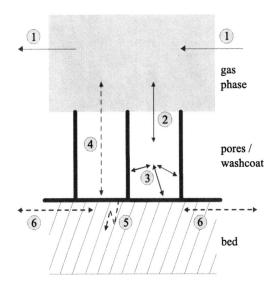

mass balance
1 exhaust mass-flow
2 mass transfer to/from pores
3 adsorption / desorption

energy balance
1 exhaust mass-flow
4 heat exchange between solid
   and gas phase
5 heat of reaction
6 heat conduction in solid phase

Fig. 2.67. Mass flows and heat transfers in the catalytic-converter model.

A second mass balance is formulated for species i in the washcoat pores:

$$4d_b\varepsilon_W d_W \rho_g \frac{\partial \xi_{i,pores}}{\partial t} = k_{i,g} A a_V \rho_g \cdot (\xi_{i,g} - \xi_{i,pores}) \tag{2.198}$$
$$-a_{cat} A R_i M_i$$

where d_b is the length of one side of the quadratic monolith channel, ε_W denotes the washcoat porosity and d_W its thickness, and a_{cat} is the relative catalytically active area per catalyst volume. The free volume (i.e., the volume that is accessible by the gas) per length of the washcoat is given by $4d_b \cdot \varepsilon_W \cdot d_W$. Of course, for non-quadratic monolith channels the terms have to be corrected accordingly.

The mass production term R_i expresses the transfer rate of species i per unit area between the solid and the gas phase in $[mol\, m^{-2}s^{-1}]$. It defines the total adsorption/desorption rate and is calculated from all elementary reactions j in which species i participates: The individual reaction rates r_j times the stoichiometric factors $\nu_{i,j}$ are summed up

---

[50] The Stefan-Maxwell equation would be a more accurate choice. But since p_{N_2} is dominant because of the high concentration of nitrogen in the gas, the inaccuracy may be compensated quite well by slightly adjusting p_{N_2}.

$$R_i = \sum_j (r_j \cdot \nu_{i,j}) \cdot L_t \qquad (2.199)$$

where L_t denotes the maximum capacity of adsorbed species in $[mol/m^2]$.

Energy Balances

The energy balance for the gas phase includes the convective heat flow in the exhaust-gas mass flow and the heat exchange with the solid catalyst bed, as shown in Fig. 2.67. The catalyst is assumed to be adiabatic, i.e., any heat exchange with the environment is neglected, yielding the following heat balance for the gas phase

$$\varepsilon \cdot \rho_g \cdot c_{v,g} \cdot \frac{\partial \vartheta_g}{\partial t} = -\frac{\dot{m}}{A} \cdot c_{p,g} \cdot \frac{\partial \vartheta_g}{\partial x} - \alpha \cdot a_V \cdot (\vartheta_g - \vartheta_{cat}) \qquad (2.200)$$

where $c_{p,g}$ and $c_{v,g}$ are the isobaric and isochoric specific heats of the gas, ϑ_g and ϑ_{cat} the gas and catalyst temperatures, and α is the heat-transfer coefficient between the solid and the gas phase.

The energy balance for the solid phase also includes the heat exchange with the gas phase (with a change of sign) as well as the heat conduction in the solid phase itself. As third term, namely the heat produced by the exothermic chemical reactions on the catalyst surface, has to be included, leading to

$$(1 - \varepsilon) \cdot \rho_{cat} \cdot c_{cat} \cdot \frac{\partial \vartheta_{cat}}{\partial t} = \lambda_{cat} \cdot (1 - \varepsilon) \cdot \frac{\partial^2 \vartheta_{cat}}{\partial x^2} + \alpha \cdot a_V \cdot (\vartheta_g - \vartheta_{cat})$$
$$+ a_{cat} \sum_k (-\Delta H_r)_k \cdot r_k \cdot L_t \qquad (2.201)$$

Here, ρ_{cat} and c_{cat} are the density and the specific heat capacity of the solid phase, respectively, and λ_{cat} denotes the heat conduction coefficient. The terms $(-\Delta H_r)_k$ and r_k are the reaction enthalpy and the reaction rate of reaction k.

Usually, the following simplifications do not introduce any significant errors and thus are allowable in the case of a control-oriented TWCmodel:

- Axial diffusion and heat conduction in the gas can be neglected; convection dominates this direction.
- The dynamics of the gas concentrations in the channel can be omitted and thus the gas concentrations are calculated quasi-statically. Since the gas turnover rate of the catalyst is high for realistic system configurations, changes in the gas concentrations are substantially faster than the oxygen storage dynamics (even during idling $\dot{V}_{exh}/V_{cat} \approx 7.5\frac{1}{s}$).
- Due to the high mass-transfer rates from the channel to the washcoat and the substantially smaller washcoat volume (compared to the channel), the dynamics of the washcoat concentrations can be considered to be much faster than the relevant dynamics. Gas-concentration dynamics in the washcoat are therefore neglected as well.

- Density is only a function of the temperature. Its dependence on the exhaust-gas composition is negligibly small (a few percent).

Chemical Model

The dynamics of surface reactions are often calculated using *Langmuir-Hinshelwood* (LH) or *Eley-Rideal* (ER) kinetics. The former allows only adsorbed reactants, whereas in the latter, only one of the reactants has to be adsorbed. The following assumptions are made:

- All adsorption sites are equivalently active for all species, i.e., all species compete for all sites.
- The total number of adsorption sites is constant.

In order to illustrate the calculation of the reaction rates, a simple example is discussed first. Consider the simplified reaction scheme for the oxidation of CO by O_2, comprising the three steps adsorption of the two species, reaction, and desorption of the product, CO_2. These three steps are described by

$$CO^g + v^* \rightleftharpoons CO^* \qquad O_2^g + 2v^* \rightleftharpoons 2O^* \qquad (2.202)$$

$$CO^* + O^* \rightleftharpoons CO_2^* \qquad (2.203)$$

$$CO_2^* \rightleftharpoons CO_2^g \qquad (2.204)$$

The symbol v^* denotes a free site on the catalyst surface, while the superscripts $*$ and g designate absorbed and free (i.e., in the gas) species, respectively.

According to the LH kinetics, the adsorption rate of species i depends on the fraction of vacant sites θ_V and the partial pressure p_i of the species i only. Often the concentration c_i is taken instead of the partial pressure for calculating the reaction rate. The occupancy θ_i is defined as the fraction of total active sites covered by species i. For CO, the adsorption rate is thus given by

$$r_{CO,ads} = k_{ads,CO} \cdot p_{CO} \cdot \theta_V \qquad (2.205)$$

If all vacant sites are equivalent, their fraction may be expressed as

$$\theta_V = 1 - \theta_O - \theta_{CO} \qquad (2.206)$$

The partial pressure of species i can be calculated using the ideal gas law

$$p_i = \rho_g \cdot \xi_{i,g} \cdot R_i \cdot \vartheta_g = \rho_g \cdot \xi_{i,g} \cdot \frac{\mathcal{R}}{M_i} \cdot \vartheta_g \qquad (2.207)$$

The rate constant $k_{ads,CO}$ is derived by means of statistical thermodynamics, since it is proportional to the number of impingements of gas particles on the catalyst surface per time and catalyst area. (Interested readers may find more information about this in [74] and [13].)

$$k_{ads,CO} = \frac{1}{L_t} \cdot \frac{s}{\sqrt{2\pi \cdot M_{CO} \cdot \mathcal{R} \cdot \vartheta_{gas}}} \qquad (2.208)$$

The symbol M denotes the molecular mass, $\mathcal{R}$ the universal gas constant, and $s \in [0,1]$ the sticking probability defining the fraction of collisions actually leading to an adsorption reaction. The sticking probability depends on both the surface material and the gas species. The adsorption of O_2 is calculated analogously.

The reactivity between two or more reactants is defined by their availability, i.e., their surface occupancy θ_i (or their concentration in the gas phase), and the reaction-rate constant k_{reac}

$$r_{CO,reac} = r_{O,reac} = k_{reac,CO_2} \cdot \theta_{CO} \cdot \theta_O \qquad (2.209)$$

Note that during this reaction, not only CO_2 is formed and its occupancy is increased, but also CO and O are consumed and their occupancies decreased. The rate constant is usually calculated according to the law of Arrhenius

$$k_{reac} = A_{reac} \cdot e^{-\frac{E_a}{\mathcal{R} \cdot \vartheta_{cat}}} \qquad (2.210)$$

For a detailed explanation, see [74] and [13]. The symbol E_a denotes the activation energy, which is the energy barrier that must be passed in order to start a reaction, whereas A_{reac} is the pre-exponential factor with units $[s^{-1}]$. Both parameters depend on both the species and the surface material. According to this law, the reactivity increases exponentially with the catalyst temperature.

The desorption rate of a species is defined by its occupancy and a rate constant k_{des} that follows Arrhenius' law as well

$$r_{CO,des} = k_{des,CO} \cdot \theta_{CO} \qquad (2.211)$$

The desorption rates of O_2 can be derived analogously.

Now, the total reaction rate of adsorbed CO can be expressed as follows

$$\frac{\partial \theta_{CO}}{\partial t} = r_{CO,ads} - r_{CO,des} - r_{CO,reac}$$
$$= k_{ads,CO}\, p_{CO}\, \theta_V - k_{des,CO}\, \theta_{CO} - k_{reac,CO_2}\, \theta_{CO}\, \theta_O \qquad (2.212)$$

The mass production term in (2.198) only includes the adsorption and desorption terms

$$R_{CO} = (r_{CO,ads} - r_{CO,des}) \cdot L_t$$
$$= (k_{ads,CO} \cdot p_{CO} \cdot \theta_V - k_{des,CO} \cdot \theta_{CO}) \cdot L_t \qquad (2.213)$$

Considering the Chemical Equilibrium

The reaction rate of an arbitrary reaction is usually calculated by the following general formula

$$r_i = k_i^f \cdot \prod_{j,\nu_{i,j}<0}^{N_g} c_j^{-\nu_{i,j}} \cdot \prod_{j,\nu_{i,j}<0}^{N_s} \theta_j^{-\nu_{i,j}} - k_i^b \cdot \prod_{j,\nu_{i,j}>0}^{N_g} c_j^{\nu_{i,j}} \cdot \prod_{j,\nu_{i,j}>0}^{N_s} \theta_j^{\nu_{i,j}} \quad (2.214)$$

The molar concentration c_j can be calculated from ξ_j with

$$\xi_j = \frac{c_j \cdot M_j}{\overline{M}} \quad (2.215)$$

where $\overline{M}$ is the average molar mass of the gas.

The symbols k_i^f and k_i^b represent the rate constants of the forward and the reverse reaction, respectively, whereas N_g is the number of gas species participating in the reaction and N_s the number of surface species involved. The reaction orders are defined by the corresponding stoichiometric coefficients $\nu_{i,j}$ of species i in reaction j. For the stoichiometric coefficients $\nu_{i,j}$, negative values have to be used for species that are consumed and positive values for species that are produced during the forward reaction. Note that the reaction orders may be different for non-elementary reactions. If the chemical equilibrium has to be considered (see below), the reaction order may only be changed by multiplying the forward and reverse reaction with the same term.

The forward rate constants are calculated using the Arrhenius equation (for reactions between two species) or the sticking equation (for adsorption reactions) as discussed in the above example

$$k_i^f = A_i \cdot e^{-\frac{E_i}{R \cdot \vartheta_{cat}}} \quad \text{or} \quad k_i^f = \frac{1}{L_t} \cdot \frac{s}{\sqrt{2\pi \cdot M_i \cdot R \cdot \vartheta_{gas}}} \quad (2.216)$$

The reverse-reaction rate-constants are then calculated from the forward ones utilizing the chemical equilibrium

$$k_i^b = \frac{k_i^f}{K_i} \cdot \prod_j^{N_g} \left(\frac{p_{exh}}{R \cdot \vartheta_{gas}} \right)^{-\nu_{i,j}} \quad (2.217)$$

The reverse reaction could also be parametrized separately, but doing so can violate the chemical equilibrium. Certain chemical reactions are limited by the chemical equilibrium, as it is the case for the water-gas shift reaction. Therefore, a separate parametrization of the reverse-reaction rate-constants is not recommended for a three-way catalyst.

The equilibrium constant K_i is calculated based on the change of the Gibbs free energy of the reaction

$$K_i = e^{-\frac{\Delta G_i}{R \cdot \vartheta_{cat}}} \quad (2.218)$$

The Gibbs free energy is defined as

$$\Delta G_i = \Delta H_i - \vartheta_{cat} \cdot \Delta S_i = \sum_j (\nu_{i,j} \cdot H_j) - \vartheta_{cat} \cdot \sum_j (\nu_{i,j} \cdot S_j), \qquad (2.219)$$

where ΔH_i and ΔS_i are the enthalpy and entropy changes of reaction i, respectively. They are calculated from the enthalpy and entropy of each species.

If all reverse reactions are included and calculated following this approach, the thermodynamic consistency of the mechanism is guaranteed not only for each reaction, but, according to [138], also for the complete reaction system. The advantages of this formulation are obvious:

- The complete reaction mechanism is thermodynamically consistent.
- The chemical equilibrium is never violated. This is true for any global reaction even if the enthalpy or entropy of any of the involved surface species in a required reaction step is estimated with a certain error.
- The number of parameters that have to be estimated (or identified by measurements) decreases. This is especially important for complex reaction mechanisms comprising many elementary reactions.

More information and a code generator which helps to define and solve such models can be found in [25].

The enthalpy and entropy values of the gas species can be obtained using measured data from databases such as the NIST Chemistry WebBook (http://webbook.nist.gov/chemistry/). In order to approximate the dependence on temperature, often second-order polynomials are accurate enough. The values for the surface species usually cannot be taken from a database since they depend on the specific materials used in the catalyst. They need to be identified using appropriate dynamic measurements.

Finally, the mass balances of the adsorbed species on the catalyst surface are calculated as follows

$$\frac{\partial \theta_i}{\partial t} = \sum_j \nu_{i,j} \cdot r_j \qquad (2.220)$$

Example: Oxygen Storage

As an example, the calculation of the reaction rates is demonstrated using a simple reaction mechanism describing the oxygen storage dynamics. The mechanism is discussed in detail in Section 2.8.3. The three reactions considered and the corresponding pre-exponential factors and activation energies are presented in Table 2.6.

The reaction rates for the three reactions are calculated using (2.214)

Table 2.6. Reactions and kinetic constants for the oxygen-storage mechanism.

no.	reaction	pre-exp. factor $\left[\frac{m^3}{mol \cdot s}\right]$	activ. energy $\left[\frac{J}{mol}\right]$
1	$O_2 + 2\,Ce_2O_3 \rightleftharpoons 2\,Ce_2O_4$	$A_1 = 1.61 \cdot 10^4$	$E_1 = 10{,}000$
2	$H_2O + Ce_2O_3 \rightleftharpoons H_2 + Ce_2O_4$	$A_2 = 1.99 \cdot 10^9$	$E_2 = 160{,}000$
3	$CO_2 + Ce_2O_3 \rightleftharpoons CO + Ce_2O_4$	$A_3 = 4.10 \cdot 10^{14}$	$E_3 = 230{,}000$

$$r_1 = k_1^f \cdot c_{O_2} \cdot \theta_{Ce_2O_3}^2 - k_1^b \cdot \theta_{Ce_2O_4}^2$$

$$= k_1^f \cdot c_{O_2} \cdot \theta_{Ce_2O_3}^2 - \frac{k_1^f}{K_1} \cdot \theta_{Ce_2O_4}^2 \tag{2.221}$$

$$r_2 = k_2^f \cdot c_{H_2O} \cdot \theta_{Ce_2O_3} - \frac{k_2^f}{K_2} \cdot \theta_{Ce_2O_4} \cdot c_{H_2} \tag{2.222}$$

$$r_3 = k_3^f \cdot c_{CO_2} \cdot \theta_{Ce_2O_3} - \frac{k_3^f}{K_3} \cdot \theta_{Ce_2O_4} \cdot c_{CO} \tag{2.223}$$

The rate constants are calculated using the Arrhenius equation (2.216) and the equilibrium constant K_i according to (2.218). For the latter, the change of the Gibbs free energy is needed for every reaction

$$\Delta G_1 = \Delta H_1 - \vartheta_{cat} \cdot \Delta S_1 = -H_{O_2} - 2 \cdot H_{Ce_2O_3} + 2 \cdot H_{Ce_2O_4}$$
$$- \vartheta_{cat} \cdot (-S_{O_2} - 2 \cdot S_{Ce_2O_3} + 2 \cdot S_{Ce_2O_4}) \tag{2.224}$$

$$\Delta G_2 = -H_{H_2O} - H_{Ce_2O_3} + H_{H_2} + H_{Ce_2O_4}$$
$$- \vartheta_{cat} \cdot (-S_{H_2O} - S_{Ce_2O_3} + S_{H_2} + S_{Ce_2O_4}) \tag{2.225}$$

$$\Delta G_3 = -H_{CO_2} - H_{Ce_2O_3} + H_{CO} + H_{Ce_2O_4}$$
$$- \vartheta_{cat} \cdot (-S_{CO_2} - S_{Ce_2O_3} + S_{CO} + S_{Ce_2O_4}) \tag{2.226}$$

Finally, the dynamics of the oxygen storage level $\theta_{Ce_2O_4}$ is calculated by summing up the three reaction rates (2.221) to (2.223)

$$\frac{\partial \theta_{Ce_2O_4}}{\partial t} = 2 \cdot r_1 + r_2 + r_3 \tag{2.227}$$

Table 2.7 lists the thermodynamic properties of the species involved in this mechanism.

The storage capacity per catalyst volume used in this model is $a_{cat} \cdot L_t = 76.7$ mol/m^3.

Reaction Mechanism

First the real dynamic behavior of the air-to-fuel ratio (AFR) and the emissions downstream of the catalyst are discussed. Fig. 2.68 shows the results

Table 2.7. Thermodynamic properties of the species participating in the oxygen-storage reaction-mechanism. The temperature dependencies are valid for temperatures ranging from 500 K to 1100 K. The data for Ce_2O_4 and the rate parameters are estimations.

Enthalpies $\left[\frac{J}{mol}\right]$	
H_{O_2}	$= -1.12 \cdot 10^4 + 33.98 \cdot \vartheta_{cat}$
H_{H_2}	$= -9.2 \cdot 10^3 + 29.93 \cdot \vartheta_{cat}$
H_{CO}	$= -1.213 \cdot 10^5 + 32.51 \cdot \vartheta_{cat}$
H_{H_2O}	$= -2.557 \cdot 10^5 + 40.00 \cdot \vartheta_{cat}$
H_{CO_2}	$= -4.128 \cdot 10^5 + 52.77 \cdot \vartheta_{cat}$
$H_{Ce_2O_4} - H_{Ce_2O_3}$	$= -288\,600 + 54\,300 \cdot \theta^2_{Ce_2O_4}$

Entropies $\left[\frac{J}{mol \cdot K}\right]$	
S_{O_2}	$= 189.8 + 0.07301 \cdot \vartheta_{cat} - 1.915 \cdot 10^{-5} \cdot \vartheta^2_{cat}$
S_{H_2}	$= 118.1 + 0.06602 \cdot \vartheta_{cat} - 1.780 \cdot 10^{-5} \cdot \vartheta^2_{cat}$
S_{CO}	$= 183.9 + 0.06799 \cdot \vartheta_{cat} - 1.732 \cdot 10^{-5} \cdot \vartheta^2_{cat}$
S_{H_2O}	$= 172.7 + 0.07873 \cdot \vartheta_{cat} - 1.863 \cdot 10^{-5} \cdot \vartheta^2_{cat}$
S_{CO_2}	$= 189.3 + 0.10563 \cdot \vartheta_{cat} - 1.732 \cdot 10^{-5} \cdot \vartheta^2_{cat}$
$S_{Ce_2O_4} - S_{Ce_2O_3}$	$= 5.03$

obtained from measurements taken on an SI engine equipped with a standard TWC. The engine is held at constant speed and manifold pressure, resulting in a constant air mass-flow. The nominal control signal for the injection duration (i.e., the fuel mass-flow) is modulated by a square wave with an amplitude of approximately $\pm 10\%$.

The top plot of Fig. 2.68 shows the signals of the two wide-range λ sensors upstream and downstream of the TWC. The fuel path dynamics can be neglected for this investigation because they are much faster than the filling and emptying of the TWC.

In the second plot of Fig. 2.68, signals obtained by fast CO and NO sensors are depicted. Note that the NO concentration drops immediately when the upstream AFR switches from lean to rich, whereas the CO concentration remains low until the oxygen storage of the TWC is depleted and oxidation ceases. When switching the input AFR from rich to lean, the CO concentration drops with a certain time constant due ceria oxidation caused by H_2O and CO_2. In contrast, the NO concentration drops immediately after switching the AFR from lean to rich. Again, when the oxygen storage is full, the NO concentration rises again. A model which can describe this effect correctly is given in table (2.6).

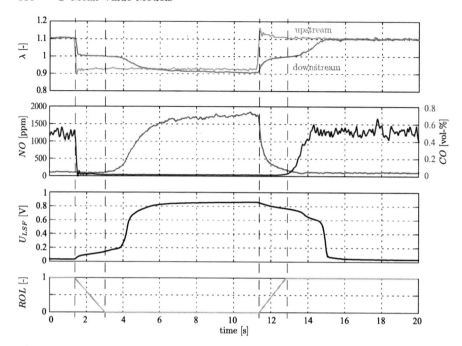

Fig. 2.68. Measurements of λ and emissions downstream of a TWC while performing step changes in the upstream air-to-fuel ratio. Engine speed: $3000rpm$, air-mass per cylinder and cycle: $0.3g$

The third plot in Fig. 2.68 shows the voltage produced by the switch-type λ sensor downstream of the TWC. The sensor switches whenever the gas exiting the TWC turns either rich or lean, which is usually attributed to the oxygen storage being either emptied or completely filled again. From the emission measurement data, the capacity of the oxygen storage can be estimated and a relative oxygen level calculated. This variable is shown in the last (bottom) plot of Fig. 2.68.

Finding a suitable reaction mechanism is not trivial. The reactions to be included and the dynamic phenomena to be reproduced strongly depend on the purpose of the model. For example, if the downstream switch-type λ sensor is used for control purposes, the dynamics of hydrogen have to be considered as well due to the strong corresponding cross-sensitivity of this sensor.

Figure 2.69 illustrates a selection of chemical reactions on a TWC, while Table 2.8 shows the reaction scheme of a relatively simple model (the numbering of the reactions is not matched). The latter was designed to accurately simulate the conversion rates of the species that are restricted by legislation (NO, CO, and HC) and the ones required for the calculation of the downstream λ sensor signal (O_2, H_2). The model is able to reproduce the dynamic behavior of the catalyst during a driving cycle. Figure 2.70 shows the simu-

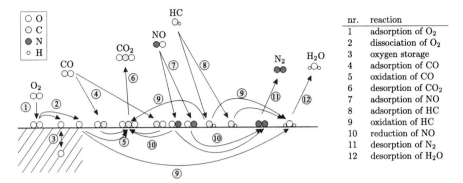

Fig. 2.69. Overall chemical reactions in a TWC.

lated and the measured cumulative emissions for NO, CO, and hydrocarbons during the last part of an FTP75 driving cycle.

Table 2.8. Reactions considered in the TWC model.

no.	reaction
1	$O_2 + 2\,Pt \rightleftharpoons 2\,PtO$
2	$CO + Pt \rightleftharpoons PtCO$
3	$H_2 + 2\,Pt \rightleftharpoons 2\,PtH$
4	$NO + Pt \rightleftharpoons PtNO$
5	$PtCO + PtO \rightleftharpoons CO_2 + 2\,Pt$
6	$2\,PtH + PtO \rightleftharpoons H_2O + 3\,Pt$
7	$PtC + PtO \rightleftharpoons PtCO + Pt$
8	$C_3H_6 + 3\,PtO + 6\,Pt \rightleftharpoons 3\,PtCO + 6\,PtH$
9	$PtNO + Pt \rightleftharpoons PtO + PtN$
10	$PtNO + PtO \rightleftharpoons PtNO_2 + Pt$
11	$2\,PtN \rightleftharpoons N_2 + 2\,Pt$
12	$PtO + Ce_2O_3 \rightleftharpoons Ce_2O_4 + Pt$
13	$CO + Ce_2O_4 \rightleftharpoons CO_2 + Ce_2O_3$
14	$2\,H_2 + Ce_2O_4 \rightleftharpoons H_2O + Ce_2O_3$

Oxygen Storage Dynamics

Usually, reaction mechanisms of TWC models are quite complex. They have to include a large number of different reaction steps, especially if the splitting and step-by-step oxidation of hydrocarbons is modeled. The reaction mechanism presented in this section is kept very simple, comprising only three global reactions. This helps understanding the dynamic behavior and avoids large numbers of unknown parameters. The concentrations of nitric oxides and hydrocarbons are typically one order of magnitude lower than those of CO,

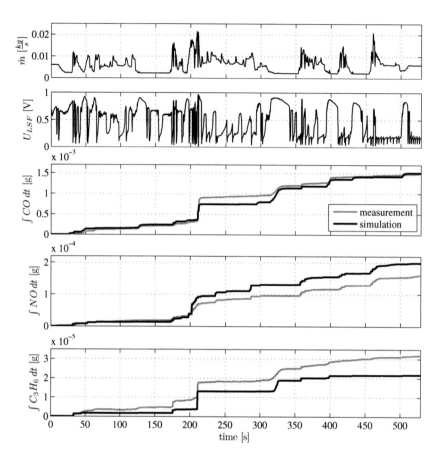

Fig. 2.70. Simulated and measured cumulative emissions during the last part of an FTP75 cycle, where U_{LSF} is the voltage of the downstream switch-type λ-sensor.

H_2, and O_2. Therefore, the conversion of hydrocarbons and nitric oxides is not modeled, but their influence on the oxygen storage dynamics is assumed to not differ substantially from that of CO, H_2 and O_2. The reaction mechanism used is the one discussed above as an example, while the list of reactions can be found in Table 2.6.

Due to the presence of H_2O and CO_2 in the exhaust gas, the filling state of the oxygen storage under rich conditions is defined by the equilibrium between the reduction of ceria by CO and H_2 and its oxidation by H_2O and CO_2. Note that the exhaust gas of an SI engine always contains about 13% H_2O and CO_2.

Two examples are presented to demonstrate the influence of H_2O and CO_2 on the oxygen storage dynamics and to emphasize the importance of the equilibrium.

The first example shows the behavior of the AFR downstream of the catalyst for lean-to-rich and rich-to-lean steps. Figure 2.71 shows the pre- and post-catalyst AFR. The downstream AFR is very critical because most controllers and on-board-diagnosis algorithms rely on it. The real λ is calculated from measured exhaust-gas concentrations in order to avoid any disturbances occurring with λ sensors downstream of a catalyst (see e.g. [15]). If the downstream sensor voltage is of interest, the model has to be expanded by a sensor model. The post-catalyst AFR exhibits an asymmetric behavior. In the case of a lean-to-rich step it switches to a value of 1 first. During this period the rich gases entering the catalyst are oxidized by the stored oxygen. When all available oxygen is used, the downstream AFR drops further until it reaches the inlet value.

After a rich-to-lean step in the upstream AFR, there is a more or less continuous rise of the outlet λ. Right after the step, rich exhaust gas is leaving the catalyst even though a lean mixture is entering. This means that during this phase more oxygen is stored than is entering the catalyst. The additional oxygen storage after the rich-to-lean step is related to the fact that H_2O and CO_2 are able to oxidize reduced ceria, producing CO and H_2 until the chemical equilibrium between ceria and the gas species is established. This observation confirms the aforementioned importance of the backward reaction paths of reactions 2 and 3 in Table 2.6. The thermodynamically consistent parametrization ensures that neither the chemical equilibrium nor the energy balance is ever violated by this reaction scheme.

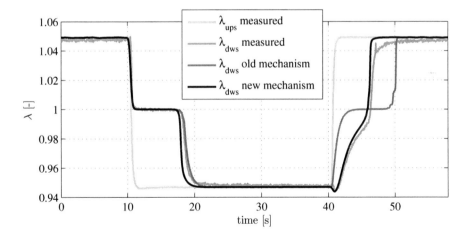

Fig. 2.71. Dynamic behavior of the downstream (dws) air-to-fuel ratio for the new storage mechanism compared to an existing storage mechanism (old). For reference, the upstream AFR is also shown (ups). Inlet temperature is 870 K and the space velocity amounts to 26,000 h^{-1}. Inlet gas composition: 10% H_2O, 10% CO_2, CO and O_2 variable, N_2 balance.

An example for a very simple old storage mechanism is given in (2.202).

An important aspect of AFR control is the behavior of TWCs during fuel cutoffs in deceleration phases, which is performed to reduce fuel consumption. During a fuel cutoff, the exhaust gas consists of air only, which causes the catalyst to completely fill its oxygen storage. Given this situation when re-enabling the fuel injection, an accidental lean excursion of the λ controller causes an immediate NO_x breakthrough. Therefore the oxygen storage must be depleted immediately at the end of every fuel cutoff period, which can be achieved by driving with rich mixture for a certain time.

The following experiment shows the catalyst's behavior after a fuel cutoff period. In a first step, the engine is run lean (about 10% from the stoichiometric AFR) for at least 20 seconds. This completely fills the oxygen storage and therefore the same conditions are established as those present after a fuel cutoff period. Subsequently, a short rich pulse in the upstream λ is generated to empty the oxygen storage. Finally, it is set to and kept at stoichiometric conditions ($\lambda = 1$).

The measured concentrations and the simulation results of the experiment are shown in Figure 2.72. During the rich pulse, no CO and H_2 emissions are present downstream of the catalyst. About $10s$ after switching to stoichiometric conditions, the catalyst starts to emit CO and H_2 for about $100s$. Thus the downstream lambda sensor signal indicates a rich mixture and the PID controller typically used tries to compensate for this signal by sustainedly decreasing the amount of fuel injected. This renders the mixture lean again and thus, eventually, leads to a breakthrough of NO.

Figure 2.73 shows profiles of oxygen storage levels computed for different times of the experiment presented in Figure 2.72. At $10s$, all the ceria is completely oxidized (not shown in the plot). When the rich pulse enters the catalyst, ceria is reduced at the inlet. During the $\lambda = 1$ phase, this reduced zone again is partially oxidized by CO_2 and H_2O. This oxidation produces CO and H_2, which in turn reduces the ceria further downstream in the TWC. As soon as the reduced zone reaches the end of the catalyst, the rich components begin to exit the catalyst, and the CO and H_2 excursions shown in Fig. 2.72 arise. The ROL profile then slowly converges towards its steady-state distribution, while less and less CO and H_2 are formed by reactions 2 and 3 shown in Table 2.6.

The dynamic behavior of a three-way catalyst strongly correlates with its oxygen-storage capacity, making the latter an important design parameter. Adapting the controller to the dynamics of the catalyst is crucial when high conversion rates under transient conditions have to be attained. Since the storage capacity decreases over the catalyst's lifetime, it can be used as an indicator for the aging level of the TWC. In fact, most onboard-diagnosis strategies aim at estimating the available oxygen-storage capacity to deduce the aging level of the catalyst therefrom.

Empirical data show a dependency of the oxygen-storage capacity on several other parameters defined by the operating conditions. This is due to the

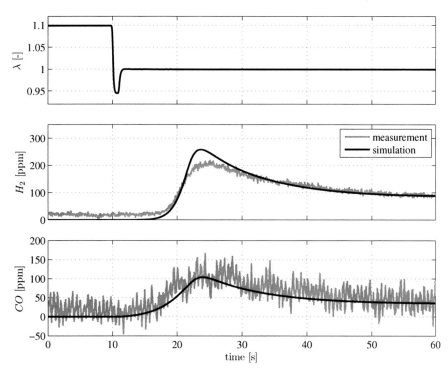

Fig. 2.72. Delayed emissions after a rich pulse. The inlet temperature is 770 K and the space velocity amounts to 78 000 h^{-1}. Inlet gas composition: 10% H_2O, 10% CO_2, O_2 and CO variable, N_2 balance.

filling state being controlled (under rich conditions) by the chemical equilibrium between the reduction of ceria by CO and H_2 and the oxidation of ceria by H_2O and CO_2, respectively. The most important relations are listed below:

- The oxygen storage capacity is significantly smaller if H_2O and/or CO_2 is present in the exhaust gas. Due to ceria being oxidized by these species, the oxygen storage is emptied only to the level where the reduction and the oxidation of ceria reach an equilibrium. This influence was also documented in [159] for H_2O.
- If the gas upstream of the catalyst does not contain any H_2O or CO_2 at all, the dynamics of the oxygen storage are symmetric for lean-to-rich and inverse steps.
- The storage capacity varies for increasingly rich mixtures. This variation is due to a shift of the equilibrium caused by the higher concentrations of CO and H_2.
- No significant changes for lean conditions are observed.

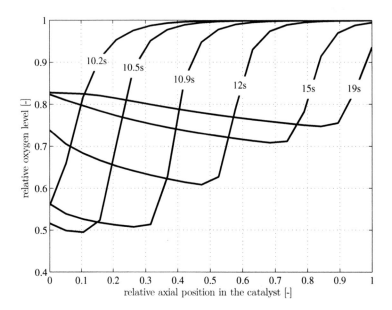

Fig. 2.73. Profiles of the relative oxygen-level for different times after the start of the rich pulse (at $10s$). The timescale matches that of Figure 2.72.

Due to lack of space, only the most important facts are summarized here. For more detailed information, [141] and [142] are recommended.

Buffer Model for the Oxygen Storage in a TWC

A simple, but commonly used model for the oxygen storage was presented in [152] (several subsequent papers further discuss this topic, see e.g. [120] or [30]).

From a control-engineering point of view, the concentrations of the individual species are of minor interest. However, it is important to keep the oxygen storage of the catalytic converter at a level allowing both the reduction of the nitrogen oxide and the oxidation of the hydrocarbons and carbon monoxides. As Fig. 2.68 shows, this is the case only if the relative oxygen level (ROL) is between 0 (empty) and 1 (full). The ROL is calculated following

$$ROL(t) = \lim_{0,1} \int 0.21 \cdot \frac{(\dot{m}_\beta(t) - \sigma_0 \cdot \dot{m}_\varphi(t))}{C_{cat}} \, dt$$

$$= \lim_{0,1} \int 0.21 \cdot \frac{\dot{m}_\beta(t)}{C_{cat}} \cdot \left(1 - \frac{1}{\lambda(t)}\right) \, dt \qquad (2.228)$$

where C_{cat} represents the total oxygen storage capacity of the catalytic converter. Since only oxygen has to be considered, the air mass is multiplied by 0.21 (assuming an oxygen mass fraction of 21% in the air). For improved clarity, all delays have been omitted; further details can be found in [162], [103] and [16].

2.9 Pollution Abatement Systems for Diesel Engines

Motivation

It is anticipated that future limits imposed on the pollutant emissions of Diesel engines will no longer be met by reducing only engine-out pollutants. Most probably, an aftertreatment system will become necessary to reduce the nitrogen oxide and the particulate matter emissions to the future legislation limits shown in Fig. 2.74.

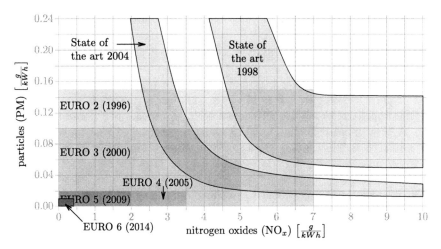

Fig. 2.74. Legislated limits and trade-off of raw emissions for heavy-duty Diesel engines: particulate matter (PM) versus nitrogen oxide (NO_x), adopted from [58].

The type of aftertreatment system selected will depend on the engine application (light/medium/heavy duty) and on the engine calibration chosen. With a high-efficiency NO_x reduction system in place, for instance, the engine can be operated close to its optimal bsfc and at low PM conditions.[51]

Three types of pollution abatement systems for Diesel engines are under consideration at the moment:

---

[51] If, however, the targeted emission limits are very low, a combination of engine tuning, in-engine measurements, and an exhaust gas aftertreatment system will have to be set up, and the engine may *not* be retuned fuel-efficiently.

- particulate filters;
- NO_x traps; and
- selective catalytic reduction (SCR) systems.

The first two systems have the drawback that for their regeneration, high temperatures (particulate filters) or rich exhaust gases (NO_x traps) are necessary. These regeneration phases increase fuel consumption. The SCR approach requires an additional agent, most probably urea, in a separate storage system. Fortunately, this substance is innocuous and not expensive. A comprehensive preliminary assessment of various exhaust aftertreatment systems is presented in [148], and a state-of-the-art overview can be found in [107].

An SCR exhaust gas aftertreatment system using urea solution as a reductant has a high NO_x reduction potential and is a well-known technique for stationary applications [26]. In mobile applications, however, the volume of the catalytic converter is limited, and the engine is operated under transient conditions most of the time. Due to the fast changes in the exhaust gas temperature and the high gas hourly space velocity (GHSV), ammonia slip is likely to occur unless special control actions are applied. Model-based control concepts, as presented for instance in [178], are therefore a must in this case.

Structure and Working Principle of SCR Systems

A catalytic converter for Diesel engines cannot work the same way as a TWC for SI engines because an excess of oxygen is always present in the exhaust gas. One approach to resolve this problem is to use an additional reducing agent. A convenient reducing agent showing a good selectivity toward NO_x is ammonia (NH_3). For dosage and safety reasons, ammonia is often injected as a water-urea solution that decomposes in the catalytic converter to form ammonia and CO_2.[52]

The primary reactions occurring in this catalytic system are:

- Evaporation of the urea solution and decomposition of the urea molecules (where s denotes solid)

$$((NH_2)_2CO)_s + H_2O \rightarrow 2NH_3 + CO_2 \qquad (2.229)$$

- Adsorption and desorption of ammonia on the surface of the catalytic converter (where * denotes an adsorbed species)

$$NH_3 \leftrightarrow NH_3^* \qquad (2.230)$$

- Selective catalytic reduction consuming adsorbed ammonia and gaseous NO_x (Eley-Rideal mechanism)

$$4NH_3^* + 2NO + 2NO_2 \rightarrow 4N_2 + 6H_2O \qquad (2.231)$$

---

[52] For the engine under investigation, 1 liter urea solution is needed for every 20 liter of Diesel fuel.

$$4NH_3^* + 4NO + O_2 \rightarrow 4N_2 + 6H_2O \qquad (2.232)$$

The first reaction is faster than the second. Up to 95% of the NO_x leaving the engine is NO. It is useful to mount an oxidation catalytic converter in front of the SCR catalyst. This device changes the ratio of NO to NO_2 (but it does not affect the amount of NO_x because of excess oxygen), which is favorable for the SCR reaction.[53] Additionally, this catalytic converter transforms hydrocarbons and soot into CO_2 and, simultaneously transforming NO to NO_2, prevents the SCR catalyst's being poisoned.

- Oxidation of adsorbed ammonia

$$4NH_3^* + 3O_2 \rightarrow 2N_2 + 6H_2O \qquad (2.233)$$

For reasonably sized catalytic converters, the temperature window where an SCR system works best is 250-400 °C. Below 200 °C, the reactions inside the SCR are too slow, whereas above 450 °C an increasing amount of ammonia is converted directly to nitrogen.

Modeling an SCR System

As shown in Fig. 2.75, the complete SCR system consists of an oxidation catalyst, a device for injecting urea solution, and the SCR catalytic converter.

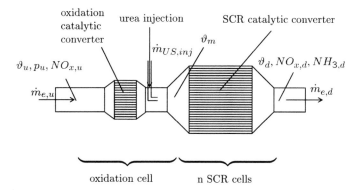

Fig. 2.75. Complete SCR system consisting of the oxidation catalytic converter, the urea injection device, and the SCR catalytic converter.

The subscripts u, m, and d stand for upstream, middle, and downstream of the SCR system. The plant is modeled by an oxidation cell and n identical SCR cells. Notice that the oxidation cell includes the oxidation catalytic converter *and*, immediately downstream of that converter, the urea injection system.

---

[53] In order to avoid additional very slow reactions, the ratio of NO to NO_2 should not fall below a value of 1.

The models derived below disregard the hydrocarbon and soot present in the exhaust gases. This simplification is based on the assumption that these species do not affect the main NO_x reduction pathways.

Oxidation Cell

The oxidation catalytic converter and the urea injection system are modeled first. This entire component is referred to as the *oxidation cell*. In general, the oxidation cell has a different coating than the SCR cell.

The assumptions made to simplify the modeling process of the oxidation cell are:

- The injected urea solution mass flow $\dot{m}_{US,inj}$ (a water-urea mixture with a urea concentration $\xi_U S$) is heated to the temperature of the exhaust gas. However, no urea decomposition takes place.
- The molar flow of NO_x is not affected.
- The oxidation catalyst is assumed to be a perfect heat exchanger, and the exhaust gas leaving the catalyst is assumed to have the same temperature as the catalyst ($\vartheta_{OC} = \vartheta_m$).
- The oxidation catalyst can be modeled accurately as a perfectly stirred reactor with concentrated model parameters.

Note that only the total flow of NO_x is taken as the input to the system and that the primary effects modeled are those causingthe thermal behavior of the oxidation cell.

Modeling the thermal behavior of the system using two thermal energy reservoirs often yields satisfactory results. These two reservoirs represent the behavior of the converter and its housing. For the first level variable, the temperature of the converter ϑ_{OC}, the following equation holds

$$\frac{d}{dt}\vartheta_{OC} \cdot c_{OC} \cdot m_{OC} = (\vartheta_u - \vartheta_{OC}) \cdot c_{p,g} \cdot \dot{m}_{e,u} - \dot{Q}_{US} - \dot{Q}_{cond} \qquad (2.234)$$

where the two heat flows $\dot{Q}_{US}$ and $\dot{Q}_{cond}$ are modeled by the following equations

$$\dot{Q}_{US} = \Big[\big((\vartheta_b - \vartheta_a) \cdot c_{p,H_2O(l)} + r_{v,H_2O} + (\vartheta_{OC} - \vartheta_b) \cdot c_{p,H_2O(g)}\big) \cdot (1 - \xi_{US})$$

$$+ (\vartheta_{OC} - \vartheta_a) \cdot c_{p,U} \cdot \xi_{US}\Big] \cdot \dot{m}_{US,inj} \qquad (2.235)$$

$$\dot{Q}_{cond} = \alpha_{cond} \cdot A_{cond} \cdot (\vartheta_{OC} - \vartheta_{OCh}) \qquad (2.236)$$

The specific heat capacity of the oxidation converter is given by the parameter c_{OC} and m_{OC} is its mass. The first term on the right-hand side of (2.234) describes the enthalpy increase by the mass flow through the oxidation converter. The specific heat capacity of the exhaust gas is given by $c_{p,g}$ and by

$\dot{m}_{e,u}$, which is the exhaust gas mass flow at the inlet of the oxidation cell. The variable $\dot{Q}_{US}$ denotes the energy flow necessary to heat and vaporize the urea solution injected. The urea solution is an eutectic solution ($\xi_{US} = 0.325$).

Urea is assumed to enter the system as a solid substance. The vaporization of the urea solution must be divided into three steps: heating from the ambient temperature ϑ_a to the boiling point ϑ_b, vaporization, and finally, heating from the boiling point to the gas temperature.

The variable $\dot{Q}_{cond}$ describes the conductive heat losses to the converter housing. The housing of the oxidation converter is modeled according to the description given in Sect. 2.6.3. The temperature of the oxidation converter housing, ϑ_{OCh}, is the second level variable. Its dynamics are described by

$$\frac{d}{dt}\vartheta_{OCh} \cdot c_{OCh} \cdot m_{OCh} = \dot{Q}_{cond} - \dot{Q}_{out,OCh} \tag{2.237}$$

where

$$\dot{Q}_{out,OCh} = k_{out,OCh} \cdot A_{out,OCh} \cdot (\vartheta_{OCh}^4 - \vartheta_a^4)$$
$$+ \alpha_{out,OCh} \cdot A_{out,OCh} \cdot (\vartheta_{OCh} - \vartheta_a) \tag{2.238}$$

The heat losses $\dot{Q}_{cond}$ in (2.234) represent the heat increase in (2.237). The second term models the heat losses of the housing to the ambient by radiation and convection.

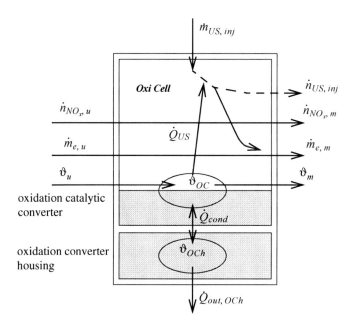

Fig. 2.76. Mass and heat flows within the oxidation cell.

Figure 2.76 shows the mass and heat flows as well as the inlet and outlet temperatures in the oxidation converter, where $\dot{n}_i$ denotes the molar flow of the species i. The level variables, i.e., ϑ_{OC} and ϑ_{OCh}, are encircled.

The liquid urea solution is vaporized, and a molar flow $\dot{n}_{U,m}$ of urea is produced. The NO_x mass flow remains unchanged, while the total mass flow $\dot{m}_{e,m}$ is increased by the mass of the urea solution injected ($\dot{m}_{e,m} = \dot{m}_{e,u} + \dot{m}_{US,inj}$).

As mentioned, the expressions (2.235) and (2.236) describe only the heating of the urea solution and the heat losses to the oxidation converter housing. The NO/NO_2 ratio change is not modeled and therefore not visible in these equations. The main justification for this simplification is that both NO and NO_2 "consume" one ammonia molecule when reduced to molecular nitrogen. Moreover, the kinetics of both reactions are faster than the ammonia adsorption such that the latter dominates the overall dynamics. Of course, these global reaction parameters must be experimentally identified and are valid only for the specific configuration used in the experiments.

SCR Cell

In this part, a model of the SCR cell is introduced. A distributed-parameter formulation using partial differential equations must be used for this subsystem. Of course, this model has to be discretized for use in a later control system. The derivation of the PDEs follows the approach used in Sect. 2.8.3.

The assumptions made when deriving the model of the SCR cell are:

- Only one solid component, urea, and only two gases, NO_x and NH_3, are considered in the model.
- The reaction for the urea decomposition is modeled using (2.229).
- Adsorption and desorption of ammonia on the surface of the catalytic converter are described using (2.230).
- The oxidation of the ammonia adsorbed on the surface of the catalytic converter is modeled using (2.233).
- Eley-Rideal reaction kinetics are assumed to be applicable for the SCR component. The gas NO_2 is considered to react the same way as NO, consuming one NH_3 molecule per NO_x molecule, see (2.232) and (2.231). Depending on the activity of the oxidation cell, a certain average value of the NO/NO_2 ratio is present at the input of the SCR converter. This results in an average SCR reduction rate which has to be identified by measurements.
- No adsorption of reaction products is considered.
- No mass storage effects are assumed to take place in the *gas* phase, i.e, the exhaust mass flow is assumed to be large compared to the mass flow of the reacting species.
- The influence of any washcoat diffusion effects is neglected.

- Compared to the oxidation catalytic converter, the SCR converter has a much smaller surface-to-volume ratio. Therefore, only one thermal energy reservoir must be considered.
- The reaction enthalpies are neglected since they do not have any significant influence on the temperature of the system.
- Regarding the thermal effects, the SCR converter is assumed to be a perfect heat exchanger that satisfies the lumped-parameter assumptions. Accordingly, the exhaust gas temperature at the outlet, ϑ_d is assumed to be equal to the converter temperature ϑ_C. The dynamics of this variable can be approximated well using an ODE.

Therefore, the simplified thermal energy dynamics for the SCR catalytic converter can be described by

$$c_{SCR} \cdot m_{SCR} \cdot \frac{d}{dt} \vartheta_C(t) = c_{p,g} \cdot \dot{m}_{e,m} \cdot (\vartheta_m - \vartheta_C) - \dot{Q}_{out,SCR} \qquad (2.239)$$

The heat losses to ambient are calculated analogously to (2.238)

$$\dot{Q}_{out,SCR} = k_{out,SCR} \cdot A_{out,SCR} \cdot (\vartheta_C^4 - \vartheta_a^4)$$
$$+ \alpha_{out,SCR} \cdot A_{out,SCR} \cdot (\vartheta_C - \vartheta_a) \qquad (2.240)$$

Applying mass balances, kinetic gas theory for the adsorption, and Arrhenius reaction kinetics for the urea decomposition, the desorption, the SCR reaction, and the ammonia oxidation, a system of coupled PDE's is obtained. These equations describe the dynamics of the various gas concentrations and of the surface coverage. Note the similarities to the corresponding set of equations presented in Sect. 2.8.3. The urea decomposition is described by

$$\varepsilon \cdot A \cdot \rho_g \cdot \frac{\partial \xi_U}{\partial t} = -\dot{m}_{e,m} \cdot \frac{\partial \xi_U}{\partial x} - \varepsilon \cdot A \cdot \rho_g \cdot r_{UD}$$

$$r_{UD} = A_{UD} \cdot e^{\left(-\frac{E_{a,UD}}{\mathcal{R} \vartheta_C(t)}\right)} \cdot \xi_U \qquad (2.241)$$

The first term on the right-hand side models the increase in concentration by the flow through the converter. The second term describes the urea decomposition. The modeling is very similar to that of the TWC.

The PDE for the ammonia concentration is given next

$$\varepsilon \cdot A \cdot \rho_g \cdot \frac{\partial \xi_{NH_3}}{\partial t} = -\dot{m}_{e,m} \frac{\partial \xi_{NH_3}}{\partial x} + \varepsilon \cdot A \cdot \rho_g \cdot 2 \cdot r_{UD}$$
$$+ \varepsilon \cdot A \cdot a_{cat} \cdot L_t \cdot (r_{des} - r_{ads}) \cdot M_{NH_3} \qquad (2.242)$$

$$r_{ads} = \frac{1}{L_t} \cdot \frac{s}{\sqrt{2\pi \cdot M_{NH_3} \cdot \mathcal{R} \cdot \vartheta_C}} \cdot p_{NH_3} \cdot (1 - \theta_{NH_3})$$

$$= \rho_g \cdot \frac{s}{L_t \cdot M_{NH_3}} \cdot \sqrt{\frac{\mathcal{R} \cdot \vartheta_C}{2\pi \cdot M_{NH_3}}} \cdot \xi_{NH_3} \cdot (1 - \theta_{NH_3})$$

$$r_{des} = A_{des} \cdot e^{\left(-\frac{E_{a,des}}{\mathcal{R} \vartheta_C}\right)} \cdot \theta_{NH_3}$$

In addition to the terms that are already known, the adsorption and desorption of urea on the surface of the catalytic converter must be considered. The adsorption rate is primarily dependent on the number of free sites on the surface $1 - \theta_{NH_3}$ and upon the ammonia concentration in the gas. The desorption rate is assumed not to be dependent upon the gas concentration. The PDE for the NO_x concentration contains the SCR reaction

$$\varepsilon \cdot A \cdot \rho_g \cdot \frac{\partial \xi_{NO_x}}{\partial t} = -\dot{m}_{e,m} \cdot \frac{\partial \xi_{NO_x}}{\partial x} - \varepsilon \cdot A \cdot a_{cat} \cdot L_t \cdot r_{SCR} \cdot M_{NO_x}$$

$$r_{SCR} = \frac{1}{L_t} \cdot \frac{s}{\sqrt{2\pi \cdot M_{NO_x} \cdot \mathcal{R} \cdot \vartheta_C}} \cdot p_{NO_x} \cdot \theta_{NH_3}$$

$$= \rho_{gas} \cdot \frac{s}{L_t \cdot M_{NO_x}} \cdot \sqrt{\frac{\mathcal{R} \cdot \vartheta_C}{2\pi \cdot M_{NO_x}}} \cdot \theta_{NH_3} \cdot \xi_{NO_x} \qquad (2.243)$$

Finally, the surface coverage can be calculated

$$\frac{\partial}{\partial t}\theta_{NH_3} = r_{ads} - r_{des} - r_{SCR} - r_{ox} \qquad (2.244)$$

$$r_{ox} = A_{ox} \cdot e^{(-\frac{E_{a,ox}}{\mathcal{R}\vartheta_C})} \cdot \theta_{NH_3} \qquad (2.245)$$

It is assumed that there is always excess oxygen in the exhaust gas and that thus no dependency on the partial pressure of oxygen arises. The oxidation of ammonia on the surface of the catalytic converter has been taken into account as well. The mass and energy flows within the SCR cell are depicted in Fig. 2.77.

The level variables are the concentration of urea, the concentrations of ammonia and NO_x in the gas phase, the surface coverage of the catalytic converter with ammonia, and the temperature of the catalytic converter. Since the listed components are indicated, the products N_2 and H_2O of the ammonia oxidation are not shown. The same is valid for the SCR reaction, which, again, produces only N_2 and H_2O.

Usually, these equations are not solved as partial differential equations. They are approximated by ordinary differential equations by discretizing the converter into n_{cells} cells along its flow axis. The temperature distribution and the gas concentrations are assumed to be homogenous in each cell (lumped-parameter assumption). The order of the discretized complete SCR system is then $2 + 5n_{cells}$, two states for the oxidation cell, and five states for each SCR cell. Even for a small number of cells, the solution of these equations requires a substantial number of numerical computations. Therefore, such a model is not useful in real-time applications.

Simplified Model for the SCR System

In order to obtain a model that is useful for the design of model-based control algorithms, additional simplifying assumptions must be made:

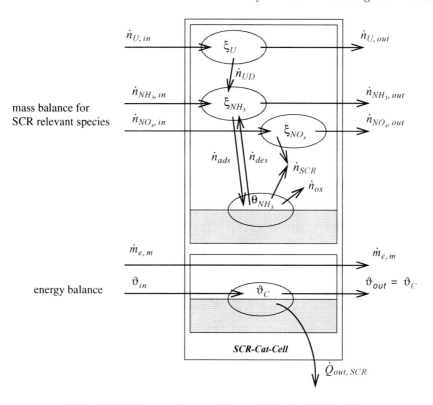

Fig. 2.77. Mass and energy flows within the SCR cell.

- The urea decomposition is fast and therefore the ammonia flow can be statically calculated from the amount of urea at the entrance of the SCR catalytic converter.
- The time constants of the storage effects in the gas phase are much smaller than those of the storage of thermal energy and of ammonia on the surface of the catalyst. Thus, the NO_x concentration as well as the gaseous ammonia concentration are calculated statically.
- The reaction enthalpies can be neglected as opposed to thermal convection and radiation.

This results in the following two ordinary differential equations for an SCR cell:

$$\frac{d\theta_{NH_3,i}}{dt} = \frac{s \cdot \rho_g}{L_t \cdot M_{NH_3}} \cdot \sqrt{\frac{\mathcal{R} \cdot \vartheta_{C,i}}{2\pi \cdot M_{NH_3}}} \cdot (1 - \theta_{NH_3,i}) \cdot \xi_{NH_3,i}$$

$$- \theta_{NH_3,i} \cdot \left[A_{des} \cdot e^{\left(\frac{-E_{a,des}}{\mathcal{R} \cdot \vartheta_{C,i}}\right)} + A_{ox} \cdot e^{\left(\frac{-E_{a,ox}}{\mathcal{R} \cdot \vartheta_{C,i}}\right)} \right. \tag{2.246}$$

$$\left. + \rho_g \cdot \frac{s}{L_t \cdot M_{NO_x}} \cdot \sqrt{\frac{\mathcal{R} \cdot \vartheta_{C,i}}{2\pi \cdot M_{NO_x}}} \cdot \xi_{NO_x,i} \right]$$

$$\frac{d\vartheta_{C,i}}{dt} = \frac{n_{cells} \cdot c_{p,g}}{c_{SCR} \cdot m_{SCR}} \cdot \dot{m}_{e,m} \cdot (\vartheta_{C,i-1} - \vartheta_{C,i}) \tag{2.247}$$

$$- \frac{A_{out,SCR}}{c_{SCR} \cdot m_{SCR}} \cdot (k_{out,SCR} \cdot (\vartheta_{C,i}^4 - \vartheta_a^4) + \alpha_{out,SCR} \cdot (\vartheta_{C,i} - \vartheta_a))$$

The gas concentrations as mass fractions must be calculated by

$$\xi_{NO_x,i} = \frac{\xi_{NO_x,i-1}}{1 + \frac{\rho_g \cdot V_C}{n_{cells} \cdot \dot{m}_{e,m}} \cdot a_{cat} \cdot \theta_{NH_3,i} \cdot s \cdot \sqrt{\frac{\mathcal{R} \cdot \vartheta_{C,i}}{2\pi \cdot M_{NO_x}}}} \tag{2.248}$$

and

$$\xi_{NH_3,i} = \frac{\xi_{NH_3,i-1} + \frac{V_C}{n_{cells} \cdot \dot{m}_{e,m}} \cdot a_{cat} \cdot L_t \cdot \theta_{NH_3,i} \cdot M_{NH_3} \cdot A_{des} \cdot e^{-\frac{E_{a,des}}{\mathcal{R} \cdot \vartheta_{C,i}}}}{1 + \frac{\rho_g \cdot V_C}{n_{cells} \cdot \dot{m}_{e,m}} \cdot a_{cat} \cdot (1 - \theta_{NH_3,i}) \cdot s \cdot \sqrt{\frac{\mathcal{R} \cdot \vartheta_{C,i}}{2\pi \cdot M_{NH_3}}}} \tag{2.249}$$

These algebraic equations can be derived from the differential equations (2.243) and (2.242) by setting the time derivative to zero and discretizing, i.e., by splitting the converter into n_{cells} identical slices. Note that $\frac{\rho_{gas} \cdot V_C}{n_{cells} \cdot \dot{m}_{e,m}}$ indicates the mean residence time of the exhaust gas in the SCR cell. The resulting system now has $2 \cdot n_{cells}$ states. Simulations with varying values of n_{cells} show that it is usually sufficient to consider two or three cells. This reduced complexity is acceptable such that this model can serve as a basis for the subsequent control system design.

3

Discrete-Event Models

Discrete-event models (DEM) of selected engine subsystems are the subject of in this chapter. The input and output signals of these systems either are not defined or are not relevant continuously, but only at certain discrete instances. Thus, the system behavior associated with the sampling of the signals, the delays caused by the timing relations, and the interaction with the ECU (which operates in a discrete-time way as well), become important.

Crank-angle based representations are often used for IC engine models. This change in the independent variable from time t to crank angle ϕ has advantages for certain control problems. Specifically, the feedforward action of the air/fuel ratio control system needs this special representation because it must be realized with the maximum bandwidth achievable. It is thus very sensitive to any errors in modeling the delays and the timing relations.

The following items are discussed in this chapter:

- *When are DEM required?*
- *Various ECU operation modes and time scales.*
- *Precise timing of injection and ignition commands.*
- *DEM of torque production in IC engines.*
- *DEM of the air flow in IC engines.*
- *DEM of the fuel flow in IC engines.*
- *DEM of back-flow dynamics of CNG engines.*
- *DEM of residual gas dynamics.*
- *DEM of exhaust pipe dynamics.*
- *DEM based on measurements data of the cylinder pressure.*

Most of these models are control-oriented *models. Only the last model represents a first step toward thermodynamically more detailed ones.*

One fundamental assumption that underlies the derivation of all of these models is that the engine speed does not significantly change during one engine cycle. This assumption has been justified in Sect. 2.5.2. Therefore, the models, although being defined in the crank-angle domain, are formulated on the basis of constant sampling times.

3.1 Introduction to DEM

3.1.1 When are DEM Required?

A model of an engine system always includes continuous-time and discrete-event subsystems. Examples of the continuous-time subsystems are the intake manifold dynamics, the acceleration of the crankshaft, the turbocharger speed dynamics. Subsystems that often are modeled as discrete-event systems are the torque production, the gas exchange of the individual cylinders, the injection and the ignition processes, etc.

The ECU is another part of the complete system that is best described using a discrete-time approach. In fact, modern ECUs are always microprocessor systems running various tasks with different sampling times. Section 3.1.3 provides a short overview of the software structure implemented in such ECUs.

The derivation of continuous-time mean-value models for the various engine subsystems is shown in Chap. 2. The discrete working principles of the subsystems must be considered in the following cases:

- When the system representation, and therefore the subsequent controller realization, is simpler in the crank-angle domain.
- When the control system has to achieve bandwidths that make it necessary to take into account the synchronization problems. This is often the case for feedforward control action, since this type of controller usually requires maximum bandwidth attainable.
- When cylinder-individual effects have to be analyzed, e.g., single-cylinder air/fuel control, misfire detection with engine speed measurements, the interpretation of cylinder-pressure measurement data, etc.

Note that it is usually not possible to know at the outset which approach is better suited. A robustness analysis can help to decide whether a time or a crank-angle discretization achieves the best performance. Moreover, the available CPU power does not allow the calculation of control tasks in the so-called "synchro mode" (synchronized with the engine's crank angle, i.e., using a discrete-event approach) unless it is mandatory to do so.

3.1.2 Discrete-Time Effects of the Combustion

Due to its reciprocating action, an internal combustion engine by its very nature is a discrete-event system. The focus of this chapter is on four-stroke spark ignited (SI) engines. Almost identical models can be derived for two-stroke gasoline engines and Diesel engines.

A cylinder-pressure trajectory and the scaled valve-lift curves are depicted in Fig. 3.1. The timing of the different strokes is indicated as well. As the diagram clearly shows, a single-cylinder engine produces torque every two engine revolutions. This torque acts on the crankshaft during approximately one-third of a revolution. In the remaining portions of the two revolutions, the engine rotates due to the inertia of the flywheel. This process leads to very strong engine speed oscillations and rough engine behavior. Having more cylinders and distributing their torque peaks evenly over the two revolutions results in a much smoother behavior. Accordingly, in a six-cylinder engine, a cylinder fires every one-third of a revolution, resulting in a rather constant engine torque over one complete engine cycle.

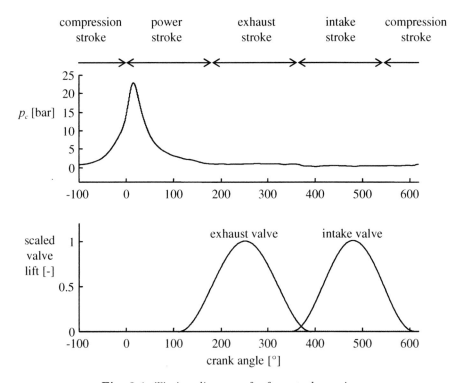

Fig. 3.1. Timing diagram of a four-stroke engine.

At a certain crank angle (for port-fuel injected Otto engines this is the crank angle where the ignition occurs) all boundary conditions for the combustion are fixed and can no longer be changed. Hence, the torque production acts as a discrete system with a sampling period of

$$\phi_{seg} = \frac{2\pi \cdot N}{n_{cyl}} \tag{3.1}$$

$$\tau_{seg} = \frac{2\pi \cdot N}{\omega_e \cdot n_{cyl}} \tag{3.2}$$

where N indicates a two-stroke ($N = 1$) or a four-stroke engine ($N = 2$), n_{cyl} is the number of cylinders of the engine, ϕ_{seg} is the sampling angle in radians, and τ_{seg} is the sampling time in seconds, usually referred to as the *segment time*.

The sampling interval is constant in the crank-angle domain and depends upon the engine speed in the time domain. In discrete-time control theory, the sampling of the signals is usually assumed to take place with constant sampling times. However, this assumption does not hold for crank-angle-based sampling. As shown in Sect. 2.5.2, changes in engine speed normally are at least a factor of ten slower than the changes in manifold pressure, which set the boundary conditions for the combustion in the cylinder. Thus, crank-angle-based sampling can be assumed to take place at evenly spaced points in time.

In four-stroke engine applications, the synchronization of any device (control system, measurement equipment, or actuator) with the sampling behavior of the engine requires information on the crank angle as well as on the camshaft position. Usually, this is accomplished by using two sensors that detect the variations in a magnetic field caused by the teeth on the outer rim of the flywheel and by the camshaft.

3.1.3 Discrete Action of the ECU

Before the advent of electronic control units, the ignition was changed continuously by a centrifugal advance mechanism (for speed adjustment) and a vacuum unit (connected to the manifold for load adaptation). Therefore, the influences of the engine speed and of the manifold pressure on the ignition angle were separate, and the characteristics of that control action were given by the shape of the mechanical cam mechanisms. The time during which the ignition current was on, thus building up energy in the ignition coil, was given by the setting of the ignition contact breaker and was constant in the crank-angle domain, yielding a constant dwell angle, but a varying dwell time. No correction for low battery voltage was possible and the coils had to be short-circuit resistant, because at start-up, whenever the ignition was on, the full current was flowing in the coil for an extended period of time.

Fuel was brought into the cylinder by means of carburetors that continuously evaporated fuel into the air stream. The resulting air/fuel ratio was not constant over the entire operating region, which caused drivability problems, high pollutant emissions, and fuel economy problems. Mechanical devices, i.e., accelerator pumps, various adjustable nozzles, etc., were added in an attempt to deal with these problems. However, they just made carburetors costly, difficult to adjust, and unreliable.

The introduction of emission legislation required the introduction of three-way catalytic converters. As shown in Sect. 1.3.2, for these devices to work properly, a precise air/fuel ratio control system is mandatory. For that reason, carburetors with some electronically controllable elements were introduced that were able to utilize the signal of the air/fuel ratio sensor to compensate for non-stoichiometric mixtures. However, those systems turned out to be more expensive than the new injection systems, which injected the fuel using electronically controlled solenoid valves. The first devices of that type were central fuel injection systems that injected fuel at an upstream position in the manifold from where it was distributed to several cylinders. Such a system is simple to realize but difficult to control, due to the large wall-wetting effects and the long transport time delays. A port fuel injection system proved to be the better solution, i.e., a configuration in which the fuel is injected just upstream of the intake valve.

Since that approach required electronic control units with electronic power amplifiers, the next step was the integration of a solid state ignition system, which made it possible to control injection and ignition timing simultaneously. Around the year 1980, the first of these SI engine control units with microprocessors appeared on the market [75]. They combined the control of ignition and injection in one control unit, and were able to implement more complex control algorithms.

The core of all ECUs is a microcontroller. This device runs in a discrete-event mode, i.e., it performs some of its tasks repetitively in constant crank angle intervals. In addition, the ECU runs several tasks using fixed sampling time intervals. A real-time operating systems coordinates these various tasks and resolves computation-time conflicts on the basis of predefined priorities (see below).

Two operating modes are possible for an ECU:

- *Fully Synchronized Operating Mode*: The ECU processor continuously counts pulses from a sensor that picks up the variations in the magnetic field caused by the teeth on the outer rim of the flywheel (where 60 teeth is a common number) and compares the demanded values for injection and ignition crank angle with the actual values of the crank angle. This operation mode guarantees the additional delays to be minimal. On the other hand, a great deal of precious processor time is spent on pulse counting. Therefore, modern ECUs avoid this fully synchronized operation mode.
- *Asynchronous Operation Mode*: The ECU processor has a dedicated slave processor, often referred to as a time-processing unit (TPU, see Sec. 1.2.2), which relieves the ECU from the synchronization tasks mentioned [78]. In fact, the microprocessor sends the new values for injection and ignition timing asynchronously to the TPU. The TPU obtains crank angle and cam shaft pulses from the engine and triggers the injection and ignition events at the correct crank angle values. This operation mode frees up more time for "intelligent" tasks performed by the processor. In production-type ECUs, these interface chips are directly mounted on the ECU board or, in more recent systems, are integrated on the same chip as the microcontroller. For research purposes, such TPUs with a very high degree of flexibility have been designed using field programmable gate arrays [69].

Even the processors of the most modern engine ECUs have clock rates that are much slower than those of modern desktop computers. This is due to the harsh environmental conditions in which they must work reliably, e.g., temperatures ranging from $-40°C$ to $+100°C$, together with high humidity and electromagnetic fields (radio signals, GSM, ignition system, etc.). One consequence of these slow clock rates is that the computing resources must be carefully managed. This is achieved by utilizing several clock rates:

- 1 ms: very fast tasks, sensor signal post processing, etc.;
- 20 ms: normal time scale for control functions;
- 100 ms: slow (adaptive) functions, temperature effects, etc;
- 1 segment (3.1): control functions which have to be synchronized with the engine's reciprocating behavior; and
- 1/2 segment or even faster: data processing for special tasks, i.e., cylinder-individual air/fuel control, interpretation of cylinder-pressure data, etc.

Computational power is limited such that every task must be run in exactly the time frame that is required by the physical constraints, but not faster. Every task is provided with a priority level, which enables the operating system to resolve conflicts in case the computational resources are not sufficient. At high engine speeds, due to the large amount of calculation time spent in the high-priority segment tasks, the low-priority tasks are hardly ever executed.

This aspect can be illustrated by considering the available computation intervals at different engine speeds and for various crank-angle intervals. Table 3.1 lists the corresponding values for engine speeds ranging from 600 rpm to 6000 rpm. These values show that at low engine speeds there is sufficient time for many calculations, but the time delays associated with the crank-angle domain becoming very large render the feedback control problems hard to solve. On the other hand, while at high speeds these time delays are short, the time available for computations is so small that not many tasks can be run simultaneously.

Table 3.1. Time-domain values for engine events (4 stroke).

	crank angle	corresponding time-intervals at:	
		$n_e = 600\ rpm$	$n_e = 6000\ rpm$
cycle	4π	200 ms	20.0 ms
revolution	2π	100 ms	10.0 ms
segment (6-cylinder)	$\frac{2\pi}{3}$	33 ms	3.3 ms
segment (4-cylinder)	π	50 ms	5.0 ms

3.1.4 DEM for Injection and Ignition

The primary task of the ECU is to guarantee the injection of the correct amount of fuel into the manifold and the ignition of the mixture in the cylinder at the desired crank-angle position with sufficient energy.

The fuel-injection module is analyzed in a first step. The timing of the injection is derived from the data shown in Fig. 3.2. The lift curves of the intake and the exhaust valves of an engine are assumed to be constant in the crank-angle domain (this is not the case with variable-valve timing, of course). Both valves are open for approximately two-thirds of an engine crankshaft revolution. It is advantageous to inject the fuel on a closed valve[1] such that sufficient time for fuel evaporation is available. Moreover, since the intake valve is very hot compared to the intake runners, the evaporation of the injected fuel is further improved.

Following this approach, the time available for fuel injection at 6000 *rpm* amounts to approximately 13 ms. This value should not be exceeded except for very harsh transients. At idling, which is the other extreme, a minimal amount of fuel must be injected. Opening and closing the fuel injector takes approximately 1 ms, a quantity that strongly depends upon manufacturing tolerances, wear, bouncing, and fuel quality. To prevent large variations in the amount of fuel injected, the minimum injection time should not be shorter than approximately 2 ms. These considerations determine the dimensions of the injection valve. Obviously, a trade-off has to be made between low amounts of hydrocarbons emitted at full load and fueling precision at idle, i.e., the more fuel that needs to be injected, the more critical the manufacturing tolerances become at idle.

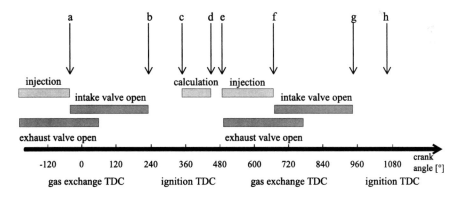

Fig. 3.2. Timing diagram for the injection.

---

[1] Injecting fuel directly into the cylinder through the open valve would increase hydrocarbon emissions.

The timing diagram for the injection depicted in Fig. 3.2 will now be discussed in detail.

- In normal operation, the injection is completed at point **a**. In the ECU, this event is fixed. The start of injection is set at the necessary crank angle, which is computed using the air mass flow information obtained at an earlier crank angle.
- Between the points **a** and **b**, the air mass and the fuel mass flow into the cylinder. Changes in these mass flows into the cylinder during that period can no longer be corrected with a regular injection.
- In normal operation, point **b** is the earliest possible start of injection for the next cycle. For high loads, the resulting injection duration must be sufficient.
- At point **c**, the measurements for the calculation of the duration of the injection are taken. This event is usually triggered at all crank-angle intervals ϕ_{seg} (3.1). This is the relevant event for the calculation of the injection-to-torque delay, because the measurement signals valid at **c** are responsible for the combustion at **h**. Between **c** and **d**, all calculations associated with the injection control algorithms are performed. This event will be referred to as the *update injection event*.
- At **d**, new values for the duration of the injection duration (as well as for the ignition timing) are sent to the time processing unit (TPU).
- Until **e** is reached, no new information for the injection timing is available. An average delay between **d** and **e** of one-half sample interval τ_{sample} can be assumed, reflecting the fact that an even distribution of the time delays between zero (the calculations finished just in time and the injection starts immediately after the update event) and one sample interval (the calculations required too much time such that the result is transferred to the TPU only in the next update injection event) may be expected.
- At point **e**, the injection with the most recent air mass information starts.
- Depending on the type of ECU, between **e** and **f**, the injection pulse can be extended if necessary, but only if critical air/fuel ratios are to be expected (i.e., misfire). Most ECUs typically allow an immediate closing of the injection valve.
- At point **f**, the regular injection pulse is finished. During very heavy transients, some ECUs offer the possibility of injecting a small amount of fuel onto the open intake valve. Of course, fuel can only be added, thus if the air mass is reduced, very rich conditions must be accepted.
- After the point **g**, the amounts of air and fuel in the cylinder are fixed, and the combustion starts at the point **h**.

Summarizing these considerations, the time delay between the start of the calculation (update injection event) and the injection start event can be approximated by

$$\tau_{c)-e)} = \tau_{calculation} + \frac{1}{2} \cdot \tau_{sample} \tag{3.3}$$

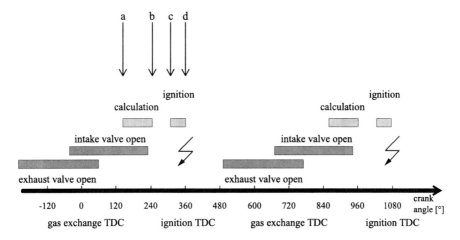

Fig. 3.3. Timing diagram for the ignition.

For the ignition, similar considerations exist, as shown in Fig. 3.3. Modern ignition coils must be connected to the full battery voltage for approximately 4 ms to build up enough energy for a sufficient spark. However, if the coil is connected to the battery voltage for more than 10 ms, it may suffer irreversible damage.

- At point **a**, the measurements of the engine variables necessary for the determination of the ignition timing are triggered by the ECU. This point is now referred to as the *update ignition event.* In general, this is not be the same crank angle as that at which the update injection event takes place.
- Calculations are performed up to point **b**, at which the calculated values are sent to the TPU.
- Again, until ignition starts, an additional time delay occurs. For this delay, a value of one-half of the sample time can be assumed. At **c**, the coil is connected to the battery terminals.
- At **d**, the ignition (or spark advance) angle, the circuit breaker is opened causing the spark to start. The time between c) and d) is the dwell angle. Both variables are chosen such that the thermodynamic conditions of combustion are optimized.

Changing the spark advance between **c** and **d** is possible within small limits. The case of missing a spark event must be strictly avoided because such misfires can cause severe damage to the TWC. Accordingly, increasing the spark advance is always more critical.

The calculation of the spark advance must be done considering the available information on engine speed, air mass in cylinder, knock onset during the last combustion event, etc. Since the dwell-time is rather short and the ignition takes place during the compression of the gases in the cylinder, the ignition has a smaller inherent delay than the injection. The ignition is thus the fastest actuator for a port-fuel injected SI engine.

3.2 The Most Important DEM in Engine Systems

3.2.1 DEM of the Mean Torque Production

In a mean-value setting, all the delays present in the torque production were lumped into the IPS time delay (2.1). This section presents a more detailed analysis of the reciprocating torque generation. The main idea is to identify for each relevant subsystem x a characteristic crank angle ϕ_x, at which the actual boundary conditions (manifold pressure, air mass flow, etc.) must be sampled. If this information permits an accurate prediction of the torque production, a crank-angle-based DEM of the corresponding subsystem can be formulated.

Note that this torque is still assumed to be constant over one engine segment, i.e., the models presented in this section are only able to predict mean-torque values. If the variations in engine speed due to the pulsating combustion are of interest, the crank-angle-resolved cylinder pressure and its effect on the cranking mechanism must be taken into account. For this purpose, some simple models are introduced in Sect. 3.3.

To model the mean torque taking into account the various delays present in an engine system, a characteristic crank-angle center position is defined for each stroke:

- Intake Stroke: The intake center corresponds to the crank angle at which the relevant air mass flow enters the cylinder.
- Compression Stroke: The events relevant to the ignition process take place, as discussed in Sect. 3.1.4.
- Power Stroke: The torque center represents the crank angle of the relevant torque production.
- Exhaust Stroke: Similar to the intake center, the exhaust center is defined as that crank angle at which the relevant exhaust gas mass flow leaves the cylinder.

The crank-angle position of these events is not easy to determine. For the intake and exhaust center, a first approximation of $\pm 90°$ crank angle before or after TDC can be given (roughly where the piston moves with maximum speed).[2]

The location of the intake and exhaust center is influenced by the cylinder pressure and therefore by the operating point, by the geometry of the cranking mechanism, and by the engine speed. Accordingly, the positions of the intake, the exhaust, and the torque center in the crank-angle domain must be determined by experiments. Since only time delays can be measured, but not

---

[2] The experiments described below show that a value of $\pm 110°$ is more realistic.

crank-angle delays, the experiments must be conducted on a test bench with a high-bandwidth engine-speed controller. Once the time delays are identified, they can be converted to crank angles by considering the engine speed.

Measuring Delays

To determine the torque-center crank angle, the input variables influencing the torque production must first be analyzed. In Sect. 2.5.1 it was shown that the following equation can be used to approximate the engine torque

$$T_e = \frac{H_l}{4\pi} \cdot m_\varphi \cdot \eta_{e0}(m_\varphi, \omega_e) \cdot \eta_\lambda(\lambda) \cdot \eta_\zeta(\zeta) \cdot \eta_{egr}(x_{egr}) \qquad (3.4)$$

Accordingly, the following "inputs" can be used to trigger changes in the engine torque:

- the air mass in the cylinder;
- the fuel in the cylinder;
- the spark advance; and
- the exhaust gas recirculation rate (internal and external).

The signal flowchart of the corresponding system is depicted in Fig. 3.4. The dynamics of the various blocks can be determined by keeping all inputs constant, with the exception of one. As discussed in Appendix A, frequency response measurements are a convenient tool to determine the time delays for linear or linearized systems.

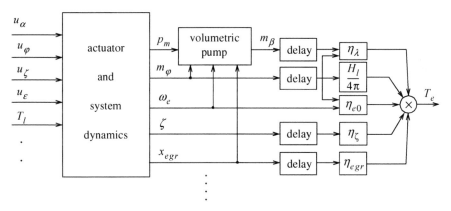

Fig. 3.4. Block diagram of the torque production taking into account the delays in the signal paths.

As mentioned above, the only input which can be assumed to influence the torque production almost instantaneously is the spark signal. Measured frequency responses with the spark signal as the input and the engine torque as the output are therefore the best choice to analyze the time delay between the ignition update event (chosen by the ECU, as explained in Sect. 3.1.4) and the torque center. Of course, the engine speed, the throttle position, the EGR valve position, and the injection duration must be kept constant for this measurement.

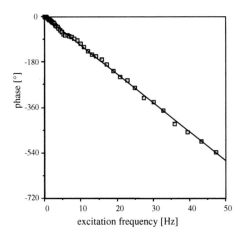

Fig. 3.5. Phase of the frequency response measurement from ignition input u_ζ to engine torque output T_e. Simulation (solid) and measurement results (squares indicating the excitation frequencies at which measurements were taken).

Figure 3.5 shows the phase plot of such a frequency response measurement. The spark angle is changed sinusoidally while the update ignition event is kept constant. Thus, all variations in engine torque phasing are due to the delay between update ignition event and torque center. The results shown in Fig. 3.5 confirm that the dynamics of this subsystem can be approximated well by a single delay element. In fact, only a pure time delay shows the constant phase roll-off over the excitation frequency ($\varphi = -\omega \cdot \tau_{delay}$) shown in that plot. From this data, the time delay, and, with the knowledge of the engine speed, the crank-angle delay, can be determined using the following relationship

$$\tau_{delay} = \frac{\Delta\varphi(\Delta f)}{360 \cdot \Delta f} \tag{3.5}$$

where the phase lag φ is assumed to be given in degrees and the excitation frequency f of the spark angle modulation in Hz.

For the example shown in Fig. 3.5, an update ignition event at $90°$ crank angle before TDC and an engine speed of 900 rpm is chosen. Using (3.5), a delay $\tau_{delay} \approx 31$ ms is identified, which, for the engine speed mentioned, is equivalent to a crank angle of $170°$. Accordingly, for the torque center, a characteristic crank angle of $80°$ after TDC is obtained.

Once the position of the torque center is known, the intake center crank-angle position can be determined with a frequency response measurement that modulates the throttle position and utilizes the manifold pressure as the input signal and the engine torque as the output signal. Obviously, this approach relies on the assumption that, for constant engine speeds, the air mass in the cylinder is a static function of the manifold pressure (see Sect. 2.3). Analogously, mounting a fast exhaust gas sensor as close as possible to the exhaust valve and exciting the spark advance, the exhaust center can be determined.

Table 3.2 summarizes a set of measured values for these engine parameters. The values given in that table are valid at idle conditions. They are influenced by the cylinder pressure and valve timing and, therefore, by the operating point. Additionally, they can be altered by the algorithms implemented in the ECU by changing the ignition timing and, in variable-cam systems, the cam phasing.

Table 3.2. Timing events of a four-stroke engine at idle speed.

engine event	position in ° crank angle after ignition TDC
intake center (IC)	470° (110° after TDC)
torque center (TC)	80°
exhaust center (EC)	250° (110° before TDC)
update ignition (U_{ign}) update injection (U_{inj})	defined in the ECU $\phi_{update} = \phi_{offset} + \phi_{seg} \cdot k$ $k = 0, 1, 2, 3, 4, \ldots$

Using the numerical values given in Table 3.2 to localize the events discussed above yields the final timing diagram shown in Fig. 3.6. For the chosen engine speed (idling), an induction-to-power-stroke delay of 330° and an induction-to-exhaust delay of 500° result, which corresponds well to the values introduced in Sect. 2.1, where $\tau_{ips} \approx \frac{2\pi}{\omega_e}$ and $\tau_{eg} \approx \frac{3\pi}{\omega_e}$ were given as approximations.

Time Domain Description

Taking into account all delays introduced so far, the engine torque can be approximated by

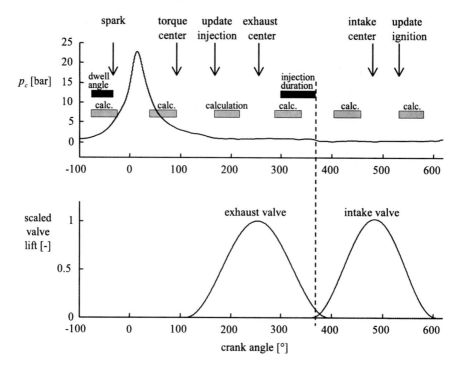

Fig. 3.6. Timing diagram of a four-stroke six-cylinder engine. Note that the calculation is performed for every segment, in this case every third of a revolution. Depending on injection duration, spark advance, and dwell angle the relevant calculation event may vary.

$$T_e(t) = \frac{H_l}{4\pi} \cdot m_\varphi(t - \tau_{Uinj \to TC}) \cdot \eta_{e0}(m_\varphi(t - \tau_{Uinj \to TC}), \omega_e(t))$$
$$\cdot \eta_\lambda \left(\frac{m_\beta(t - \tau_{IC \to TC})}{m_\varphi(t - \tau_{Uinj \to TC})} \cdot \frac{1}{\sigma_0} \right) \cdot \eta_\zeta(\zeta(t - \tau_{Uign \to TC})) \qquad (3.6)$$
$$\cdot \eta_{egr}(x_{egr}(t - \tau_{tra} - \tau_{IC \to TC}))$$

In this formulation, the two delays between the update-injection to intake-center event and the intake-center to torque-center event are lumped into a single delay $\tau_{Uinj \to TC}$.

Using the numerical values indicated in Table 3.2, the delays introduced above can be written as

$$\tau_{IC \to TC} \approx \frac{330\,^\circ}{360\,^\circ} \cdot \frac{2\pi}{\omega_e} \qquad (3.7)$$

$$\tau_{Uinj \to TC} = \tau_{calc} + \frac{1}{2}\tau_{sample} + \tau_{inj} + \tau_{IVO \to IC} + \tau_{IC \to TC} \qquad (3.8)$$

$$\tau_{Uign \to TC} = \tau_{calc} + \frac{1}{2}\tau_{sample} + \tau_{dwell} + \tau_{sp \to TC} \qquad (3.9)$$

where the following abbreviations have been used:

sp   spark

IVO   intake valve opening

calc   calculation

τ_{tra} transport delay from EGR valve to intake manifold

Crank Angle Domain Description

Changing to a crank-angle domain formulation, the corresponding delays must be divided by the segment time $\tau_{seg} = (2\pi \cdot N)/(\omega_e \cdot n_{cyl})$.

In this formulation, the advantages of a crank-angle-based representation become obvious. In fact, in this domain most delays are constant. For instance, assuming a six-cylinder engine and neglecting the calculation times, the dwell time, and the injection duration (which is the most critical assumption, but which holds well at low speed) the following discrete event formulation can be derived for the torque production at idle

$$T_e(t_k) = \frac{H_l}{4\pi} \cdot m_\varphi(t_{k-5}) \cdot \eta_{e0}(m_\varphi(t_{k-5}), \omega_e(t_k))$$

$$\cdot \eta_\lambda \left(\frac{m_\beta(t_{k-3})}{m_\varphi(t_{k-5})} \cdot \frac{1}{\sigma_0} \right) \cdot \eta_\zeta(\zeta(t_{k-2})) \qquad (3.10)$$

$$\cdot \eta_{egr}(x_{egr}(t_{k-3} - \tau_{tra}))$$

In this equation, the delays have been rounded up to the next higher integer multiple of the sampling time. Of course, for other operating points the delays are be different. Moreover, for high loads and high engine speeds it is not possible to neglect the injection duration.

3.2.2 DEM of the Air Flow Dynamics

Aside from the fuel flow, the air flow of an engine is the most important subsystem to be considered when designing an open-loop air/fuel ratio control system. The best possible drivability and the lowest possible emissions are only realized if a detailed model of the air flow dynamics is used in that design problem.

As shown in Sect. 2.3, the mean value model for the manifold dynamics can be described by the following equations (no EGR is considered here)

$$\frac{d}{dt}m_m(t) = \dot{m}_\alpha(t) - \dot{m}_\beta(t) \qquad (3.11)$$

An approximate description of the mass flow $\dot{m}_\beta(t)$ entering the cylinder is given by

$$\dot{m}_\beta(t) = \rho_m(t) \cdot \dot{V}(t) = \rho_m(t) \cdot \lambda_l(p_m(t), \omega_e(t)) \cdot \frac{V_d}{N} \cdot \frac{\omega_e(t)}{2\pi} \qquad (3.12)$$

For convenience, the definition (2.20) of the volumetric efficiency has been used here.[3]

Using (3.12) and $m_m = \rho_m \cdot V_m$, it is possible to rewrite (3.11) as follows

$$\frac{d}{dt}\left(\frac{\dot{m}_\beta(t)}{\lambda_l(p_m(t),\omega_e(t))}\cdot\frac{V_m}{V_d}\cdot N\cdot\frac{2\pi}{\omega_e(t)}\right) = \dot{m}_\alpha(t) - \dot{m}_\beta(t) \qquad (3.13)$$

This equation must be integrated over the time interval of one engine segment yielding

$$\int_{t_{k-1}}^{t_k}\frac{d}{dt}\left(\frac{\dot{m}_\beta(t)}{\lambda_l(p_m(t),\omega_e(t))}\cdot\frac{V_m}{V_d}\cdot N\cdot\frac{2\pi}{\omega_e(t)}\right)dt \qquad (3.14)$$

$$= \int_{t_{k-1}}^{t_k}\dot{m}_\alpha(t)dt - \int_{t_{k-1}}^{t_k}\dot{m}_\beta(t)dt \qquad (3.15)$$

$$(3.16)$$

where t_k is equal to $k\cdot\tau_{seg}$ and $k = 1,2,3,\ldots$ is the engine segment counter. Obviously, this equation can be reformulated as follows

$$m_{\alpha,k} - m_{\beta,k} = \frac{V_m}{V_d}\cdot N\cdot\frac{\dot{m}_\beta(t_k)}{\lambda_l(p_m(t_k),\omega_e(t_k))}\cdot\frac{2\pi}{\omega_e(t_k)} \qquad (3.17)$$

$$-\frac{V_m}{V_d}\cdot N\cdot\frac{\dot{m}_\beta(t_{k-1})}{\lambda_l(p_m(t_{k-1}),\omega_e(t_{k-1}))}\cdot\frac{2\pi}{\omega_e(t_{k-1})}$$

The symbol $m_{\alpha,k}$ represents the mass of air flowing through the throttle during one segment. Similarly, the symbol $m_{\beta,k}$ represents the mass of air entering the cylinder during the event k. The corresponding mass flows $\dot{m}_{...}(t)$ do not necessarily have to be constant. It is only assumed that they are periodic, with period τ_{seg}, such that the integration is independent of the location of the sampling event.

Since $\dot{m}_\beta$ and ω_e vary much more slowly than $\dot{m}_\alpha$, an Euler integration is usually acceptable for the computation of the mass entering the cylinder

$$m_k = \dot{m}_\beta(t_k)\cdot\tau_{seg} \qquad (3.18)$$

The variable $m_{\alpha,k}$ is the forcing term in the manifold system description (3.17). If the throttle angle command u_α is the actual input signal, the mass flow through the throttle must be calculated using (2.10) and numerical integration. However, often the mass flow $\dot{m}_\alpha$ entering the manifold is measured with a hot film anemometer. In that case, its measurement output can be used to calculate $m_{\alpha,k}$ directly using, again, a suitable numerical integration

[3] The error that has to be expected thereby is in the order of 5% for gasoline fuels. If the engine is fueled by natural gas, the error would be larger. In that case, the more precise, but also more cumbersome, approximation (2.19) is recommended.

method. Therefore, in the text following, the variable $m_{\alpha,k}$ will be taken as the input signal that drives the manifold dynamics.

Combining (3.17) and (3.18) yields

$$m_{\alpha,k} - m_{\beta,k} = \frac{(m_{\beta,k} - m_{\beta,k-1})}{\lambda_l(p_m,\omega_e)} \cdot \frac{V_m}{V_d} \cdot N \cdot \frac{2\pi}{\omega_e} \cdot \frac{1}{\tau_{seg}} \tag{3.19}$$

$$= \frac{(m_{\beta,k} - m_{\beta,k-1})}{\lambda_l(p_m,\omega_e)} \cdot \frac{V_m}{V_d} \cdot n_{cyl} \tag{3.20}$$

and introducing the parameter

$$X = \lambda_l(p_m,\omega_e) \cdot \frac{V_d}{V_m} \cdot \frac{1}{n_{cyl}} \tag{3.21}$$

a discrete event formulation for the mass m_k entering the cylinder during the k-th engine cyle is obtained

$$m_{\beta,k} = m_{\beta,k-1} \cdot \frac{1}{1+X} + m_{\alpha,k} \cdot \frac{X}{1+X} \tag{3.22}$$

Remarkably, assuming a slowly varying volumetric efficiency, the manifold dynamics become invariant in the crank-angle domain.

The last equation is the zero-order hold discrete-time equivalent of a continuous-time first-order lag element with gain equal to one (see Appendix A). The time constant τ of that continuous-time element is related to the parameter X by the following well-known relationship

$$e^{-\tau_{seg}/\tau} = \frac{1}{1+X} \tag{3.23}$$

Assuming that τ_{seg} is much smaller than the time constant τ, the following approximation is valid

$$\frac{1}{1+X} \approx \frac{1}{1+\frac{\tau_{seg}}{\tau}} \quad \Rightarrow \quad \tau \approx \frac{\tau_{seg}}{X} \tag{3.24}$$

and using the definition (3.21) of the parameter X, the following expression for the equivalent continuous-time time constant τ is obtained

$$\tau \approx \tau_{seg} \cdot n_{cyl} \cdot \frac{V_m}{V_d} \cdot \frac{1}{\lambda_l(p_m(t),\omega_e(t))} \tag{3.25}$$

Assuming $\lambda_l = 1$, this time constant corresponds to the time of one segment multiplied by the ratio of the size of the manifold volume to the cylinder displacement. Moreover, inserting the definition of the segment time (3.2) into the last equation, the following expression for τ is obtained

$$\tau \approx \frac{4\pi}{\omega_e} \cdot \frac{V_m}{V_d} \tag{3.26}$$

The result is equal to the manifold time constant $\tilde{\tau}_p$ (2.135) obtained in Sect. 2.5.2 in a rather different context.

3.2.3 DEM of the Fuel-Flow Dynamics

Introduction

The fuel injection in a sequentially injected SI engine is inherently synchronized with the engine's crank angle. The correct identification of the delays present in this loop is important for the design of the feedforward and the feedback action of the corresponding control system. A simple, but useful discrete-event formulation of the fuel dynamics is shown below.

The primary question that must be answered in this subsection is how the single command issued by the fuel-control system is branched to the n_{cyl} different cylinders of the engine. Similarly, the same cylinders produce n_{cyl} different air/fuel ratio signals that are all recorded by only one air/fuel ratio sensor. Of course, one approach would be to treat all cylinders individually. This solution offers the greatest flexibility but imposes a large computational burden. The models of the fuel-flow dynamics presented in this section offer an optimal compromise between computational efforts and the performance of the control loop designed based on these models.

Multi-Cylinder Engines as Multiplexed Systems

In a first step, a single cylinder of the engine is analyzed. The mean-value model of the injection dynamics derived in Sect. 2.4.2 assumes a continuous flow of fuel into the cylinders. A discrete-event formulation, which predicts the total amount of fuel that enters the cylinder in the k-th cycle, can be obtained by using a $\mathcal{Z}$-transformation approach. Assuming first-order fuel dynamics (2.60)

$$G(s) = (1 - \kappa) + \kappa \cdot \frac{1}{s\tau + 1} \tag{3.27}$$

and a constant engine speed, the following discrete-time system description can be found for one cylinder

$$G(z) = (1 - \kappa) + \kappa \cdot \frac{1 - e^{-\frac{n_{cyl}\tau_{seg}}{\tau}}}{z - e^{-\frac{n_{cyl}\tau_{seg}}{\tau}}} \tag{3.28}$$

Note that in this step the relevant delay is the *cycle* time defined by the expression

$$\tau_{cycl} = n_{cyl} \cdot \tau_{seg} \tag{3.29}$$

In the second step, all cylinders of the engine are analyzed simultaneously. Of course, in a port-injected engine the fuel path includes n_{cyl} separate wall-wetting dynamics. Similarly, the gas mixing dynamics of those parts of the exhaust manifold immediately after the exhaust valves evolve in parallel as well. These dynamic elements obtain their inputs in a sequential way synchronized to the crank angle, and they deliver their output signals in the same manner.

On an abstract level, such a multi-cylinder system can be modeled as a *multiplexed system*. Using this approach, a multiplexing element is assumed to act at the input and at the output of the system. Figure 3.7 shows the structure of such a system, including the necessary sample-and-hold elements.

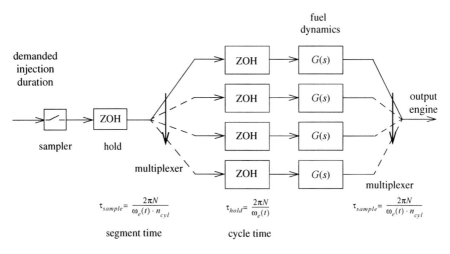

Fig. 3.7. Continuous-time fuel dynamics of a port-fuel-injected four-cylinder engine.

The input signal is sampled with the segment time τ_{seg}. Accordingly, in each segment a new value enters the multiplexer. In accordance with the chosen firing sequence, these values are distributed to the different injection valves. The individual wall-wetting dynamics are synchronized with the engine cycle time τ_{cycl} defined in (3.29). Therefore, the block diagram seen in Fig. 3.7 can be redrawn as shown in Fig. 3.8. The input and output of that system are continuous-time signals. Therefore, there must be a continuous-time transfer function $G_{total}(s)$ that describes the dynamics of that module. Again assuming the fuel dynamics for each cylinder to be of the first order, the system depicted in Fig. 3.8 is represented by the following transfer functions

$$G_{total}(s) = G_{hold}(s) \cdot G_{fd}(z)\big|_{z=e^{s\tau_{seg}}} \tag{3.30}$$

$$= \frac{1 - e^{-s\tau_{seg}}}{s\tau_{seg}} \cdot \left((1-\kappa) + \kappa \cdot \frac{1 - e^{-\frac{n_{cyl}\tau_{seg}}{\tau}}}{z^{n_{cyl}} - e^{-\frac{n_{cyl}\tau_{seg}}{\tau}}} \right)\Bigg|_{z=e^{s\tau_{seg}}}$$

In this expression, the sample-and-hold block shown in Fig. 3.7 is represented by $G_{hold}(s)$. The second term represents the wall-wetting dynamics. The only difference to the single-cylinder case is that the z-variable in the single-cylinder formulation (3.28) must be substituted with $z^{n_{cyl}}$ because the sampling in the multi-cylinder case is n_{cyl} times faster than in the single-cylinder case. Accordingly, the single cylinder, which still operates on the same cycle duration as before, must be delayed by n_{cyl} sampling intervals.

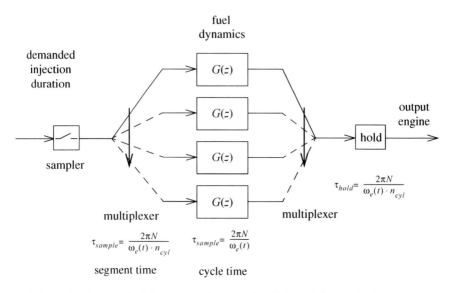

Fig. 3.8. Discrete fuel dynamics of a port-fuel-injected four-cylinder engine.

Assuming a four-cylinder engine, the state-space description of this system has the form

$$x_{k+1} = \begin{bmatrix} 0 & 0 & 0 & F \\ 1 & 0 & 0 & 0 \\ 0 & 1 & 0 & 0 \\ 0 & 0 & 1 & 0 \end{bmatrix} \cdot x_k + \begin{bmatrix} 1-F \\ 0 \\ 0 \\ 0 \end{bmatrix} \cdot u_k$$

$$y_k = \begin{bmatrix} 0 & 0 & 0 & \kappa \end{bmatrix} \cdot x_k + (1-\kappa) \cdot u_k$$

(3.31)

where $F = e^{\frac{-n_{cyl} \cdot \tau_{seg}}{\tau}}$ is the only non-trivial parameter. For each new segment, a new fueling value is calculated, discretizing the associated continuous-time description with one engine cycle. Then this value is shifted $(n_{cyl} - 1)$ times until it reaches the output. This permits each cylinder to operate with its own fuel dynamics. Therefore, defining $a = 1 - F$, the block diagram of this discrete system can be drawn as shown in Fig. 3.9.

Figure 3.10 shows the response of the system (3.31) to a step input (left plot) and to a harmonic input (right plot). The first plot suggests that the system includes a sample-and-hold element with a delay equal to $n_{cyl} \cdot \tau_{seg}$, while in the second plot the same system seems to include a phase lag and a much shorter sample-and-hold delay equal to τ_{seg}. This is easy to explain with the system description (3.31):

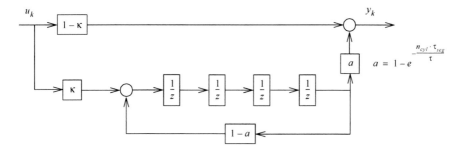

Fig. 3.9. Block diagram of the discrete-time dynamics of the fuel system.

- When a step-wise input is applied, all cylinders obtain the same new values for the injection timing such that the output is similar to the response of a single-cylinder engine.
- When a sinusoidal input is applied, the injection command changes from segment to segment. Accordingly, each of the output signals changes in the same way.

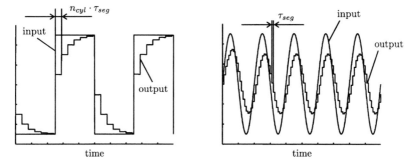

Fig. 3.10. Simulation of the multiplexed system.

Analysis of the Complete System

Many dynamic effects in the fueling system can be understood best by measuring frequency responses. The following measurements were all conducted on a 1.8-liter, four-cylinder, four-valves-per-cylinder, SI engine. Figure 3.11 shows a typical measurement setup for such an experiment.

Frequency response measurements are chosen because they permit the correct interpretation of the aliasing effects that are caused by the discrete-event behavior. Although the system is sampled in a segment mode, the individual fuel dynamics are excited only once every engine cycle. Therefore, for the identification of the fuel dynamics, only frequencies up to $\omega_e/4$ satisfy the

Shannon theorem, by which the slowest sampling defines the limits. The input modulation frequency can be chosen higher than this limit, and the resulting frequency response must be matched by the system model derived above using physical first principles, provided the system model is sufficiently close to the real system. Of course, the inverse is not true, i.e., at higher frequencies it is not possible to find a unique mathematical model using only the measured frequency response. The ambiguities introduced by the sampling process at frequencies higher than the Nyquist frequency $\omega_e/4$ preclude this.

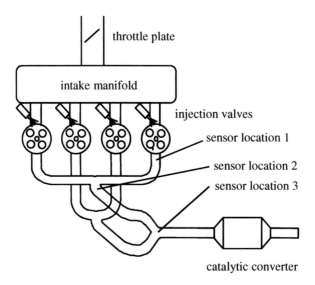

Fig. 3.11. Measurement setup for the fuel path investigations.

Depending on the output chosen the fuel dynamics incorporate different subsystems:

- *Engine Torque:* The fuel dynamics contain wall-wetting phenomena and back flow from the cylinder into the intake manifold.
- *Air/Fuel Ratio:* The fuel dynamics contain wall-wetting phenomena, back flow from the cylinder into the intake manifold, as well as gas mixing in the exhaust manifold and transport delays between the actuator and sensor locations.

The engine used in the subsequent tests is equipped with a symmetric 4-in-2-in-1 exhaust manifold, as shown in Fig. 3.11. The main advantage of this configuration is that the gas mixing dynamics may be assumed to be identical for all cylinders. Output measurements were made at three locations with a BOSCH LSU 4 wide-range air/fuel sensor, which can be modeled well as a first-order lag with a time constant of approximately 20 *ms*. All measurements were performed at 2000 *rpm* and 50% of full load. As shown in Fig. 3.12, the chosen input signal is the fuel command u_φ, while the output is the signal $y_{\lambda,i}$ of one of the three wide-range air/fuel ratio sensors.

Fig. 3.12. Main dynamic subsystems in the fuel dynamics of an SI engine.

The structure of the complete system consisting of several delays, the wall-wetting dynamics, the sensor dynamics and the gas mixing dynamics, is shown in Fig. 3.12. Such an input-output setting of course does not permit the identification of the contribution of any single module to the total frequency response of a system. Therefore, the individual sub-blocks of the fuel path of an SI engine must be identified and validated sequentially.

The dynamic behavior of the torque generation has been described in Sect. 3.2.1 such that in this section the focus is on the air/fuel ratio dynamics. As a first step, the air/fuel ratio immediately after the exhaust valves is chosen as the output signal. To avoid any gas mixing dynamics, the corresponding sensor must be placed as close as possible to one of the exhaust valves. Figure 3.13 shows the Bode plot of the frequency response from the injection quantity u_φ to the corresponding air/fuel ratio signal $y_{\lambda,1}$.

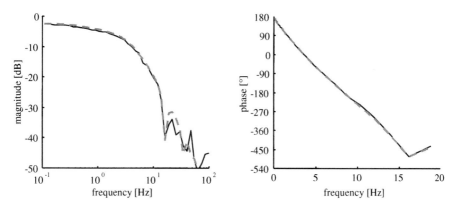

Fig. 3.13. Frequency response measurement (solid) and simulation (dashed) from fuel input u_φ to the output $y_{\lambda,1}$ of the LSU sensor at the exhaust valve.

To identify the parameters of this system, first the structure of the transfer function must be fixed. A series connection of a sample-and-hold element with sampling time $4\tau_{seg}$, an additional delay[4] τ_{delay}, the wall-wetting dynamics,

---

[4] This delay includes all delays from the update injection to the exhaust center event.

and the first-order lag with time constant τ_{LSU} caused by the sensor dynamics is proposed.

The wall-wetting subsystem is discussed first. At low engine temperatures, a first-order wall-wetting model does not sufficiently well reproduce all the dynamic effects observed. Therefore, a second parallel first-order lag is added,[5] resulting in the following continuous-time representation for the wall-wetting dynamics

$$G(s) = (1 - \kappa_1 - \kappa_2) + \frac{\kappa_1}{\tau_1 s + 1} + \frac{\kappa_2}{\tau_2 s + 1} \qquad (3.32)$$

When the engine is fully warmed up, typical values for the parameters are

$$1 - \kappa_1 - \kappa_2 \approx 0.5, \quad \kappa_1 \approx 0.15, \quad \tau_1 \approx 0.5 \ s, \quad \tau_2 \approx 0.05 \ s \qquad (3.33)$$

For a cold engine, the following values are more realistic choices

$$1 - \kappa_1 - \kappa_2 \approx 0.4, \quad \kappa_1 \approx 0.4, \quad \tau_1 \approx 1 \ s, \quad \tau_2 \approx 0.1 \ s. \qquad (3.34)$$

Of course, these values are different for each operating point and for each engine system. The numerical values given here just show the orders of magnitude that can be expected.

The discrete-time representation of the wall-wetting dynamics (3.32) of the four-cylinder engine analyzed above reads as follows

$$G_{wwd}(z) = (1 - k_1 - k_2) + k_1 \cdot \frac{1 - e^{-\frac{4\tau_{seg}}{\tau_1}}}{z - e^{-\frac{4\tau_{seg}}{\tau_1}}} + k_2 \cdot \frac{1 - e^{-\frac{4\tau_{seg}}{\tau_2}}}{z - e^{-\frac{4\tau_{seg}}{\tau_2}}} \qquad (3.35)$$

The transfer function of the complete system can be written as follows

$$G(s) = \frac{1 - e^{-4\tau_{seg}s}}{4\tau_{seg}s} \cdot e^{-\tau_{delay}s} \cdot G_{wwd}(z)\big|_{z=e^{4\tau_{seg}s}} \cdot \frac{1}{\tau_{LSU}s + 1} \qquad (3.36)$$

Note that in the case where the sensor is placed close to the exhaust valve of one cylinder, the air/fuel ratio dynamics are essentially those of a single-cylinder engine. Accordingly, in (3.36) the sampling time is equal to the cycle time $\tau_{cycl} = 4\tau_{seg}$.

If the structure, the gains, and the time constants are correct, the frequency response obtained with (3.36) should be close to that measured at frequencies even higher than the Nyquist frequency. As Fig. 3.13 shows, this is the case indeed. A crucial element in the loop is the sample-and-hold element introduced to model the reciprocating engine behavior. This element causes the strong drop in amplitude at approximately 16 Hz, which corresponds to the sampling frequency of a single cylinder at 2000 rpm. Modulating the fuel input with this frequency results in a constant injection duration for each cylinder, i.e., every cylinder operates under constant but different conditions. This also holds for integer multiples of that frequency.

---

[5] This additional element also models the back flow of combustion gases into the manifold after BDC when the intake valve is not yet completely closed.

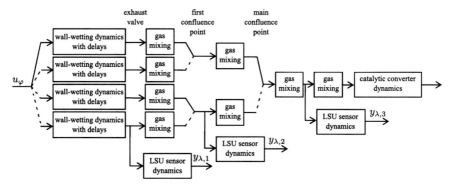

Fig. 3.14. Flowchart of the complete system investigated.

The approaches developed so far are now applied to the multi-cylinder case. The structure of that system is depicted in Fig. 3.14. Combining the multiplexing modeling approach of the wall-wetting dynamics with the gas mixing dynamics and using again the ideas of multiplexing, the structure of the complete system with input u_φ and output $y_{\lambda,3}$ is found to be

$$G(s) = \frac{1 - e^{-\tau_{seg}s}}{\tau_{seg}s} \cdot \left.\widetilde{G}_{wwd}(z)\right|_{z=e^{\tau_{seg}s}} \cdot \left.\left(\frac{z^4(b_1 z^4 + b_2)}{z^8 - a_1 z^4 - a_2}\right)\right|_{z=e^{\tau_{seg}s}} \quad (3.37)$$

$$\cdot \left.\left(z^2 \frac{1 - e^{-\frac{2\tau_{seg}}{\tau_3}}}{z^2 - e^{-\frac{2\tau_{seg}}{\tau_3}}}\right)\right|_{z=e^{\tau_{seg}s}} \cdot \left(\frac{\omega_0^2}{s^2 + 2\zeta\omega_0 s + \omega_0^2}\right) \cdot \frac{e^{-\tau_{delay}s}}{1 + \tau_{LSU}s}$$

where

$$\widetilde{G}_{wwd}(z) = \left((1 - k_1 - k_2) + k_1 \frac{1 - e^{-\frac{4\tau_{seg}}{\tau_1}}}{z^4 - e^{-\frac{4\tau_{seg}}{\tau_1}}} + k_2 \frac{1 - e^{-\frac{4\tau_{seg}}{\tau_2}}}{z^4 - e^{-\frac{4\tau_{seg}}{\tau_2}}}\right) \quad (3.38)$$

The various parts in this expression model the following effects:

1. The sample-and-hold element is represented in its usual form.
2. The wall-wetting dynamics (in fact all dynamic effects taking place between injection location and exhaust valve) are approximated by a second-order element.
3. The gas mixing phenomena in the exhaust tubes from the exhaust valve to the first confluence point are described by a second-order element that is multiplexed. The structure of this element is chosen such that it reflects the underlying phenomena caused by the multiplexed operation of four exhaust parts. This choice reduces to the bare minimum the number of parameters that must be identified experimentally.
4. Similarly, the exhaust pipes between first confluence point and main confluence point can be approximated by a first-order element, multiplexed with one-half of the segment frequency.

5. The gas mixing in the main confluence point then again is a continuous element, which can be described by two additional states. Interestingly, these dynamics are best described by a second-order system. This approach permits the combination of the two extreme situations of *plug flow* and *perfectly stirred reactor* conditions, as described in [128] and [216].

6. The overall delay and the LSU sensor dynamics follow as the last elements.

The corresponding parameters must be identified using a sequential approach. Starting with the model description (3.36), obtained for the case where the sensor is mounted close to one of the exhaust valves, the subsequent dynamic elements can be identified by using intermediate measurement signals. Figure 3.15 shows the measured and the predicted frequency response of the full system with the air/fuel ratio sensor at position 3.

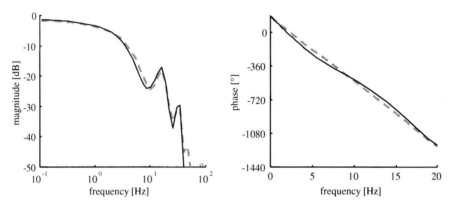

Fig. 3.15. Frequency response measurement (solid) and simulation (dashed) from fuel excitation to the output $y_{\lambda,3}$ of the LSU sensor at the main confluence point.

As Fig. 3.15 shows, the model represented by (3.37) well predicts the measured frequency response. The system is a low-pass element with a first corner frequency at approximately 2 Hz. At 16.66 Hz a local maximum is attained because at that frequency the fuel path of all cylinders operates at steady-state conditions,[6] which eliminates the influence of the wall-wetting dynamics. The same observation is true at $2 \cdot 16.66$ Hz and $3 \cdot 16.66$ Hz. At $4 \cdot 16.66$ Hz (not shown in Fig. 3.15) the sample-and-hold element introduces a zero such that at that frequency no local maximum is attained. The continuous-time elements in (3.37) cause a steady drop in magnitude and phase. The latter is dominated by the delays, as the left part of Fig. 3.15 shows (notice that the frequency in this diagram is linear).

---

[6] The fuel modulation and the multiplexing frequency are identical. Hence, all cylinders obtain different, but constant fuel quantities.

3.2.4 DEM of the Back-Flow Dynamics of CNG Engines

For several reasons, compressed natural gas (CNG) is an interesting fuel. For instance, it permits a clean and quiet combustion, produces low non-methane emissions even at cold start, features low CO_2 emission, and its high octane numbers are well suited to supercharged engine applications. Moreover, it is readily available in many places at relatively low cost. Compared to energy carriers based on liquid hydrocarbons, its main disadvantages are its smaller energy density, the demand for additional energy to compress the gas, leakage losses in production and transportation, and the currently small number of operational fueling stations.

The majority of all CNG engines are spark-ignited, port-injected engines that operate in an Otto cycle. In their dynamic behavior, such CNG engines are similar to SI engines. The cause and effect diagram shown in Fig. 2.5 remains valid in principle for such CNG engines as well.

The only three major differences that must be considered are:

- Compared to gasoline, natural gas occupies a substantial part of the aspirated volume. This requires that the detailed model of the engine mass flow dynamics, as expressed in equation (2.27), must be used in simulations.
- Although with a gaseous fuel no wall-wetting phenomena occur, the fuel path exhibits a similar dynamic effect due to a substantial back flow of aspirated mixture.
- Since the fuel is gaseous, it can be injected while the intake valves are open without having to accept an associated penalty in hydrocarbon emissions. This permits a certain amount of charge stratification. Therefore, this effect must be included in the model of the back-flow dynamics.

Few authors have discussed the control-oriented modeling of CNG engines, e.g., [168]. In the following, a model of the back-flow dynamics is presented which was first introduced in [54]. The discrete-event formulation of this model, as shown below, requires the following variables to be defined first:

m_e Mass of NG entering the cylinder.

m_b Mass of NG flowing from the cylinder back to the intake manifold.

m_c Mass of NG remaining in the cylinder when the intake valve closes.

m_l Mass of NG injected after back flow has started.

m_t Total mass of NG injected into the intake manifold.

V_a Cylinder volume filled with air.[7]

V_m Cylinder volume filled with air and NG mixture.

V_b Cylinder volume that flows back to the intake manifold.

Figure 3.16 shows the status of the cylinder at its bottom dead center (BDC). Simplifying the complex interactions that take place in a real gas-exchange process, it is assumed that the cylinder charge consists of a first volume, V_a, filled with air and a second volume, V_m, filled with a mixture of NG and air. It is assumed that the maximum amount of NG entering the cylinder m_e is achieved at BDC, that the back flow takes place after that point, and that no mixing between V_m and V_a must be considered.

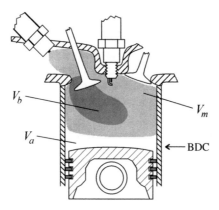

Fig. 3.16. Simplified engine stratification model used in the derivation of the back-flow model.

The following three dimensionless variables will be used to compactly formulate the model. The *back-flow* parameter α defines the relative amount of NG flowing back into the intake manifold with respect to the maximum aspirated amount of NG. Assuming a homogeneous mixture yields

$$\alpha = \frac{V_b}{V_m + V_a} \tag{3.39}$$

Obviously, a value of $\alpha = 0$ indicates no back flow.

The *stratification* parameter β describes the ratio between the total charge and the mixture

$$\beta = \frac{V_m + V_a}{V_m} \tag{3.40}$$

In the case of an unrealistically large stratification factor $\beta = 1/\alpha$, all of the aspirated NG would flow back again.

Finally, the *overlap* factor ξ indicates the relative amount of fuel that is injected after back flow has started, i.e., that part of the fuel that will not enter the cylinder in the actual cycle. Obviously, a value of $\xi = 0$ indicates that all injected fuel enters the cylinder (but does not necessarily stay there due to back flow), while a value of $\xi = 1$ indicates that all of the fuel injected in the current cycle will be aspirated into the cylinder in the next cycle.

Using the definitions introduced above, the dynamic behavior of one cylinder of an n-cylinder engine is described in the k-th segment by the following set of equations

$$m_b(k) = \alpha(k) \cdot \beta(k) \cdot m_e(k) \qquad (3.41)$$

$$m_l(k) = \xi(k) \cdot m_t(k) \qquad (3.42)$$

$$m_e(k) = m_b(k-n) + m_l(k-n) + m_t(k) - m_l(k) \qquad (3.43)$$

$$m_c(k) = m_e(k) - m_b(k) \qquad (3.44)$$

These equations essentially represent a mass balance with one reservoir one input, and one output, where the reservoir is intake manifold , the input is the total mass $m_t(k)$ of NG injected, and the output is the mass $m_c(k)$ of NG taking part in the combustion in the k-th cycle. The mass $m_b(k-n) + m_l(k-n)$ is stored in the intake manifold between two cycles.

The variables $\{\alpha(k), \beta(k), \xi(k)\}$ completely describe the intrinsic dynamic properties of the back flow dynamics. As is the case in the wall-wetting dynamics in port-injected gasoline engines, these parameters generally are non-linear functions of the engine's load and speed and of many other influencing variables.

For small changes around a nominal operating point, the parameters $\{\alpha(k), \beta(k), \xi(k)\}$ may be assumed to be constant such that the following DEM transfer function formulation may be derived

$$m_c(z) = P(z) \cdot m_t(z) \qquad (3.45)$$

with

$$P(z) = (1 - \alpha \cdot \beta) \cdot (1 - \xi) + \frac{(1 - \alpha \cdot \beta) \cdot (\xi + \alpha \cdot \beta \cdot (1 - \xi))}{z^n - \alpha \cdot \beta} \qquad (3.46)$$

As in the wall-wetting dynamics, a feedthrough element and a "memory" element can be seen in this expression. As expected, the static gain $P(1)$ of the transfer function (3.46) is equal to 1.

The parameters $\{\alpha, \beta, \xi\}$ have to be identified using experiments that are essentially the same as those used in Sect. 3.2.3 in the case of gasoline engines. The well-known correlation between air/fuel ratio and NO_x emissions and the availability of very fast NO_x sensors permit the identification of the unknown parameters using step responses.

Figure 3.17, taken from [54], shows the results for one specific case in which the values identified were $\{\alpha = 0.16, \beta = 1.6, \xi = 0\}$.

3.2.5 DEM of the Residual Gas Dynamics

The residual gas is the burnt gas remaining in the cylinder after the exhaust valve has closed. Especially for varying air/fuel ratios, this mass and its corresponding air/fuel ratio may not be neglected. In this section, a DEM of the residual gas dynamics is introduced that is useful for an understanding of the main effects.

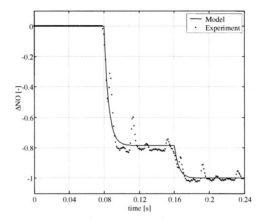

Fig. 3.17. Model/experiment validation of the back flow dynamics for a 1.8-liter, four-cylinder SI engine. Normalized step response of the transfer function (3.46). Engine operating point $\omega_e = 150 \; rad/s$, $p_{me} = 3.8 \; bar$. The simulation combines the discrete-event model with a continuous-time model of the sensor dynamics.

The variable $\tilde{\lambda}(k)$ is used to denote the air/fuel ratio of the mixture that enters the cylinder in the k-th cycle. According to Sect. 2.4.3, this variable is defined as

$$\tilde{\lambda}(k) = \frac{m_\beta(k)}{m_\varphi(k) \cdot \sigma_0} \tag{3.47}$$

Using this definition, the total mass of fresh mixture in the cylinder in the k-th cycle can be expressed as follows

$$m_{fc}(k) = m_\beta(k) + m_\varphi(k) \tag{3.48}$$

$$= m_\beta(k) \left(1 + \frac{1}{\tilde{\lambda}(k) \cdot \sigma_0} \right) \tag{3.49}$$

$$= m_\varphi(k) \cdot (\tilde{\lambda}(k) \cdot \sigma_0 + 1) \tag{3.50}$$

Similar equations describe the dynamics of the mass of residual gas $m_{rg}(k)$. The air/fuel ratio of the residual gas is equal to the air/fuel ratio of the total mass[8] at event t_{k-1}, assuming a sampling time of one engine cycle. Therefore, formulating a mass balance of both the fresh charge and the burnt gases present in the cylinder after the intake valve closed yields the following equation for the air/fuel ratio $\lambda(k)$ of the cylinder contents

---

[8] This definition is based on a carbon-to-hydrogen balance, which is valid even after the combustion. This approach is used in [97], for instance, to derive the air/fuel ratio by analyzing the exhaust gases.

$$\lambda(k) = \frac{m_{fc}(k) \cdot \frac{\tilde{\lambda}(k) \cdot \sigma_0}{1 + \tilde{\lambda}(k) \cdot \sigma_0} + m_{rg}(k) \cdot \frac{\lambda(k-1) \cdot \sigma_0}{1 + \lambda(k-1) \cdot \sigma_0}}{m_{fc}(k) \cdot \frac{1}{1 + \tilde{\lambda}(k) \cdot \sigma_0} + m_{rg}(k) \cdot \frac{1}{1 + \lambda(k-1) \cdot \sigma_0}} \cdot \frac{1}{\sigma_0} \tag{3.51}$$

This equation can be transformed into the following (nonlinear) equation

$$\lambda(k) = \frac{m_{fc}(k)\tilde{\lambda}(k)(1 + \lambda(k-1)\sigma_0) + m_{rg}(k)\lambda(k-1)(1 + \tilde{\lambda}(k)\sigma_0)}{m_{fc}(k)(1 + \lambda(k-1)\sigma_0) + m_{rg}(k)(1 + \tilde{\lambda}(k)\sigma_0)} \tag{3.52}$$

As expected, for steady-state conditions, i.e., for $\lambda(k) = \lambda(k-1)$, the identity $\lambda = \tilde{\lambda}$ results. Moreover, since in most situations $\lambda(k) \cdot \sigma_0 \gg 1$, the last equation may be simplified

$$\lambda(k) \approx \frac{m_{fc}(k) \cdot \tilde{\lambda}(k) \cdot \lambda(k-1) + m_{rg}(k) \cdot \lambda(k-1) \cdot \tilde{\lambda}(k)}{m_{fc}(k) \cdot \lambda(k-1) + m_{rg}(k) \cdot \tilde{\lambda}(k)} \tag{3.53}$$

$$\approx [m_{fc}(k) + m_{rg}(k)] \cdot \left[\frac{m_{fc}(k)}{\tilde{\lambda}(k)} + \frac{m_{rg}(k)}{\lambda(k-1)} \right]^{-1} \tag{3.54}$$

This equation can be reformulated as

$$\frac{m_{fc}(k) + m_{rg}(k)}{\lambda(k)} \approx \frac{m_{fc}(k)}{\tilde{\lambda}(k)} + \frac{m_{rg}(k)}{\lambda(k-1)} \tag{3.55}$$

which can be interpreted as a weighted sum of the corresponding amounts of gases. If, in addition, it can be assumed that the air/fuel ratio does not deviate too much from its stoichiometric value of 1, then the following simplification can be made

$$\frac{1}{\lambda} \approx 1 - (\lambda - 1) = 2 - \lambda \tag{3.56}$$

This leads to the formulation that has already been used in Sect. 2.4.3

$$\lambda(k) \approx \frac{m_{fc}(k) \cdot \tilde{\lambda}(k) + m_{rg}(k) \cdot \lambda(k-1)}{m_{fc}(k) + m_{rg}(k)} \tag{3.57}$$

Clearly, (3.57) is still a nonlinear equation because m_{fc} and m_{rg} are varying quantities. However, if the fuel quantity is varied only, m_{fc} and m_{rg} can be approximated well by constant values. In this case, the air/fuel ratio dynamics are well approximated by a discrete first-order lag element.

It is usually not easy to obtain information on the precise amount of burnt gas that remains inside the cylinder. A very rough estimation would be a value for m_{rg} of 5% of the full-load total mass in the cylinder. A better approximation is obtained if the mass of residual gas is estimated using an affine relationship of the form

$$m_{rg} = a_{rg} \cdot \frac{1}{\omega_e} + b_{rg} \tag{3.58}$$

Figure 3.18 shows as an example the mass of burnt gas that remains in the cylinder for a six-cylinder 3.2-liter SI engine, which at full load has a total cylinder charge mass of approximately 450 mg.

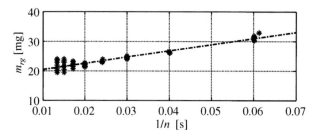

Fig. 3.18. Residual burnt-gas mass as a function of the time per revolution of a six-cylinder 3.2-liter SI engine. Stoichiometric combustion, engine speed $1000 - 4500$ rpm, engine load $p_{me} = 0 - 6$ bar.

3.2.6 DEM of the Exhaust System

In this section, a DEM of the effects taking place in the exhaust system is presented. One motivation for doing this is the objective of estimating the air/fuel ratios of each cylinder[9] using one air/fuel ratio sensor only. The main simplifications assumed in this section are steady-state operating conditions, i.e., each cylinder operating at constant but different air/fuel ratios, and a symmetric layout of the exhaust manifold. Under these conditions a compact model formulation can be derived as first proposed in [88].

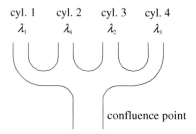

Fig. 3.19. Idealized exhaust manifold model.

The symmetric exhaust layout assumed in this section is shown in Fig. 3.19. Since all cylinders are operated in constant conditions, such a symmetric layout results in periodic air/fuel ratio signals at the confluence point where the sensor is assumed to be placed. Accordingly, the air/fuel ratio λ_{cfp} of the exhaust gas that the sensor observes can be approximated by the following

---

[9] Manufacturing tolerances for the injection valves and intake ports, as well as different flow characteristics in the intake runners, result in different air/fuel ratios for each cylinder. Variations in the air/fuel ratios can cause drivability and emission problems, especially when the engine is cold.

expression[10]

$$\lambda_{cfp}(k) = c_1(k) \cdot \lambda_1 + c_2(k) \cdot \lambda_2 + c_3(k) \cdot \lambda_3 + c_4(k) \cdot \lambda_4 \qquad (3.59)$$

where k is the segment counter and λ_i the air/fuel ratio of the i-th cylinder. The coefficients $c_i(k)$ are functions of the segment counter. They are periodic and their sum is equal to 1 for all k. Figure 3.20 shows an example of these variables. The specific form of the curves depends on the engine operating condition and on the manifold layout.

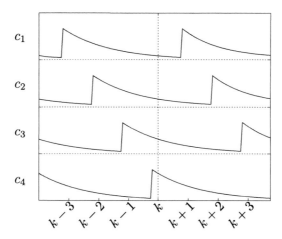

Fig. 3.20. Trajectories of the weighting functions $c_i(k)$.

Instead of using periodic coefficients $c_i(k)$, a cyclic permutation of the four λ_i signals can be used to describe the same behavior. In this case, the following permutation law must be adopted

$$\begin{bmatrix} \widetilde{\lambda}_1(k+1) \\ \widetilde{\lambda}_2(k+1) \\ \widetilde{\lambda}_3(k+1) \\ \widetilde{\lambda}_4(k+1) \end{bmatrix} = \begin{bmatrix} 0 & 1 & 0 & 0 \\ 0 & 0 & 1 & 0 \\ 0 & 0 & 0 & 1 \\ 1 & 0 & 0 & 0 \end{bmatrix} \cdot \begin{bmatrix} \widetilde{\lambda}_1(k) \\ \widetilde{\lambda}_2(k) \\ \widetilde{\lambda}_3(k) \\ \widetilde{\lambda}_4(k) \end{bmatrix} \qquad \begin{bmatrix} \widetilde{\lambda}_1(0) \\ \widetilde{\lambda}_2(0) \\ \widetilde{\lambda}_3(0) \\ \widetilde{\lambda}_4(0) \end{bmatrix} = \begin{bmatrix} \lambda_1 \\ \lambda_2 \\ \lambda_3 \\ \lambda_4 \end{bmatrix} \qquad (3.60)$$

where λ_i is the air/fuel ratio of the cylinder i and $\widetilde{\lambda}_i(k)$ is the i-th fictitious air/fuel ratio signal that is used to define the output signal

$$\lambda_{cfp}(k) = \begin{bmatrix} \widetilde{c}_1 & \widetilde{c}_2 & \widetilde{c}_3 & \widetilde{c}_4 \end{bmatrix} \cdot \begin{bmatrix} \widetilde{\lambda}_1(k) \\ \widetilde{\lambda}_2(k) \\ \widetilde{\lambda}_3(k) \\ \widetilde{\lambda}_4(k) \end{bmatrix} \qquad (3.61)$$

---

[10] This formulation is valid for a four-cylinder engine. The model can be easily adapted to an arbitrary number of cylinders.

where the *constants* $\tilde{c}_i$ are obtained by taking the explicit numerical values of $c_i(k)$ for any k. The variable $\lambda_{cfp}(k)$ is the air/fuel ratio at the confluence point at event k. Of course, the sensor dynamics must be taken into account as well. Assuming the usual first-order time lag model with linearized static behavior, the following result is obtained

$$\lambda_{meas}(k+1) = a \cdot \lambda_{meas}(k) + (1-a) \cdot \lambda_{cfp}(k) \qquad (3.62)$$

where

$$a = e^{-\frac{\tau_{seg}}{\tau_{sensor}}} \qquad (3.63)$$

Since the sensor time constant is fixed in the time domain, the discrete representation of the sensor dynamics must be recalculated every time a sample is taken.

3.3 DEM Based on Cylinder Pressure Information

3.3.1 General Remarks

The pressure of the gases in the cylinder provides the most direct signal available for engine control purposes. Since the interpretation of this variable is a well-known tool in ICE research and development, numerous control algorithms based on that information have been proposed (see, e.g., [150], [151], [183] [212]). However, the equipment used in those experiments is far too expensive to be included in series production engines. Recently, new in-cylinder pressure sensors have been developed that are sufficiently robust and reasonably priced such that their inclusion in series production vehicles appears feasible (see e.g., [95] and [205]).

Since full process calculations are not yet possible in the ECU, algorithms requiring a reduced amount of computational burden are of high interest. In [140], such simplified correlations useful for a control-oriented cylinder pressure interpretation have been derived. The highlights of that approach are reviewed in this section.

The pressure in the cylinder is determined by the movement of the piston and the corresponding compression of the air/fuel mixture inside the cylinder, the heat release caused by the combustion of this mixture, and the heat transfer from the gas to the cylinder walls. The position of the piston can be described by

$$s(\phi) = r_{cs} \cdot \left(\frac{l + r_{cs}}{r_{cs}} - \sqrt{\frac{l^2}{r_{cs}^2} - \sin^2\phi} - \cos\phi \right) \qquad (3.64)$$

where $s(\phi)$ is the distance of the piston from its TDC position, ϕ is the crank angle ($\phi = 0$ at TDC), r_{cs} is the radius of the crankshaft, i.e., one-half of the stroke, and l is the length of the connecting rod. A good approximation of (3.64) is given by

$$s(\phi) \approx r_{cs} \cdot \left(1 - \cos\phi + \frac{r_{cs}}{2 \cdot l} \sin^2\phi\right) \tag{3.65}$$

The equations describing the piston speed and the piston acceleration, in particular, are easier to formulate using (3.65). Denoting the compression ratio by ε and the displaced volume by V_d, the following equation holds for the cylinder volume V_c

$$V_c(\phi) = V_d \cdot \left(\frac{1}{\varepsilon - 1} + \frac{s(\phi)}{2 \cdot r_{cs}}\right) \tag{3.66}$$

3.3.2 Estimation of Burned-Mass Fraction

In "forward" thermodynamic process calculations, the development of combustion is often described with the burnt-mass fraction $x_B(\phi)$ defined as

$$x_B(\phi) = \frac{m_{\varphi b}(\phi)}{m_{\varphi t}} \tag{3.67}$$

where $m_{\varphi b}(\phi)$ is the mass of burnt fuel at crank angle ϕ and $m_{\varphi t}$ is the total burnt fuel at the end of combustion. For rich conditions, the total mass of burnt fuel must be calculated with the available mass air, thus $m_{\varphi t} = \lambda \cdot m_\varphi$ where m_φ is the fuel mass in the cylinder at IVC. Under lean conditions, $m_{\varphi t}$ is simply m_φ.

The burnt-mass fraction (3.67) is usually parameterized using the well-known Vibe function, [208] and Appendix C

$$x_B(\phi) = 1 - e^{c \cdot \left(\frac{\phi - \phi_s - \Delta\phi_d}{\Delta\phi_{bd}}\right)^{m_v + 1}} \tag{3.68}$$

where the parameter $\Delta\phi_d$ indicates the spark delay, $\Delta\phi_{bd}$ represents the duration of the combustion, $c \approx \log(0.001)$ indicates that the end of combustion is assumed at $x_B = 0.999$, and m_v is used to shape the resulting curve for the burnt-mass fraction.

Note that if a symmetric burnt-mass fraction is acceptable, an even simpler approximation can be used as well

$$x_B(\phi) = \sin^2\left(\frac{\phi - (\phi_s + \Delta\phi_d)}{\Delta\phi_{bd}} \cdot \frac{\pi}{2}\right) \tag{3.69}$$

Using a heat release function, combined with a heat-transfer approximation and the thermodynamic conservation laws, the pressure inside the cylinder can be estimated [97]. This information can then be used to compute indicated efficiencies or, combined with a friction model such as that expressed by (2.112), to compute brake efficiencies.

In [49], functions were derived that describe the variations of the three Vibe parameters of equation (3.68) for different operating conditions. This approach allows a control-oriented description of the combustion processes and, thus, for instance, the determination of the engine inputs for optimal fuel consumption [156]. More details are given in Appendix C.

If the cylinder pressure is measured, the burnt-mass fraction can be determined by an "inverse" thermodynamic process calculation. However, this is computationally quite demanding, some details can be found in Appendix C. Therefore, an approximation that was first described in [171] is used here. The primary points of that approach are visualized in Fig. 3.21.

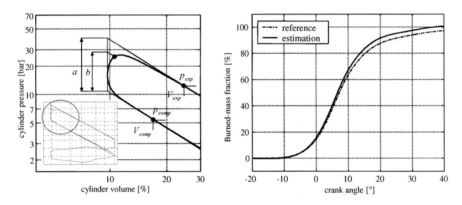

Fig. 3.21. Estimation of the burn rate.

The procedure starts with the selection of a pressure/volume pair, p_{comp} and V_{comp}, during the compression phase (before the combustion starts) and a pressure/volume pair, p_{exp} and V_{exp}, during the expansion phase (after the combustion is completed). Extrapolating these two pairs to the minimum volume value using a polytropic approximation yields the pressure difference indicated by **a** in Fig. 3.21. The actual pressure/volume data during the combustion is extrapolated in the same way, yielding the pressure difference **b**. The mass of burnt fuel is then estimated by the ratio of these two pressure differences

$$x_B(\phi) \approx \frac{b}{a} \approx \frac{p_c(\phi) \cdot \left(\frac{V_c(\phi)}{V_{TDC}}\right)^{n(\phi)} - p_{comp} \cdot \left(\frac{V_{comp}}{V_{TDC}}\right)^{n_{comp}}}{p_{exp} \cdot \left(\frac{V_{exp}}{V_{TDC}}\right)^{n_{exp}} - p_{comp} \cdot \left(\frac{V_{comp}}{V_{TDC}}\right)^{n_{comp}}} \qquad (3.70)$$

$$n(\phi) = \begin{cases} n_{comp} & \text{for} \quad \frac{\partial V_c(\phi)}{\partial \phi} \leq 0 \\ n_{exp} & \text{for} \quad \frac{\partial V_c(\phi)}{\partial \phi} > 0 \end{cases}$$

Choosing the reference points to be symmetric with respect to TDC (say at 80 ° before and after[11] TDC) simplifies the computations. In fact, as illustrated in Fig. 3.22, in this case the burn rate can be estimated using the expression

$$x_B(\phi) \approx \frac{p_c(\phi) \cdot \left(\frac{V_c(\phi)}{V_{TDC}}\right)^{n_{pol}} - p_{c1}}{\Delta p_c} \qquad (3.71)$$

To further simplify the computations, a constant value of $n_{pol} = 1.31$ can be used for the polytropic exponent.

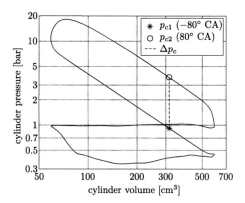

Fig. 3.22. Pressure/Volume (pV) diagram using a double-logarithmic representation including the reference points chosen for the symmetric case.

3.3.3 Cylinder Charge Estimation

Introduction

A typical application for the method introduced in the last section is the estimation of the 50% burnt-mass position. Of course, the cylinder pressure trace can be used for various other purposes. Two straightforward possibilities are the detection of knock and misfiring or the identification of the actual engine cycle.

A computationally more demanding use of the pressure trace, which is based on physical first principles, is the estimation of the fresh air, fuel, and burnt-gas mass in the cylinder. This application is discussed in this section. Assuming each cylinder to be equipped with a pressure sensor, this approach offers the possibility of cylinder-individual control, resulting in a more uniform load and air/fuel ratio distribution. Moreover, if the estimation results are sufficiently accurate, the air mass sensor can be left out.

---

[11] Note that the combustion must be completed at that point. This is the case for most regular combustion systems, but not necessarily for new, "exotic" engines.

The total mass in the cylinder, defined by $m_{tot} = m_\beta + m_\varphi + m_{bg}$, is fixed as soon as the intake valve has closed. The variable $m_{bg} = m_{rg} + m_{egr}$ is the total mass of burnt gas, i.e., the sum of the residual gas and the externally recirculated exhaust gas. Choosing a reference position during the compression stroke (indicated by an asterisk in Fig. 3.22), and using the measured cylinder pressure p_{c1} and the known cylinder volume (3.66), the ideal gas law

$$\hat{m}_{tot} = \frac{p_{c1} \cdot V_{c1}}{R \cdot \vartheta_{c1}} \qquad (3.72)$$

yields an estimation of m_{tot} by *assuming* the temperature ϑ_{c1} at the same reference point to have a value of approximately $470° K$. Note that (3.72) is valid only because the gas constant R of the air/fuel mixture and burnt gases is quite similar (a value of $287 \, J/kgK$ is recommended). Once the total mass is known, the amount of burnt gas can be estimated if the mass of aspirated mixture is estimated with any of the methods presented in either Sect. 2.3 or Sect. 3.2.2, or with the approaches shown below.

The estimation of m_{tot} using (3.72) is straightforward but does not yield sufficiently precise results that would allow for the omission of the air mass flow sensor. To achieve this goal, additional information must be obtained by analyzing the complete pressure trajectory $p_c(\phi)$. However, in this case some information on the heat release during the combustion must be taken into account as well.

Basic Model

Using the lower heating value H_l of the fuel and introducing the parameter C_c ("completeness of combustion") that indicates how much of the fuel is actually burnt until the exhaust valve starts to open, the following equation describes the total heat release

$$Q_c = m_\varphi \cdot H_l \cdot C_c = m_{tot} \cdot \frac{H_l}{1 + \lambda \cdot \sigma_0} \cdot (1 - x_{bg}) \cdot C_c \qquad (3.73)$$

At ideal conditions, i.e., for $m_{bg} = 0$, $\lambda = 1$, and $C_c = 1$, the heat released is

$$Q_{c,ideal} = m_{tot} \cdot \frac{H_l}{1 + \sigma_0} = \frac{p_{c1} \cdot V_{c1}}{R \cdot \vartheta_{c1}} \cdot \frac{H_l}{1 + \sigma_0} \qquad (3.74)$$

Thus, a new variable η_c, referred to as the *cylinder charge efficiency*, can be defined as follows

$$\eta_c = \frac{Q_c}{Q_{c,ideal}} = (1 - x_{bg}) \cdot C_c \cdot \frac{1 + \sigma_0}{1 + \lambda\sigma_0} \qquad (3.75)$$

For a given value of m_{tot}, this parameter quantifies the ratio between the actual and the maximum theoretically possible heat released in the combustion.

The relationship between heat release and pressure in the cylinder is analyzed next. For the two reference positions indicated above, the pressure difference

$$\Delta p_c = p_{c2} - p_{c1} \tag{3.76}$$

can be calculated. For the relevant operating range of a six-cylinder 3.2-liter SI engine, the pressure difference as a function of the combustion energy is depicted in Fig. 3.23. The air/fuel ratio is set to stoichiometry, and the 50% burnt-mass position, ϕ_{50}, is always kept near its optimum value of approximately $8°$ crank angle after TDC.

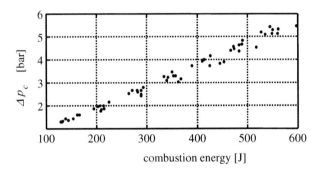

Fig. 3.23. Pressure difference as a function of the combustion energy. Engine operating range $n = 1000 - 4500 \ rpm$ and $p_{me} = 0 - 6 \ bar$. SI engine, $\lambda = 1$, $\phi_{50} \approx 8°$ crank angle after TDC.

As expected, an almost linear dependency can be observed. Note that this is just another confirmation of the remarks made in Sect. 2.5.1. In fact, the parameter e in the Willans approximation shown in Fig. 2.29 can be interpreted as the "indicated" (internal) efficiency that describes the ratio between combustion energy and pressure difference that eventually results in mechanical work on the piston.

Improved Model

The deviations of the single pressure difference point from a purely linear dependency occur mainly for two reasons:

- Variations in the engine speed influence the time available for the combustion and heat exchange with the cylinder walls.
- Variations in the combustion center, represented by the 50% burnt-mass position, influence the conversion of heat to mechanical work and, hence, the heat balance.

These observations lead to the following parametrization of the pressure difference as a function of engine speed and heat-release center

$$\Delta p_c = k(\omega_e, \phi_{50}) \cdot Q_c$$

$$= k(\omega_e, \phi_{50}) \cdot \eta_c \cdot Q_{c,ideal} \tag{3.77}$$

$$= k(\omega_e, \phi_{50}) \cdot \eta_c \cdot \frac{p_{c1} \cdot V_{c1}}{R \cdot \vartheta_{c1}} \cdot \frac{H_l}{1 + \sigma_0}$$

To investigate these influences, the effects of varying the 50% burnt-mass position and the engine speed can be estimated using thermodynamic process simulations. Assuming an ideal combustion with $\eta_c = 1$, the factor $k(\omega_e, \phi_{50})$ can be estimated by using (3.78), because all other terms on the right-hand side of this equation can be assumed to be constant

$$\frac{\Delta p_c \cdot \vartheta_{c1}}{p_{c1}} = k(\omega_e, \phi_{50}) \cdot \frac{V_{c1}}{R} \cdot \frac{H_l}{1 + \sigma_0} \tag{3.78}$$

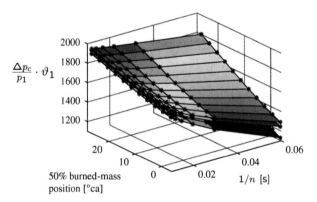

Fig. 3.24. Numerical value of $\frac{\Delta p_c}{p_{c1}}\vartheta_{c1}$ for simulated variations in engine speed n and 50% burned-mass position ϕ_{50}.

Figure 3.24 shows the calculated variation for three different total mass values (representing idle, half-load, and full-load conditions). Obviously, the time available for the combustion is an important parameter. Accordingly, in this figure the inverse engine speed is used as an independent variable. A bilinear parametrization is proposed here

$$k(\phi_{50}, \omega_e) \cdot \frac{V_{c1}}{R} \cdot \frac{H_l}{1 + \sigma_0} = \sum_{j \in corners} \left(\frac{\vartheta_{c1} \cdot \Delta p_c}{p_{c1}}\right)_j \cdot \mu_j \tag{3.79}$$

where

$$\mu_j = \left(1 - \frac{|\phi_{50,OP} - \phi_{50,j}|}{\phi_{50,max} - \phi_{50,min}}\right) \cdot \left(1 - \frac{|\omega_{OP}^{-1} - \omega_j^{-1}|}{\omega_{min}^{-1} - \omega_{max}^{-1}}\right) \tag{3.80}$$

where μ_j are the weighting parameters that allow for the calculation of the value of an arbitrary operating point OP within a rectangle defined by the corner values.

With these preparations, the cylinder charge efficiency can be calculated based on the knowledge of p_{c1} and ϕ_{50}, and an estimation for ϑ_{c1}

$$\eta_c = \frac{\Delta p_c}{k(\omega_e, \phi_{50})} \cdot \frac{R \cdot \vartheta_{c1}}{p_{c1} \cdot V_{c1}} \cdot \frac{1 + \sigma_0}{H_l} \tag{3.81}$$

Choosing $\vartheta_{c1} = 470°\ K$ as a first approximation of the temperature of the cylinder charge at the IVC position, an estimation for the charge efficiency with an accuracy of typically $\pm 5\%$ is obtained. If a better estimation of ϑ_{c1} can be obtained from a detailed thermodynamic process calculation, the approximation of the charge efficiency can be improved. As Fig. 3.25 shows, an estimation of (3.81) with an accuracy of $\pm 3\%$ is feasible (see [139] for details).

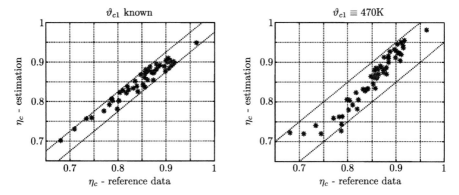

Fig. 3.25. Estimation of the charge efficiency.

Finally, assuming a minimum and a maximum completeness of combustion of 0.93 and 0.98, respectively, the second-order effects of the amount of fuel injected, of the 50% burned-mass position, of the combustion duration, etc., may all be collected in a linear function depending on p_{c2}

$$C_c = 0.93 + \frac{0.05(p_{c2} - p_{c2,min})}{p_{c2,max} - p_{c2,min}} \tag{3.82}$$

Reasonable values for a naturally aspirated engine for $p_{c2,max}$ and $p_{c2,min}$ are

$$p_{c2,min} = 1.5 \text{ bar} \quad \text{and} \quad p_{c2,max} = 15 \text{ bar}$$

Now, using (3.75), if λ is known from sensor information, the burned-gas fraction at IVC can be estimated. The estimated values plotted against the reference values (obtained with a two-zone thermodynamic process calculation) are shown in Fig. 3.26.

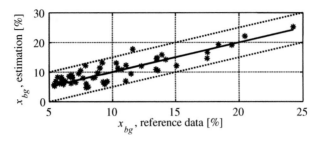

Fig. 3.26. Estimation of the burned-gas fraction at IVC, comparison between the detailed process calculations and the estimations using the cylinder charge efficiency.

3.3.4 Torque Variations Due to Pressure Pulsations

The varying pressure in the cylinder produces a non-uniform torque on the crankshaft. If these variations must be modeled, the following approach can be used. The starting point for this analysis is the mass of fuel from which the combustion energy can be calculated with (3.73). Using (3.77), the pressure difference Δp_c can then be calculated.

Assuming reasonable values for the Vibe parameters and, thus, for the burned-mass fraction and inverting the relationship (3.71), an approximation of the cylinder pressure is obtained.

The pressure at reference position 1, p_{c1}, can be obtained according to (3.70) by assuming a polytropic compression starting with the manifold pressure p_m at the IVC crank angle to the chosen reference point at $\phi_1 = -80\,^\circ$ crank angle.

The pressure acts on the piston surface producing a force

$$F_p(\phi) = p_c(\phi) \cdot A_p = p_c(\phi) \cdot \frac{V_d}{2 \cdot r_{cs}} \tag{3.83}$$

The d'Alembert force F_m, which takes into account the inertia[12] of the piston and the crank-slider parts, must be considered as well

$$F_m = -m_p \cdot \ddot{s}(\phi) \tag{3.84}$$

Neglecting the friction forces and formulating a power balance for the crankshaft, the torque T_e on the crankshaft can be calculated

$$T_e \cdot \omega_e = T_e \cdot \dot{\phi} = (F_p + F_m) \cdot \dot{s}(\phi) \tag{3.85}$$

Using the approximation (3.65), the following expression for the engine torque can be derived

---

[12] One-third of the connecting-rod mass is usually added to the piston mass. The other two-thirds are added to the crankshaft inertia.

$$T_e = \frac{(F_p + F_m) \cdot \dot{s}(\phi)}{\dot{\phi}} = (F_p + F_m) \cdot r_{cs} \cdot \left(\sin\phi + \frac{r_{cs}}{2 \cdot l} \cdot \sin(2\phi)\right) \quad (3.86)$$

The final ODE describing the crank-shaft dynamics is obtained using a torque balance

$$T_e - T_l = \Theta_e \cdot \ddot{\phi} \quad (3.87)$$

The rotational inertia of the system, Θ_e, includes the inertia of the crankshaft, the flywheel, and two thirds of the mass of the piston rods.

Finally, assuming each cylinder to produce the same torque pattern, the total torque can be derived by shifting the single cylinder torque appropriately. In Fig. 3.27, this is shown for a four-, a six-, and an eight-cylinder engine. In these simulations, the rotational inertia has been chosen to be very large such that a constant engine speed results.

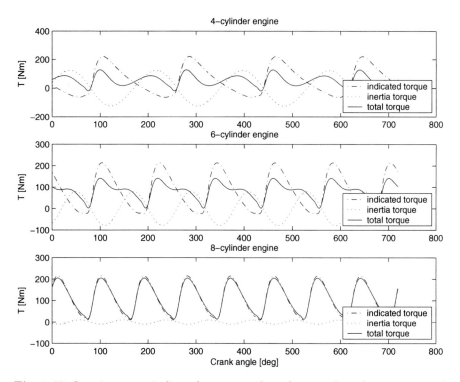

Fig. 3.27. Inertia torque, indicated torque, and total torque for a four-, a six-, and an eight-cylinder engine (note the different torque scales).

4

Control of Engine Systems

As discussed in the preceding chapters, engine systems contain a large number of control loops. For the design of these feedforward and feedback control systems, the main objectives are:

- *The driver's demands for immediate torque response, good drivability, and low fuel consumption must be met.*
- *The engine must be kept in a safe operating region where damage to or fatigue of the material is avoided. Knocking must be prevented, the catalytic converter must not overheat, etc.*
- *The emission limits must be met. In the case of SI engines, this requires a rapid light-off of the catalytic converter, precise stationary air/fuel ratio control, and good compensations of the transient phenomena.*

Since these requirements are partially contradictory, they must be fulfilled according to the priorities set by legislation and customer demands. In this optimization, control systems play a major role as enabling technology.

The complete treatment of all control problems arising in engine systems is beyond the scope of this text. The following five important case studies are presented with the intention to show how model-based methods can be used to streamline the design process:

- *Physics-based modeling and control of engine knock as an example for model-based signal processing, Sec. 4.2*
- *The air/fuel-ratio control as a typical combined feedforward and feedback control problem. The conventional approach is discussed in Sec. 4.3.2, whereas model-based design methods are presented in Sec. 4.3.3 and 4.3.4.*
- *The combined engine-speed and air/fuel-ratio control system as an example of a MIMO control problem, Sec. 4.3.5*
- *The SCR control system in Sec. 4.4 shows how sensor models can help to improve the system performance.*
- *The engine-temperature control-system or "thermomanagement" as an example of a control system that must cope with a changing system structure, Sec. 4.5*

4.1 Introduction

4.1.1 General Remarks

The general structure of automatic control systems is discussed in Appendix A. Readers not familiar with the main ideas of general control system theory may want to consult that section before starting to read this chapter.

Figure 4.1 shows the various actuators and sensors of a modern SI engine management system. The ECU must set all actuator commands according to driver demands and available sensor signals, taking into account constraints imposed by engine safety considerations and pollutant emission limitations.

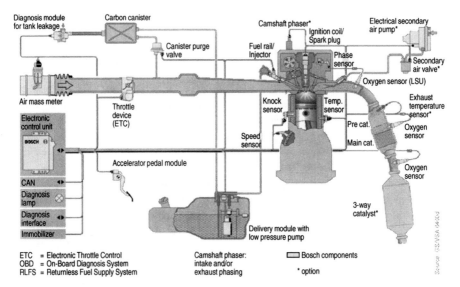

Fig. 4.1. Layout of a modern engine management system, reprinted with the permission of Robert Bosch GmbH.

In the following, it is assumed that the engine system is equipped with an electronic throttle device, i.e., the driver does not directly control the angle of the throttle plate. Instead, an accelerator pedal module captures the driver's input and produces an electronic signal which is then sent to the ECU. This signal is usually interpreted as a desired engine torque. Such a decoupling permits a coordination of all involved actuators.

As discussed in Chapter 3, the engine speed sensor and the camshaft phase sensor allow the synchronization of the ECU with the engine's reciprocating behavior. The throttle plate angle and the measured mass flow and temperature of the air entering the intake manifold provide the necessary information to calculate the amount of fuel required. Since the sensors are placed upstream of the intake manifold, the mass of air entering the cylinder must be calculated using a physical model of the intake runner. This model takes into account the temperature of the engine coolant as well.

The engine management system shown in Fig. 4.1 uses three catalytic converters: the close-coupled TWC optimized for rapid light-off at cold start, the main TWC a short distance downstream of the close-coupled TWC; and an optional underfloor catalytic converter. To control this complex exhaust aftertreatment system, three exhaust gas oxygen sensors are used: one located as close as possible to the exhaust valves; one installed between the close-coupled and the main TWC; and one attached downstream of the main TWC. An optional exhaust gas temperature sensor can help to prevent damage to the catalytic converters due to overheating and to improve the overall control system performance. This configuration is adapted to the specific pollutant emission limitations that are imposed by legislation. Obviously, the smallest number of TWC and sensors are chosen with which these limitations can be met in order to realize the configuration.

Two additional subsystems can be identified in Fig. 4.1:

- At cold start, the secondary air system pumps air into the exhaust manifold. When the engine is run under slightly rich conditions, this amount of air improves the heating of the catalytic converter system and thus shortens the light-off time.
- When the engine is not running, the fuel that evaporates in the tank cannot be directed to the intake manifold for later combustion. Nevertheless, in order to prevent excessive pressures in the tank system, these fuel vapors must be vented. Since venting fuel vapors directly into the environment is not permitted, a carbon canister is used to collect them. Since this canister has a limited capacity, it must be purged when the engine is running. Controlling that process and detecting unwanted leakages are the goals of some of the algorithms that are implemented in the ECU.

Notice that the structure shown in Fig. 4.1 does not represent the most general case. For instance, it does not contain an external EGR device or a turbocharging system. Additional sensors and actuators will be included in these situations.

4.1.2 Software Structure

The most important task of an engine is to deliver the desired amount of torque. As mentioned earlier, the demanded torque is generally defined by the driver. However, several subsystems of the engine influence that request, e.g., the cruise-control module, the catalyst heating system, the control system that avoids oscillations of the drive-line, etc.

Figure 4.2 shows the often contradictory signal paths used in conventional engine management systems. For such systems, the design of the control algorithms is very difficult. Many control loops must be run in parallel and, due to frequent changes in the system structure, a correct multivariable control system design is not possible. Moreover, software structures for control systems like the one shown in Fig. 4.2 are difficult to parameterize, and each new safety or monitoring task included at a later stage adds new and potentially disrupting signal couplings to the control system.

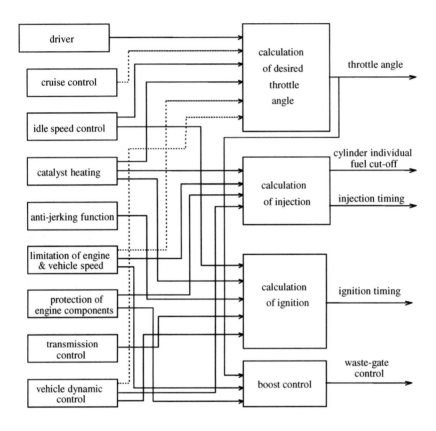

Fig. 4.2. Engine torque control structure in a conventional engine control unit (ECU).

In contrast to the engine control structure shown in Fig. 4.1.2, a torque-based control structure as shown in Fig. 4.3 is structurally much simpler. This structure decouples the strategic decisions from the low-level control tasks. In fact, the *torque demand manager* collects and evaluates all demands for engine torque. Then, only one signal is transferred to the *torque conversion manager* that issues the appropriate commands to the actuators such that the demanded torque can be realized as best as possible.[1]

---

[1] Of course, such an approach requires an electronic throttle valve.

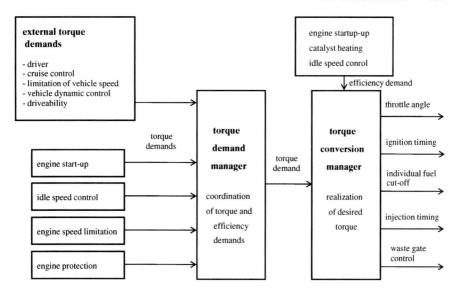

Fig. 4.3. Engine torque control structure in a torque-based engine control unit.

Since the torque-based structure reflects the underlying physics of the engine, the control algorithms implemented in the ECU reflect much more closely the actual processes. This makes the control system structure easier to understand and, thus, the corresponding software easier to maintain. For the design of the control system, only the blocks following the torque demand signal are important. The other blocks implement the algorithms for strategic decisions which are very important for issues concerning fuel economy, pollutant emission, drivability and comfort. Thus, the set points for the various actuator input signals must be chosen mainly as a function of the desired torque at a defined engine speed. The corresponding block diagram is shown in Fig. 4.4, where the block designated as torque conversion contains the inversion of the torque model that is introduced in Sec. 2.5.1 above.

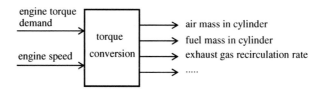

Fig. 4.4. Relevant subsystem for the control system design.

4.1.3 Engine Operating Point

Engine speed and engine torque represent the most important inputs to the engine control systems. These two variables define the engine's *operating point*. The operating point in turn characterizes the engine's main variables (air mass flows, pressures, pollutant formation, etc.). If an engine is fully warmed, most actuator inputs remain constant if the operating point is not changed.

To calculate actuator set points, instead of the actual torque, the *relative load* is often chosen as an independent variable. For each fixed engine speed, this variable indicates the actual air charge in the cylinder with respect to its maximum value. For a known full-load torque curve, the relative load can be used to derive the actual torque at a later stage.

When the engine is fully warmed up, its actuator inputs are determined using steady-state optimizations (to be discussed further in Sect. 4.1.4). The results of these optimizations are stored in *maps* that parameterize the dependency of each actuator input on the engine operating point. It is not convenient to use maps with dimensions greater than two. Therefore, other influencing engine variables (e.g., coolant temperature, battery voltage, etc.) must be taken into account by adding one- or two-dimensional correction maps.

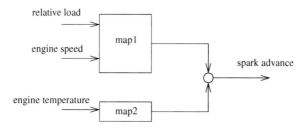

Fig. 4.5. Calculation of spark advance; example of a simple correction map structure.

Using the ignition control loop as an example, Fig. 4.5 shows the most basic structure of such a map-based algorithm. Based on the measured engine speed and the relative load, a nominal spark advance is chosen (block "map1" in Fig. 4.5). Then the corrections are applied; in this simplified case a correction for engine temperature only is considered.[2] For such a separation of effects, it is important to understand the underlying thermodynamic principles. In the case of the spark advance, for instance, the effects shown in Fig. 2.34 and explained in [70] have to be considered.

Figure 4.6 shows an example of a spark advance map as a function of engine speed and load (torque) as implemented in a modern ECU. The map

---

[2] In a real ignition control system, additional corrections would be applied for varying battery voltages, to avoid engine overheating and knock, etc. In addition, the idle-speed control system also influences the ignition control loop.

is mainly defined by the best fuel-efficiency spark advance, but factors such as emission limits, knock avoidance (see section 4.2 below), etc., are included as well.

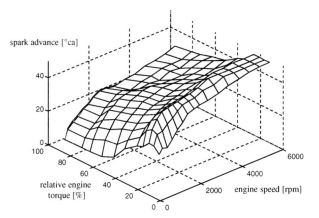

Fig. 4.6. Example of a "map1" as used in the example shown in Fig. 4.5.

Notice that, despite the fact that these maps are often referred to as feed-forward control maps, they introduce feedback action in the control system. In fact, by using measured values for the actual engine speed and load to compute the actuator signals, the inherent dynamic properties of the controlled system are changed.

4.1.4 Engine Calibration

Although significant progress has been made over the years to reduce engine-out emissions, in most cases these reductions are not sufficient for the engine to comply with actual and future emission regulations. From a control engineering perspective, these regulations reduce the degrees of freedom for the optimization of the engine control system by introducing additional boundary conditions. These constraints lead to increased fuel consumption since the engine control system cannot be further optimized without violating the regulation limits.

The best possible efficiency must be guaranteed in each operating condition while, simultaneously, the emission limitations, defined for a specific test cycle, must be met. This optimization can be performed using a Lagrange multiplier approach.

The cost function of the optimization is the fuel consumption over one driving cycle

$$J = \int_0^{t_{dc}} \dot{m}_\varphi \cdot dt \tag{4.1}$$

where t_{dc} is the duration of the driving cycle. The boundary conditions are defined by the pollutant emission constraints that must be met:

$$m_{HC} = \int_0^{t_{dc}} \dot{m}_{HC} \cdot dt \leq \overline{m}_{HC} \qquad (4.2)$$

$$m_{CO} = \int_0^{t_{dc}} \dot{m}_{CO} \cdot dt \leq \overline{m}_{CO} \qquad (4.3)$$

$$m_{NO_x} = \int_0^{t_{dc}} \dot{m}_{NO_x} \cdot dt \leq \overline{m}_{NO_x} \qquad (4.4)$$

Of course, the pollutant mass flows $\dot{m}_{...}$ are tail-pipe emissions, i.e., the dynamics of the aftertreatment systems must be included in the system analysis.

Equations (4.1) and (4.4) describe the system using a continuous-time approach. However, especially in the European test cycle, the engine system can often be assumed to operate in a series of quasi-stationary points. Therefore, by splitting up the driving cycle into a number of *constant* operating points with associated retention periods $t_{op,i}$, the optimization problem can be reformulated as follows

$$J = \sum_{op.points} \dot{m}_{f,i} \cdot t_{op,i} - \lambda_{HC} \cdot \left(\overline{m}_{HC} - \sum_{op.points} \dot{m}_{HC,i} \cdot t_{op,i} \right) \qquad (4.5)$$

$$- \lambda_{CO} \cdot \left(\overline{m}_{CO} - \sum_{op.points} \dot{m}_{CO,i} \cdot t_{op,i} \right)$$

$$- \lambda_{NO_x} \cdot \left(\overline{m}_{NO_x} - \sum_{op.points} \dot{m}_{NO_x,i} \cdot t_{op,i} \right)$$

or in the form

$$J = \sum_{op.points} (\dot{m}_{f,i} + \lambda_{HC} \cdot \dot{m}_{HC,i} + \lambda_{CO} \cdot \dot{m}_{CO,i} + \lambda_{NO_x} \cdot \dot{m}_{NO_x,i}) \cdot t_{op,i}$$

$$- \lambda_{HC} \cdot \overline{m}_{HC} - \lambda_{CO} \cdot \overline{m}_{CO} - \lambda_{NO_x} \cdot \overline{m}_{NO_x} \qquad (4.6)$$

For a given set of Lagrange multipliers λ, the sum of all operating points is then optimal, when each summand in each operating point is optimal. Of course, this requires the single operating points to be mutually independent. Especially if cold-start situations are excluded, this assumption is well satisfied for the engine-out emissions and for fuel consumption. Engine optimization methods based on this assumptions have been presented in [36], [200], [198], and [156]. The dynamic phenomena are considered by the feedforward and feedback loops discussed below.

Thus, in each operating point

$$J = \dot{m}_{f,i} + \lambda_{HC} \cdot \dot{m}_{HC,i} + \lambda_{CO} \cdot \dot{m}_{CO,i} + \lambda_{NO_x} \cdot \dot{m}_{NO_x,i} \qquad (4.7)$$

must be minimized using the following procedure:

- Set all Lagrange multipliers to zero. The resulting optimum is the fuel-optimal solution.
- Determine the sum of the masses of the limited components over the complete test cycle.
- If all limits are met, the optimization is finished. If one or more limits are exceeded, the corresponding Lagrange multipliers must be increased.
- Repeat the optimization process with the new cost function.

The optimization process consists of changing the parameters of the engine control system.[3] This can be accomplished using a purely experimental, or a partially model-based approach [154], [196]. In either case, the iterations can be performed using heuristics based on experience with previous engine calibration tasks or using systematic approaches, e.g., forming the gradients of the cost function (4.6) and using these vectors in a steepest-descent algorithm as discussed in [170], [110], [179] and [180].

4.2 Engine Knock

The phenomenon of knock has been a major limitation for SI engines since the beginning of their evolution. Engine knock has its name from the audible noise that results from autoignitions in the unburned part of the gas in the cylinder. The most probable locations for harmful self-ignitions lie in proximity of hot surfaces, i.e., piston and cylinder walls, and in the largest possible distance from the spark plug. This can be explained by the concept of the pre-reaction level. In this notion, the autoignition is a result of the chemical state of the unburned gas exceeding a critical level in which enough highly reactive radicals are formed, leading to a spontaneous ignition. This pre-reaction level, being proportional to the concentration of radicals, increases over time, primarily under the influence of high temperatures and, secondarily, high pressures. The pressure in the cylinder can be assumed to be spatially constant (it varies with time) since the speed of sound, at which the pressure is equalized, is several orders of magnitude larger than the speed of the flame propagation. In contrast, the temperature varies significantly within the cylinder volume. In the unburned gas, regions of the highest temperature levels are located in the boundary layers close to hot surfaces. In those regions the gas flow is slow and therefore the heat from the walls is transferred to a small volume during a long period of time. This time is available since, due to the large distance from the spark plug, the flame reaches these regions at the end of the combustion process.

If the mass fraction of unburned gas at the time of autoignition is large and its pre-reaction level is high (i.e., close to critical), several adjacent hot

---

[3] This process is often referred to as the *engine calibration*.

spots are ignited and merge to a fast expanding "reaction region" such that all of the highly reactive unburned gas burns almost at once. Under theses conditions the chemical reactions spread[4] faster than the speed of sound, resulting in insufficient pressure equalization. This in turn leads to shock waves and consequently to harmful pressure peaks in the cylinder.

Conventional SI engines are limited by engine knock mainly at operating points characterized by low speed and high load, where both enough time is available and high pressure and temperature levels in the end gas (i.e., the gas reached last by the flame front) are present. Whether critical pre-ignition levels are attained or not depends on several engine-specific and also external factors: mainly the layout and geometry of the combustion chamber, temperatures of different surfaces as well as of the fresh air, recirculated and residual exhaust gases, compression ratio, intensity of both global and local turbulence, and fuel composition all play a role.

In order to avoid knock, the engine has to be run in a sub-optimal way with respect to efficiency: The global knock tendency is reduced by limiting the compression ratio and therefore lowering pressure and temperature levels, which unfortunately goes along with a loss in thermodynamic efficiency. In addition, preventing knock in the critical operating regions requires the spark timing to be delayed from the position where it provides maximum brake torque (MBT). This reduces the peak pressure and temperature in the cylinder by moving the peak of the burn-rate trajectory beyond the top dead center. This causes the combustion to burn into the expansion stroke leading to a further sacrifice of efficiency.

4.2.1 Autoignition Process

Knock is closely related to the process of autoignition. Particularly, if the occurrence (not the intensity) of knock is of interest only, spontaneous self-ignition can be seen as the onset of knock. As described above, this happens as soon as a critical concentration of reactive radicals is attained. A vast number of mathematical models describing the chemical reaction kinetics in the unburned mixture exists. The complexity of the models ranges from extremely detailed (incorporating up to 1972 elementary reactions with 380 species [47]) over simplified (32 elementary reactions and 14 species [112]) to simple (19 elementary reactions [34]).

The models that are probably used most ofen are of the Shell type [84, 98, 177], consisting of eight elementary reactions, and the model developed by the authors of [52], which lumps all elementary reactions into one Arrhenius-type function (see next section). Although concerns questioning the accuracy

---

[4] There is no regularly propagating "flame front" in the classical sense during knocking combustion of the remaining unburnt gases. Also, the notion of engine knock being merely an uncontrolled acceleration of the main flame front has been disproven experimentally.

and applicability have been raised recently [109], this simple approach and its extensions are still widely used in models designed for parametric studies, where many simulations have to be carried out in a reasonable time frame.

The faster the pre-ignition reactions take place, i.e., the faster the reactive radicals are formed, the sooner self-ignition occurs. Based on this notion, the concept of the "ignition delay" $\tau_{ID}(\vartheta, p)$ can be formulated. This parameter is roughly inversely proportional to the rate of formation of the radicals (i.e., the pre-ignition reactions) and therefore depends strongly on the temperature and pressure. It defines the time interval after which an air-fuel mixture spontaneously ignites at certain given conditions. The aforementioned models for the reaction kinetics may be used to estimate this ignition delay with varying degrees of accuracy (which of course correlates with the complexity of the model).

One important characteristic phenomenon concerning the combustion of hydrocarbons is the so-called "negative temperature-coefficient" (NTC) region as described, for instance, in [211]. Usually, the ignition delay decreases with increasing temperature. At higher temperatures, the pre-ignition reactions take place at higher rates, also forming more radicals. In the NTC region, however, this characteristic is switched; the ignition delay increases for higher temperatures. This phenomenon typically occurs in a temperature band around $900K$ (depending on the pressure and on the fuel used), a temperature region often reached by the unburned gas-fraction at an advanced stadium of the combustion. This is relevant for engine knock, as will become clear in the following sections, and a model capable of reproducing this behavior is therefore desirable.

One significant drawback of simple models such as the "one-Arrhenius" approach (see next section) is their incapability of reproducing the NTC behavior. A qualitative sketch showing the ignition delay around the NTC region and the (insufficient) prediction of a simple model, i.e., the one-Arrhenius model, is shown in Figure 4.7.

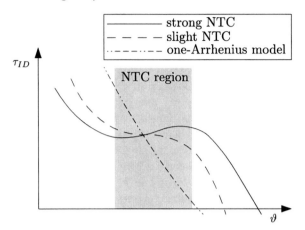

Fig. 4.7. Qualitative sketch of the NTC behavior of the ignition delay of hydrocarbons.

The "One-Arrhenius" Approach

In [52], the authors presented an autoignition model based on a single Arrhenius-type function and identified the corresponding constants for primary reference fuels by experimental data. In this approach, the ignition delay is given by

$$\tau_{ID} = A \cdot p^b \cdot e^{\frac{B}{\vartheta}} \tag{4.8}$$

with

$$A = 0.01869 \left(\frac{ON}{100} \right)^{3.4017}$$

$$b = -1.7$$

$$B = 3800$$

where ON is the octane number of the fuel. This model yields good results in the range of $80 \le ON \le 100$.

Another approach is to measure or simulate the ignition delay by very detailed models in a grid spanned over the relevant region of the pressure p and the temperature ϑ. This data is then stored in a lookup table, and the knock criteria described in the next section draw the values of $\tau_{ID}(p, \vartheta)$ from this table by interpolation.

4.2.2 Knock Criteria

In [129] the authors introduced an integral-based knock criterion. Their idea was to adapt the static calculation of the ignition delay τ_{ID}, usually based on simple models such as (4.8), to the highly transient situation in the cylinder of a reciprocating engine. The pressure, temperature, and other operational values relevant for the calculation of the ignition delay τ_{ID} are usually obtained from an engine process model. In a time-invariant environment, the condition for autoignition can be written as

$$t = t_{knock} = \tau_{ID} \iff \frac{t_{knock}}{\tau_{ID}} = 1. \qquad (4.9)$$

This can be extended to a partially constant situation, consisting of n time-invariant intervals of duration Δt_i

$$\sum_{i=1}^{n} \frac{\Delta t_i}{\tau_{ID,i}} = 1 \qquad (4.10)$$

With $n \longrightarrow \infty$ it follows that $\Delta t_i \longrightarrow dt$ and $\tau_{ID,i} \longrightarrow \tau_{ID}(t)$, such that the above sum becomes an integration over the time from the start of the compression stroke t_{SOCpr} until the occurrence of knock

$$\int_{t_{SOCpr}}^{t_{knock}} \frac{1}{\tau_{ID}(t)} dt = \int_{t_{SOCpr}}^{t_{knock}} \frac{1}{\tau_{ID}(p(t), \vartheta(t))} dt = 1. \qquad (4.11)$$

Note that this equation has the same structure as the one for a perfect plug-flow (2.173). The distance already covered by the gas thereby corresponds to the pre-ignition level (i.e., concentration of radicals) and the velocity of the gas parallels the inverse of the ignition delay.

In Fig. 4.8, equation (4.11) is plotted versus the crank angle for variations of the engine speed (at full load) and the load (at constant speed). The integral is only evaluated up to the end of combustion, i.e., to the point where 100% of the mixture is burned. In both cases, the ignition timing was chosen such that maximal brake torque was achieved.

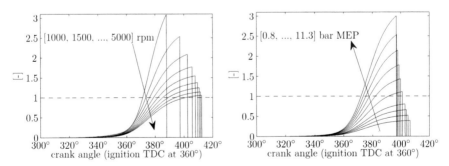

Fig. 4.8. Knock criterion for variations in engine speed at full load (left plot) and variations in load at 1500 rpm (right plot). Ignition timing tuned for MBT.

The ignition delay was modeled using the simple approach described by (4.8). For high loads, the curves would look different if a more complex model,

capable of reproducing the NTC behavior, was used. In fact, when the unburned gas reaches the NTC temperature band towards the end of combustion,
the ignition delay would increase, the slope of the integral would decrease and,
in some cases, probably would not exceed unity as it does now.

At full load, the criterion reaches the critical value of 1 during combustion,
even for very high speeds. However, it has been shown in [67] that harmful (and
audible) knock only occurs if autoignition starts before a certain amount of the
mixture is burned. Figure 4.9 depicts the regions that have been computed to
be susceptible to knock for different critical values of the burnt mass fractions
x_B.

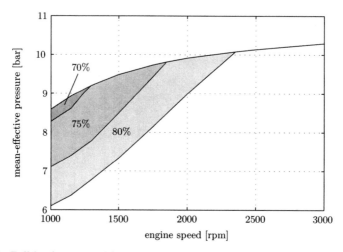

Fig. 4.9. Full-load curve and knocking operating regions under the assumption that
severe knock only occurs if autoignition starts before $x_B = 70\%$, 75%, or 80%. Note
that the plot does not show the full engine speed range.

4.2.3 Knock Detection

There is no widely recognized definition of knock since there is no standard
procedure to detect and quantify it. Most commonly, knock is defined to occur when it is audible. Most algorithms therefore detect the resulting pressure
waves propagating through the combustion chamber. This is done either by
using dedicated knock sensors, basically piezoelectric acceleration sensors attached to the engine body, resonating at the frequency of these waves, or
by sufficiently fast measurements of the cylinder pressure. For cost reasons,
the former method is widely used in the automotive industry, whereas the
latter is the most reliable and is therefore often used on test-benches. Note
that since the recorded signals of both measurement concepts contain very

similar frequency-domain information, only minor changes in the evaluating algorithm are necessary when switching from one to the other.

Cylinder Pressure Analysis

The pressure waves resulting from knocking combustion have a characteristic frequency that depends mostly on the characteristic length of the oscillation and the speed of sound in the combustion chamber [57]. Assuming that the cylinder is filled with air (modeled as an ideal gas) at a temperature ϑ_{cyl} of 2000K, the speed of sound is

$$c_{cyl} = \sqrt{\kappa \cdot R \cdot \vartheta_{cyl}} = \sqrt{1.4 \cdot 287 \frac{J}{kg \cdot K} \cdot 2000K} \approx 896 \frac{m}{s}. \qquad (4.12)$$

The characteristic frequency of the pressure waves is then calculated by

$$f_{knock} = \frac{c_{cyl} \cdot \alpha_{m,n}}{\pi \cdot B} \qquad (4.13)$$

where B is the cylinder bore and $\alpha_{m,n}$ the vibration mode factor. This parameter $\alpha_{m,n}$ can be approximated using the analytical solution of the general wave equation in a closed cylinder with flat ends. For the first circumferential mode this yields $\alpha_{1,0} = 1.841$. For an engine with a bore of $B = 85mm = 0.085m$ the frequency related to knock therefore is

$$f_{knock} = \frac{c_{cyl} \cdot \alpha_{1,0}}{\pi \cdot B} = \frac{896 \frac{m}{s} \cdot 1.841}{\pi \cdot 0.085m} \approx 6.2kHz \qquad (4.14)$$

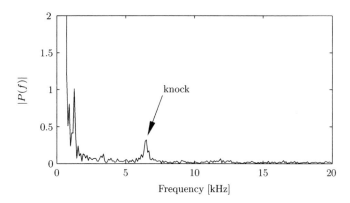

Fig. 4.10. Frequency spectrum of the cylinder-pressure signal in a operating point where knocking occurs.

Note that this frequency does not depend on engine speed. Figure 4.10 shows the frequency decomposition of the cylinder-pressure signal produced by such an engine while it runs under knocking conditions.[5] The peak made up by the knock-induced pressure waves at approximately 6.5 kHz correlates well with the frequency calculated previously.

A bandpass filter windowing the frequency band of the first circumferential mode is able to separate the superposed knock pressure-waves from the cylinder-pressure bias. Their amplitude is a good measure for the knock intensity. Analyzing the spectrum in Fig. 4.10, the specifications for the filter may be chosen as:

- block band: −60 dB for frequencies between 0 and 1 kHz;
- pass band: 0 dB for frequencies between 4 and 8 kHz; and
- block band: −20 dB for frequencies higher than 11 kHz.

The gain rise of 60 dB within less than a decade necessitates a large order of the filter. Figure 4.11 depicts the magnitude response of a filter of order 46 that meets the specifications.[6] Figure 4.12 presents the distinctive results

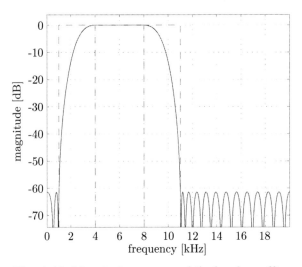

Fig. 4.11. Magnitude response of the bandpass filter.

of the bandpass filtering for a knocking and a non-knocking cycle. A slightly positive offset can be observed in the filtered signal. This is due to the low

---

[5] The engine was run at full load with additional supercharging, and the ignition angle was changed towards earlier ignition until knocking occurred.

[6] If needed, the order of the filter can be reduced by decreasing the attenuation of the high-frequency block band.

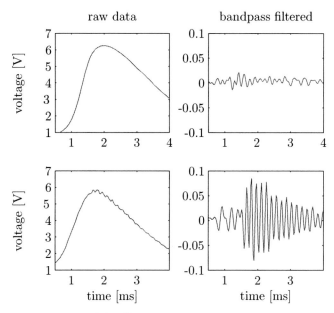

Fig. 4.12. Comparison of the filtered cylinder pressure of a knocking and a non-knocking cycle (filtered off-line).

frequencies not being fully attenuated by the filter. However, an offset does not impair the performance of the knock-detection algorithm described below.

Whether knock is present in an individual cycle or not is determined by defining a threshold for the amplitude of the filtered signal. Since this threshold is a limit for knock intensity, it can be determined directly on the engine by increasing the load or spark advance is increased until the slightest audible knock occurs. The threshold is then determined to be at that value where a significant percentage of the cycles are classified as knocking ones.[7]

By defining this threshold to be variable and using it as a parameter for the knock controller (see next section), a desired amount of knocking[8] can be chosen (e.g. for various applications of the same engine or even depending on the operating point or other parameters). Note that when such a controller is applied, differently knock-resistant fuels (i.e., fuels with higher or lower octane numbers) may be used without the drawbacks arising otherwise, namely the risk of damaging the engine by knock (when using low-octane fuel) or

---

[7] What percentage is "significant" may be determined by statistical off-line calculations using measurements around those critical conditions and/or by engineering experience and empirical considerations. An extensive description of the former can be found in [218].

[8] For improved efficiency, very slight knocking is acceptable as long as it does not impair the desired life cycle of the engine.

sacrificing efficiency (when using knock-resistant fuel) such as by lowering the compression or choosing a generally later ignition.

4.2.4 Knock Controller

If knock is detected it can either be reduced by delaying the ignition or by reducing the load, i.e., closing of the intake throttle.[9] Combining both in one controller is possible (and even advantageous) since these effects act on different time scales. Figure 4.13 presents an example of a possible knock-control strategy. The individual loops are explained in the subsequent sections.

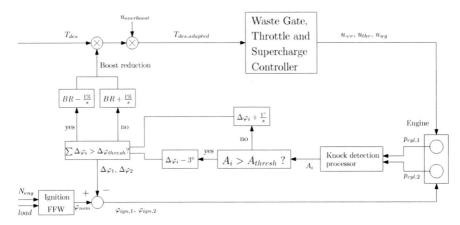

Fig. 4.13. Signal flow of a knock-controller

Another possibility of reducing the engine's susceptibility to knock is to cool the cylinder charge. This can be achieved by either cooling the aspirated air (in turbocharged systems, often an intercooler is applied after the compressor anyway) or by fuel enrichment, i.e., driving at air-to-fuel ratios lower than stoichiometry. The latter introduces additional mass into the cylinder, increasing the total heat capacity and therefore lowering the overall temperature level. In addition, the evaporation withdraws heat (the evaporation enthalpy) from the cylinder charge.

Whereas intercooling is more of a static improvement, fuel enrichment is already used to increase the full-load power and cool the engine at high speeds and loads. This may be extended to the knock-relevant low speeds by simply adjusting the injection map and is not further discussed here.

---

[9] In turbocharged engines, the waste gate of the turbine could be used to reduce the boost pressure and thus reduce the load.

Delaying the Ignition

The ignition angle is the fastest control input to an engine because changes take effect already in the next cycle. The downside of delaying the ignition is the delayed combustion leading to a reduction of torque (and therefore efficiency since the same amount of fuel is actually burnt) as well as increased exhaust-gas temperatures. The ignition angle correction $\Delta\varphi$ is defined as:

$$\varphi_{ign} = \varphi_{nom} - \Delta\varphi \tag{4.15}$$

Note that $\Delta\varphi$ is limited to negative values since only delaying the ignition leads to a reduction of knock intensity. For every cycle of any cylinder i, the maximum amplitude A_i of the bandpass-filtered pressure signal is calculated and passed to the knock controller. If it exceeds the threshold value A_{thresh}, the ignition for the considered cylinder is retarded by $3°CA$. Afterwards, the ignition angle is advanced again at the rate of $1°CA$ per second as long as the knock amplitude remains lower than the threshold. The ignition correction is subtracted from the nominal ignition angle given by the feedforward ignition controller. For an operating point at which the torque-optimal ignition angle "inherently" leads to knocking combustion, this controller exhibits a limit cycle behavior: The ignition angle is retarded until the knock intensity falls below the threshold and then advanced again, as shown in Fig. 4.14.

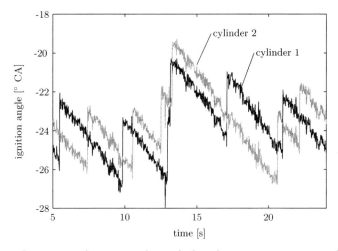

Fig. 4.14. Ignition angles at an inherently knocking operating point of an engine with two cylinders.

Load Reduction

Reducing the load is subject to slower dynamics since it depends on the dynamic state variables intake pressure and, if the engine is turbo-charged, turbocharger speed. Therefore, load reduction is used only if knock cannot be suppressed by delaying the ignition only. This is the case when ignition had to be extremely late and the accordingly increased exhaust-gas temperatures would damage the components in the exhaust system (e.g. turbine or after-treatment systems). While excessively high temperatures are acceptable during short transients, the load is reduced whenever the ignition is delayed over too long a period of time.

On one hand, the algorithm reduces the desired torque by one percent per second as soon as the sum of the ignition delays in all cylinders exceeds $\Delta\varphi_{thresh}$. On the other hand, the desired torque is increased at the same rate as long as the total spark retardation is lower than this threshold. To prevent limit-cycle behavior, a hysteresis for the threshold value $\Delta\varphi_{thresh}$ is usually included.

4.3 Air/Fuel-Ratio Control

The air/fuel-ratio control-system is one of the most important control loops. As discussed in detail in Sect. 2.8.3, the TWC achieves its best efficiency only if the engine is operated within a narrow band around the stoichiometric air/fuel ratio. Due to the TWC's ability to store oxygen and carbon monoxide on its surface, short excursions of the air/fuel ratio can be tolerated as long as they do not exceed the remaining storage capacity and the mean deviation is kept below 0.1%. These two observations require the air/fuel-ratio control-system to comprise both an approximate but fast feedforward control to handle transients as well as a slow but precise feedback control-loop to ensure the required high steady-state accuracy. The former is discussed in the subsequent section and different approaches to solve the feedback-control problem, ranging from the classical to highly sophisticated design methods, are presented in Sections 4.3.2 to 4.3.4.

4.3.1 Feedforward Control System

Without an appropriate feedforward component the performance of the air/fuel ratio control system would be much too slow. This slowness is due to the many delays and lags between the input (fuel injection) and the output (air/fuel ratio sensor) of this dynamic system.

Under transient operating conditions, the correct amount of fuel to be injected must be determined as quickly as possible. This is only possible if the intake manifold dynamics are considered. The information available on either the throttle plate angle, the manifold pressure, or the air mass flow into the intake manifold must be used to predict the air mass flow into the cylinder. This information is then used to estimate the amount of fuel to be injected in the next cycle. Since the fuel path has its own "wall-wetting" dynamics, that effect has to be compensated as well. These two parts will be discussed separately below.

Air Mass Estimation

The mean value model of the manifold dynamics was introduced in Sect. 2.3 and Sect. 3.2.2. The main continuous-time domain equations describing the air flows in an SI engine are 1) the equation that models the air mass flow entering the intake manifold

$$\dot{m}_\alpha(t) = c_d \cdot A(t) \cdot \frac{p_a}{\sqrt{R \cdot \vartheta_a}} \cdot \Psi(p_a, p_{in}(t)) \tag{4.16}$$

and 2) the equation that models the air mass flow leaving the intake manifold and entering the cylinders

$$\dot{m}(t) = \rho_m(t) \cdot \dot{V}(t) = \frac{p_m(t)}{R \cdot \vartheta_m} \cdot \lambda_l(p_m(t), \omega_e(t)) \cdot \frac{V_d}{N} \cdot \frac{\omega_e(t)}{2\pi} \tag{4.17}$$

The amount of fuel must be calculated according to the amount of air entering the cylinder. Since a mass flow measurement at the intake of the cylinder usually is not possible, a model-based estimator must be used to obtain that information.

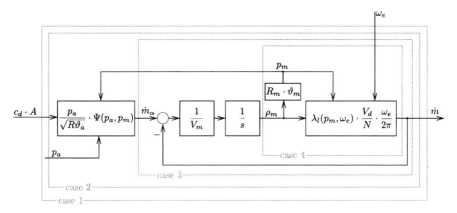

Fig. 4.15. Possible approaches to estimate the air mass in the cylinder, continuous-time formulation.

Depending upon the sensors installed, one of the following four structures is typically used for such an air mass flow estimator (see also Fig. 4.15):

1. The throttle angle is measured. The corresponding throttle area $A(t)$ multiplied with the discharge coefficient c_d can be determined by a-priori measurements. The rest of the manifold dynamics must be simulated with an open-loop observer. A correct steady-state operation cannot be guaranteed, since the ambient pressure p_a can vary, ord manufacturing tolerances result in wrong throttle area estimates. This is the cheapest solution, but obviously, also the least precise.

2. Same as case 1, except that the ambient pressure p_a also is measured. Therefore, a correct altitude compensation is possible. Ambient pressure sensors are not expensive. Because they can be mounted directly on the ECU board, this improvement is usually adopted.
3. The air mass flow $\dot{m}_\alpha$ entering the intake manifold is measured with an air mass flow sensor (typically a hot-film anemometer). This guarantees a stationary correct estimation of the air mass flow into the cylinder. The compensation of the transient effects is realized using an open-loop observer. This is the preferred sensor configuration since only the precision of the air mass flow measurement is relevant under steady-state conditions.
4. The intake manifold pressure p_m is measured. At first, this configuration seems to be optimal since the dynamically correct signal is used. Nevertheless, a relevant intake manifold temperature has to be estimated, and the volumetric efficiency must be known with high precision. An additional disadvantage is that the information about transient changes in the throttle area $A(t)$ is perceived with substantial delay. Therefore, predicting the required amount of fuel (which is necessary due to the timing delays between fuel and air) is rather difficult.

So far, continuous-time models have been discussed. However, to realize high-precision feedforward control systems a discrete-event formulation will be necessary. Using the discrete-event representations introduced in Sect. 3.2.2 the discrete-event signal flowchart of the cylinder air mass estimation shown in Fig. 4.16 can be derived.

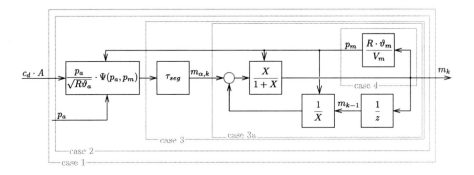

Fig. 4.16. Possible approaches for the estimation of the air mass in the cylinder, discrete-time formulation.

In the discrete-event formulation, the same three cases as those discussed in the continuous-time formulation are recognized. The only difference is that case 3a was not present before. In fact, in the discrete-event formulation of case 3, either the mass flow $\dot{m}_{\alpha,k}$ or the total mass $m_{\alpha,k}$ flowing into the manifold, is chosen as the input signal to the estimator. In the first case, the integration over one segment is accomplished in the ECU using a Euler forward integration (multiplication of the mass flow by τ_{seg}). In the second case, the output signal of the hot-film anemometer is integrated using a dedicated hardware component. Usually the second approach yields the more precise result.

Compensation of the Fuel Dynamics

General Approach

The discrete-event structure of the air mass dynamics has the advantage of being easily combined with the discrete-event representation of the fuel dynamics. Assuming first-order wall-wetting dynamics and an additional timing delay of two segments, the block diagram shown in Fig. 4.17 represents the model used to determine the duration of the fuel injection.

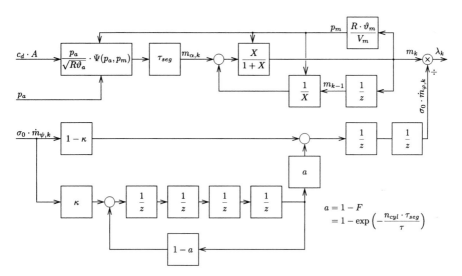

Fig. 4.17. Discrete-time air/fuel-ratio model.

The mass of air in the cylinder can be estimated using one of the approaches introduced above. The wall-wetting dynamics can be compensated as explained in Sect. 2.4.2, although the additional two-segment time delay cannot be eliminated. One way of coping with this problem is to artificially delay the input to the throttle actuator by two segments and to predict the corresponding air mass in the cylinder using a case 1 or case 2 estimator. Using this air mass information, the amount of fuel to be injected is calculated (inverting the wall-wetting dynamics) and the corresponding signal is sent to the power amplifiers of the injectors. If the air mass flow is measured, the air mass estimator can be corrected by using the measured value and taking into account the time delays.

The drawback of this solution is that at low engine speeds, the driver has to accept a significant time delay when issuing an acceleration request (for 900 rpm and a four-cylinder engine this delay amounts to 60 ms). If this is not acceptable, the air mass must be predicted using a suitable driver command prediction algorithm.

Combining these methods (delaying the driver pedal, predicting the actuation) a good drivability and a close-to-stoichiometric air/fuel ratio mixture can be guaranteed in most transients. An additional late injection into the open intake valve may become necessary in some extreme tip-in situations.

Limitations

It is interesting to note that the inversion of the wall-wetting dynamics is only possible for physically reasonable parameters. The discrete-time transfer function $G(z)$ of the wall-wetting dynamics for one cylinder can be expressed according to (3.28) as follows

$$G(z) = (1 - \kappa) + \kappa \cdot \frac{1 - F}{z - F} \qquad (4.18)$$

The inverse wall-wetting dynamics can thus be expressed with the following discrete transfer function

$$G^{-1}(z) = \frac{1}{1 - \kappa} - \frac{\kappa}{1 - \kappa} \cdot \frac{1 - \frac{F - \kappa}{1 - \kappa}}{z - \frac{F - \kappa}{1 - \kappa}} \qquad (4.19)$$

where the variable κ describes how much of the injected fuel adheres on the wall. For $\kappa = 1$ all of the injected fuel accumulates on the manifold wall, whereas for $\kappa = 0$ all of the injected fuel enters the cylinder in the same cycle. The wall film evaporation rate is expressed by F, where a small amount of F indicates a quick evaporation. Accordingly, these parameters satisfy the inequalities

$$0 \leq F \leq 1 \qquad \text{and} \qquad 0 \leq \kappa \leq 1 \qquad (4.20)$$

Obviously, the only pole p of the discrete time representation of the inverted wall wetting dynamics (4.19) lies between -1 and 1 to guarantee that the inverted system is stable. If mere stability is not enough, but non-oscillatory behavior is requested as well, then the pole must be confined to the restricted interval $0 \leq p \leq 1$. These two requirements yield the following conditions for the parameters κ and a:

condition for stability: $2 \cdot \kappa - 1 \leq F \leq 1$

and condition for non-oscillating behavior: $\kappa \leq F \leq 1$

These limits are illustrated in Fig. 4.18. Obviously, for a gain $\kappa > 0.5$ the evaporation rate may not exceed certain values (F may not be too small) if unstable transients are to be avoided. If oscillations must be avoided as well, the evaporation rate may not be higher (F may not be smaller) than the deposition rate κ. This fact can be explained as follows. For a gain κ close to one (almost all fuel adheres on the wall), the term $\frac{1}{1-\kappa}$ is relatively large, i.e., in this situation a full compensation of the wall-wetting dynamics forces the feedforward control system to inject a relatively large amount of fuel. Since a large part of this fuel is deposited on the manifold walls, the size of the wall film increases substantially. Therefore, if the evaporation process is fast (F is small), after one engine cycle a large amount of evaporated fuel is ready for induction into the cylinder. This amount can exceed the amount of fuel needed to achieve a stoichiometric air/fuel ratio, resulting in an overshooting behavior.

With a well-designed injection system (injector position, intake runner geometry, etc.), this phenomenon can be avoided. In addition, large gains κ are usually encountered at cold start. Fortunately, in this situation the evaporation rates are slow, i.e., large values of the parameter F are to be expected.

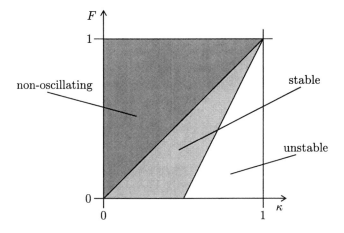

Fig. 4.18. Conditions for κ and F for stability and non-oscillating behavior of the wall-wetting compensation.

4.3.2 Feedback Control: Conventional Approach

One straightforward approach to realize an air/fuel ratio feedback control system is to use a single switch-type air/fuel ratio sensor upstream of the TWC. The voltage function of a typical switch-type sensor is depicted in Fig. 4.19.

Typically, the sensor output is converted to a binary signal, lean for sensor voltages U_λ smaller than $450\ mV$, rich for U_λ greater than $450\ mV$. Using this

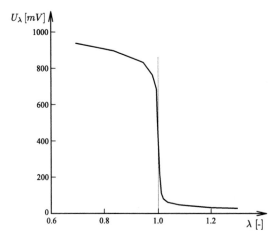

Fig. 4.19. Steady-state voltage characteristic of the Bosch LSF switch-type air/fuel ratio sensor.

binary signal, a control structure similar to a PI system is chosen. Figure 4.20 shows the block diagram of such a control system.

Instead of the usual additive approach, the correction of the nominal injection duration τ_{inj} is achieved by *multiplying* this value with the control system output F_λ. This approach is advantageous because of the multiplicative definition of the air/fuel ratio signal and because most errors in the air/fuel path are of a multiplicative nature.[10]

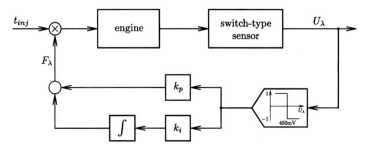

Fig. 4.20. Block diagram of the air/fuel ratio control loop with a switch-type sensor.

Due to the asymmetric behavior of the switch-type sensor with respect to steps from rich to lean compared to those from lean to rich, the control

---

[10] For instance, slightly blocking an injection valve with soot can produce a 10% smaller aperture area and, therefore, a 10% smaller amount of fuel injected. The correction of that error requires the nominal injection duration to be increased by 11%. If no changes in the valve aperture occur, all subsequent injection durations must be multiplied by the *same* correcting value.

system shown in Fig. 4.20 produces a bias of the mean value of the air/fuel ratio toward lean conditions. To compensate this bias and to achieve a shift of the mean air/fuel ratio to slightly richer conditions,[11] the switch from rich to lean is delayed by a time interval t_v.

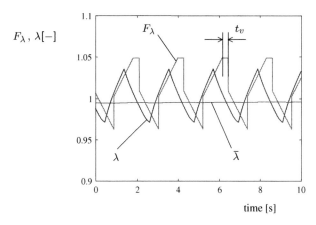

Fig. 4.21. Simulated steady-state behavior of a switch-type control system.

Figure 4.21 shows a simulation of the resulting trajectories for λ and for the control input F_λ. The fuel path of the engine is modeled as a first-order lag with a time delay. The dynamics of the switch-type sensor are neglected in that simulation. This figure also shows the mean value $\bar{\lambda}$ of λ that is obtained by averaging λ using a moving window whose size is compatible with the oxygen storage capacity of the TWC.

In summary, this type of controller has three parameters $\{k_p, k_i, t_v\}$ that must be tuned for every operating point. Since these parameters must be experimentally identified, the parametrization becomes rather time consuming.

The above approach works well as long as the emission levels that have to be attained are not too demanding. Note that with this control structure, the TWC is operated in an open-loop approach, i.e., its dynamics are not taken into account.

4.3.3 Feedback Control: H_∞

The rather complex dynamic behavior of a TWC has been described in Sect. 2.8.3. It is advantageous to include these storage effects in the development of the fueling control system. In particular, the excursions of the air/fuel ratio in one direction must be compensated with equivalent corrections in the other.

---

[11] As discussed in Sect. 2.8.3, the optimal catalytic converter efficiency is achieved for $\lambda = 0.99 \dots 1.00$.

The block diagram of the plant model used to develop the feedback air/fuel-ratio control-system is depicted in Fig. 4.22. Only the fuel path from the injector valves to the air/fuel ratio sensor is considered. The air path is not included, and all influences of that component are treated as disturbances.

The sensor, which is placed immediately upstream of the TWC, is assumed to be a wide-range sensor, giving not only lean-rich binary information on the air/fuel ratio, but also quantitative values. This sensor is used to form a first, fast air/fuel ratio control loop. In a second stage, an additional air/fuel ratio sensor downstream of the TWC is used to close an outer, slower control loop. This loop takes the oxygen storage capacity of the TWC into account and compensates for sensor drifts in the inner loop.

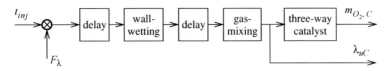

Fig. 4.22. Block diagram of the fuel path model.

As in the classical approach discussed in the last section, the output of the control system is a multiplicative factor F_λ that corrects the nominal injection duration. The error signal used as the input to the control system is defined in a multiplicative way as well.

The outputs of the plant are the air/fuel ratio value at the main confluence point λ_{uC} (upstream of the TWC) and the mass of oxygen stored in the TWC, $m_{O_2,C}$. The system up to the main confluence point has already been analyzed in detail in Chapter 3. However, that model is too complicated to be used in the design of a feedback control system. Therefore, the complete air/fuel ratio path, from the fuel injectors to the air/fuel ratio sensor at the main confluence point, is approximated by a serial connection of a first-order lag with time constant τ and a time delay δ

$$P(s) = \frac{\Delta \lambda_{uC}}{\Delta F_\lambda} = \frac{-1}{s \cdot \tau + 1} \cdot e^{-s \cdot \delta} \qquad (4.21)$$

In this expression, the error is defined by $\Delta \lambda_{uC} = \lambda_{uC} - \lambda_{uCd}$, where λ_{uCd} is the demanded value for the upstream catalytic converter air/fuel ratio, and the linearized input is calculated according to $\Delta F_\lambda = F_\lambda - 1$.

A typical measured and simulated step response is shown in Fig. 4.23. The engine used in this experiment is a 1.8-liter four-cylinder SI engine. The approach presented below is applicable to any port-injected SI engine. In this section, the numerical and experimental data will better illustrate that the proposed design methodology is always valid for that engine.

The parameters of this model, i.e., the time constant τ of the lag element and the time delay δ, depend on the operating point of the engine. An on-line identification for these two parameters is possible using the approach

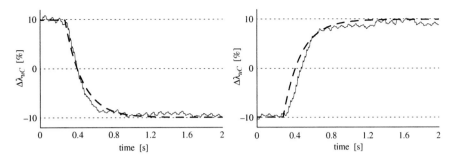

Fig. 4.23. Simulated (dashed) and measured step response of the engine mentioned in the text; operating point 1500 rpm and 0.1 g air per cylinder per cycle (approximately 25% of fullload).

presented in [186]. A complete map of these two parameters can easily be obtained off-line as well. Figure 4.24 shows an example of such maps for the full load range and the relevant speed range.[12]

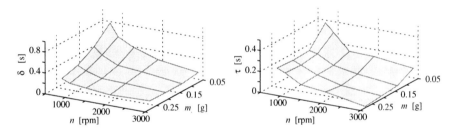

Fig. 4.24. Time delay δ and time constant τ of the fuel path of a 1.8-liter four-cylinder engine for various engine speeds n and air masses in the cylinder m.

The actual design of the feedback controller can be accomplished in many ways. Here, an approach is shown that is based on the H_∞ frequency-domain optimization technique.[13] For this control design method, a linear and finite dimensional system description is required. Therefore, the time delay δ must be approximated using a Padé allpass element

$$e^{-s \cdot \delta} \approx \frac{\sum_{j=0}^{n_{apx}} a_j \cdot (-s \cdot \delta)^j}{\sum_{j=0}^{n_{apx}} a_j \cdot (s \cdot \delta)^j} \qquad a_j = \frac{(2 \cdot n_{apx} - j)! \cdot n_{apx}!}{j! \cdot (n_{apx} - j)! \cdot (2 \cdot n_{apx})!} \qquad (4.22)$$

---

[12] For higher engine speeds, the parameters do not change substantially because the non-engine speed dependent factors such as air/fuel ratio sensor time constant, etc. become dominant.

[13] For a general discussion of this control system design method, the reader is referred to [53] and, for more information about the engineering aspects, [41] and [175] are good sources.

This approach has been found to be the best approximation method in this context. The order n_{apx} of the approximation is chosen as a function of the time delay. For delays up to 40 ms, a first-order approximation is recommended, and for every additional 40 ms, one order should be added. This results in a rather precise approximation of the delay, but also in a rather high system order. Accordingly, in order to obtain a small-order control system, an order reduction step will be necessary.

Inclusion of the Oxygen Storage Capacity

Since the three-way catalytic converter is capable of storing oxygen as well as carbon monoxide, it is not the actual value of the air/fuel ratio that is relevant for tailpipe emissions. As long as an excess or lack[14] of oxygen can be compensated for by the oxygen storage capability of the TWC, no substantial amounts of emissions are expected at the tailpipe. Only if the storage capacity of the TWC is reached, a rise in the concentrations of NO_x or CO and HC, respectively, is to be expected. Therefore, as suggested in (2.228) in Sect. 2.8.3, an integrator is added to the system dynamics described above to model the storage behavior of the TWC

$$\Delta m_{O_2,C} = \frac{1}{s} \cdot \dot{m}_{O_2,uC} \cdot \Delta\lambda_{uC} \qquad (4.23)$$

The variable $\dot{m}_{O_2,uC}$ defines the oxygen mass flow at the same location where the upstream wide-range sensor is placed. For the linear control design, a value for $\dot{m}_{O_2,uC}$ equal to 1 is assumed, i.e., a normalized system representation is used. In the controller realization, the correct mass flow will be considered.

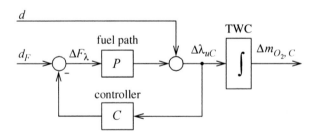

Fig. 4.25. Block diagram of the closed-loop system model used for the control system synthesis.

Inner Control Loop

The goal of the first controller's design is to minimize the impact of an air/fuel ratio deviation on the oxygen storage level in the TWC. Using a wide-range

---

[14] Missing oxygen describes rich gas mixtures which result in high levels of CO.

sensor, this must be achieved by measuring only the upstream air/fuel ratio (see Fig. 4.25).

The transfer function from the disturbance d to the oxygen mass $\Delta m_{O_2,C}$ stored in the TWC is given by 4.24

$$G_{Cd}(s) = \frac{\Delta m_{O_2,C}}{d} = \frac{1}{s} \cdot \frac{1}{1 + P(s) \cdot C(s)} \tag{4.24}$$

where $P(s)$ has been defined in (4.21) and (4.22) and $C(s)$ is the controller designed below.

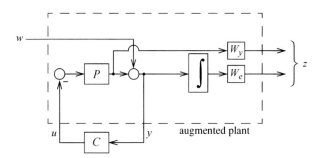

Fig. 4.26. Block diagram of the augmented control system used in the H_∞ design, where w is the generalized disturbance input and z is the generalized output whose magnitude has to be minimized.

For the design of an H_∞ control system, the plant must be augmented by appropriate weighting functions. The block diagram of the resulting closed-loop system is shown in Fig. 4.26. The associated closed-loop transfer function T_{zw} has the form

$$T_{zw}(s) = \begin{bmatrix} W_e(s) \cdot G_{Cd}(s) \\ W_y(s) \cdot T_e(s) \end{bmatrix} \qquad T_e(s) = P(s) \cdot C(s) \cdot (1 + P(s) \cdot C(s))^{-1} \tag{4.25}$$

If a stabilizing controller exists that satisfies the conditions $||T_{zw}||_\infty < 1$,[15] the following two inequalities for the disturbance transfer function G_{Cd} and for the complementary sensitivity T_e will be satisfied

$$\begin{aligned} |G_{Cd}(j\omega)| &< |W_e^{-1}(j\omega)| \\ |T_e(j\omega)| &< |W_y^{-1}(j\omega)| \end{aligned} \qquad \forall\, \omega \tag{4.26}$$

The dynamic weight $W_e(s)$ thus serves as an upper bound for the magnitude of the disturbance transfer function G_{Cd}. A possible choice for $W_e(s)$ is

---

[15] See (A.47) in the Appendix A for a definition of the H_∞ norm.

$$W_e(s) = \frac{1}{d_{max}} \cdot \frac{s + \omega_c}{s} \qquad (4.27)$$

The corresponding magnitude plot is shown in Fig. 4.27. This diagram also illustrates the influence of the two design parameters d_{max} and ω_c on the disturbance transfer function $G_{Cd}(s)$. The parameter d_{max} limits the maximum value of $|G_{Cd}(j\omega)|$. At low frequencies, the disturbance transfer function $G_{Cd}(s)$ is forced to zero. This is achieved by the integrator element included in $W_e(s)$. The frequency at which this reduction starts can be specified by the parameter ω_c.

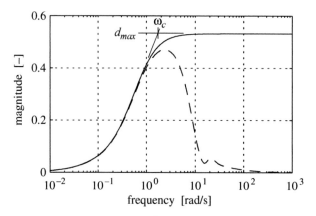

Fig. 4.27. Magnitude plot of $W_e^{-1}(j\omega)$ (solid) and $G_{Cd}(j\omega)$ (dashed).

To avoid an unnecessarily high bandwidth of the control system, the complementary sensitivity $T_e(s)$ is limited with an appropriate choice of $W_y(s)$

$$W_y(s) = \frac{s}{k_c \cdot \omega_c} \qquad (4.28)$$

where $k_c \approx 10$ is recommended. An example of the magnitude of the weighting function $W_y(s)$ is shown in Fig. 4.28.

The plant parameters τ and δ vary over quite a large range. For the engine example used in this section, the following range must be considered

$$\delta \in [0.02 , 1.0] \ s \qquad \tau \in [0.01 , 0.5] \ s$$

Since the first-order lag in the plant (4.21) can always be compensated, it has no influence on the weighting functions. The time delay cannot be compensated, and the performance specification in the weighting matrices must be adjusted to the varying operating points and, hence, to varying delays. In other words, for larger time delays it is mandatory that the demanded bandwidth of the control system is reduced. This leads to the following parametrization of the design parameters

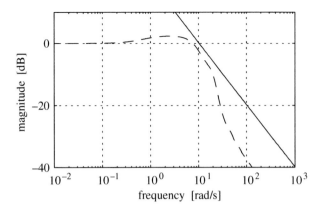

Fig. 4.28. Magnitude of $W_y^{-1}(j\omega)$ (solid) and $T_e(j\omega)$ (dashed); parameter $k_c = 13$.

$$d_{max} = f_{d_{max}} \cdot \delta, \qquad \omega_c = \frac{f_{\omega_c}}{\delta} \tag{4.29}$$

This choice of the design parameters, which has been found empirically, results in an almost identical Nyquist diagram for all G_{Cd} and, thus, in the same robustness properties over the whole operating range of the engine. The parameters $f_{d_{max}}$ and f_{ω_c}, which are the same for all operating points, allow a shaping of the Nyquist curve. The following values have been found to be a good choice for the engine discussed in this section

$$f_{d_{max}} = 1.9, \qquad f_{\omega_c} = 0.22 \tag{4.30}$$

Since the phase shift induced by the time delay cannot be avoided, a robust stabilizing controller must place a significant amount of actuator energy (high loop gain) at frequencies at the opposite of the Nyquist point -1. The solution $C(s)$ to the H_∞ problem formulated above produces one or several characteristic "bubbles" in the Nyquist diagram of the loop transfer function

$$L(s) = \frac{-1}{s \cdot \tau + 1} \cdot e^{-s \cdot \delta} \cdot C(s) \tag{4.31}$$

in the right half plane, as shown in Fig. 4.29.

Assuming that the error in the estimation of the time delay δ is small, this relatively high gain at frequencies higher than the crossover frequency causes a differential behavior of the loop that adds beneficial damping to the control system. Notice, though, that an error in the identified time delay results in an additional, usually detrimental phase shift of the Nyquist curve. Accordingly, using more than one "bubble" is only possible with a very accurate system model at hand.

The "one-bubble" form can always be realized with a fourth-order controller (the "bubble" requires a complex conjugate pair of poles). If a "no-bubble" form is chosen, then the minimal order of the controller is two, with both poles at zero in order avoid a steady-state error in $\Delta m_{O_2,C}$.

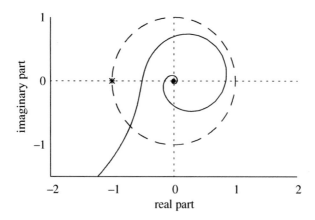

Fig. 4.29. Nyquist plot of the open-loop gain $L(j\omega)$.

For a disturbance step of 10%, the control system designed in this section yields the response depicted in Fig. 4.30. The effect of varying d_{max} and ω_c on the resulting trajectories is indicated in the same figure. Decreasing d_{max} results primarily in a faster descent of the air/fuel ratio back to the stoichiometric value, and thus in a smaller peak of the oxygen to be stored in the TWC. Increasing ω_c results in a faster descent of the total oxygen stored in the TWC to its reference value. Assuming a constant air mass flow, the return of the stored oxygen mass to its nominal value obviously can only be achieved if $\Delta\lambda_{uC}$ does not return directly to zero but overshoots or undershoots to realize the necessary compensation effects.

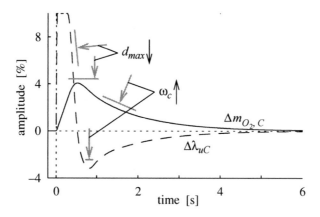

Fig. 4.30. Response of the closed-loop system to a step of 10% in the disturbance $d(t)$.

Realization Aspects

As mentioned above, with increasing delay times, the Padé approximation of the delay element causes an increase of the model order of the plant and, hence, of the control system. However, a fourth-order controller is sufficient to realize the specific form proposed in the last section. Therefore, an order-reduction step is usually added after the full-order control system has been synthesized. Any of the well-known order-reduction techniques can be applied. More details of that aspect can be found in [175].

Several state-space forms can be chosen to realize the fourth-order controller. Because of the order reduction, the states of the control system are different in each operating point and do not allow any direct physical interpretation. Since the control system designed in the last section must be gain-scheduled, a fixed, and preferably physically motivated, structure must be found. This structure must incorporate the minimum number of parameters.[16] Moreover, the chosen approach must minimize the sensitivity of the control system to parameter errors which may arise due to the gain scheduling. This can be achieved by choosing a structure that utilizes as many parallel elements as possible. For the "one-bubble" solution, the following form is recommended

$$C(s) = c_1 + c_2 \cdot \frac{1}{s} + c_3 \cdot \frac{1}{s^2} + c_4 \cdot \frac{s \cdot (c_7 \cdot s + 1)}{c_5^2 \cdot s^2 + 2 \cdot c_6 \cdot c_5 \cdot s + 1} \qquad (4.32)$$

The coefficients c_1 to c_7 are functions of the time delay, of the time constant, or of the operating point defined by engine speed and load. For each operating point, these coefficients are obtained from the result $C(s)$ of the original control system design by comparing powers of the independent variable s.

Outer Control Loop

The above control system design does not consider the drift of the sensor in front of the catalyst nor the dynamics of the TWC. In fact, this device has its own dynamic effects, the most important being oxygen storage. In this section the previously designed control system is extended to compensate for sensor drifts and to consider these dynamic effects as well. This will be achieved by using the information collected by an additional air/fuel ratio sensor downstream of the TWC.

For a realistic physical model of the storage behavior, the varying mass flow through the TWC must be taken into account. This is done by multiplying the air/fuel ratio deviation $\Delta\lambda_{uC}$ with the air mass flow $\dot{m}_{O_2,uC}$ before it is integrated. The output of the integrator is again divided by the same mass

---

[16] For a "one-bubble" solution, the number is seven. A general fourth-order controller has nine parameters, but since in this case two poles are forced to zero, two parameters are structurally fixed.

flow, yielding the structure shown in Fig. 4.31. Using this approach, a correct oxygen balance is obtained without changing the input-output behavior at a fixed operating point. An alternative but equivalent formulation of this component is possible using the concept of *relative oxygen level* introduced in (2.228).

The limits 0 and $C_{O_2,C}$ of the oxygen storage capacity of the TWC must be modeled as well. To estimate $C_{O_2,C}$, dedicated experiments similar to those whose results are shown in Fig. 2.68 must be carried out. Since the storage capacity decreases with time, an on-line identification, as proposed in [185], is necessary.

With ageing of the upstream wide-range air/fuel ratio sensor, a slowly increasing offset of its output signal is observed. This offset and the errors in the observed amount of oxygen must be corrected with an external control loop that utilizes the downstream air/fuel ratio sensor as the main input signal. Both wide-range and switch-type sensors are used for that purpose. Compared to the wide-range sensor, the switch-type sensor, in addition to being relatively cheap, has the advantage of offering higher precision in detecting stoichiometric air/fuel ratios.

In both approaches, the outer control loop is chosen to be a relatively "slow" PI controller. The signal from the downstream air/fuel ratio sensor is used to infer the actual oxygen storage level $m_{O_2,C}$. This estimation problem is not easy and its solution is the topic of ongoing research. The main difficulty is that as long as the TWC does not reach its storage capacity limits, the downstream air/fuel ratio sensor will not detect any substantial changes in the air/fuel ratio in the exhaust gases. Using a switch-type sensor, whose high sensitivity around stoichiometric air/fuel ratios is welcome in this context, an *approximation* of the oxygen content is obtained as follows

$$m_{O_2,C} \approx C_{O_2,C} \cdot \begin{cases} 1 & \text{if } U_\lambda \leq 0.2 \ V \\ 1 - (U_\lambda - 0.2)/0.5 & \text{if } 0.2 \ V < U_\lambda \leq 0.7 \ V \\ 0 & \text{if } U_\lambda > 0.7 \ V \end{cases} \qquad (4.33)$$

The difference between this measured oxygen level $m_{O_2 C}$ and the observed level $\hat{m}_{O_2,C}$ is the input to a PI controller that produces a correction signal for the oxygen storage observer. This somewhat indirect approach is chosen because of uncertainty with respect to the true oxygen storage dynamics, as mentioned.

The estimated value $\hat{m}_{O_2,C}$ of oxygen stored in the TWC is compared to the corresponding desired[17] value $\bar{m}_{O_2,C}$, and the resulting difference is used as input signal to the "buffer control" component. The resulting control system structure is shown in Fig. 4.31. Note that the gain c_3, which depends on the engine's operating point, must be placed in front of the second (rightmost)

---

[17] A value of $\bar{m}_{O_2,C} = 0.6 \cdot C_{O_2,C}$ is recommended.

integrator to avoid an excitation of the controller because of varying operating points. If no controller action is required, the input to the integrator will be zero. Therefore, varying the gain c_3 does not affect the output of the controller if it is placed before the integrator.

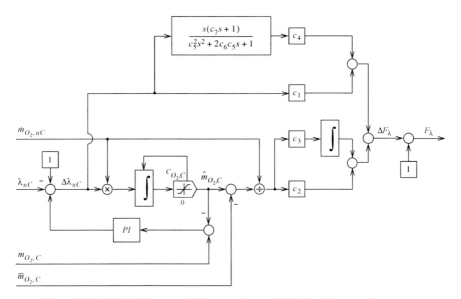

Fig. 4.31. Block diagram of the complete air/fuel ratio control system.

Experimental Results

To demonstrate the benefits of the proposed control system, especially of its capability to automatically adapt the bandwidth of the closed loop to changing operating points, the results of simulations and measurements at two operating points are shown below. The precise definition of the two cases analyzed is given in Table 4.1. The first operating point (OP 1) was also used for the simulations shown above. Note that in OP 1, the time delay δ corresponds to almost 14 segments. This large delay substantially limits the achievable bandwidth of the feedback control system.

Figure 4.32 shows the response of the upstream lambda buffer control system for a disturbance[18] step at d_F as defined in Fig. 4.25.

Because of the parametrization of the control system parameters proposed above, the two system responses have similar properties. However, because the

---

[18] The signal d_F is the more realistic choice as a disturbance source than the signal d. The latter represents the worst case situation because it acts immediately on the controller and is preferable for the design of the control system.

Table 4.1. Operating points chosen in the experiments.

operating point			OP 1	OP 2
engine speed	n	[rpm]	1500.0	1500.0
cylinder air-mass	$m_{air,cyl.}$	[g]	0.1	0.3
engine torque	T_e	[Nm]	18.0	80.0
time delay	δ	[s]	0.27	0.116
time constant	τ	[s]	0.192	0.146
segment time	τ_{seg}	[s]	0.02	0.02
cycle time	τ_{cycle}	[s]	0.08	0.08

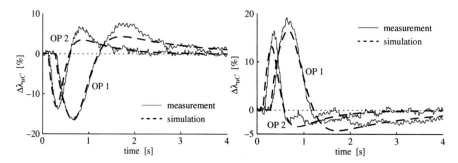

Fig. 4.32. Response of the upstream lambda buffer control to a -20% step (left) and a +20% step (right) in the disturbance d_F.

time delays in OP 1 and OP 2 are different, the bandwidths of the control systems vary accordingly.[19] The maximum deviation of the air/fuel ratio in OP 1 is slightly larger than in OP 2 because of the larger time delay. Also, the maximum value of the *integrated* λ error is larger in OP 1 than in OP 2. Nevertheless, because of the substantially smaller air mass flow, the error in stored oxygen is not as large.

The variations of the oxygen storage level in the TWC are used to assess the performance of the "buffer control system" to adapt to changing operating conditions. Since the oxygen storage level cannot be measured directly, the output of the switch-type air/fuel ratio sensor downstream of the TWC is taken as a measure of that quantity. It is reasonable to assume that if the oxygen storage level of the TWC remains constant, the composition of the exhaust gas remains nearly constant as well. Accordingly, the output voltage of the downstream sensor remains constant, too.

The transient test used to assess the performance of the proposed control approach consists in rapidly (in approximately 100 ms) opening the throttle plate from 9° to 24°. In steady-state, the mass of air entering the cylinder in each cycle due to that change is increased by a factor of three. The delayed increase in air mass visible in Fig. 4.33 is a consequence of the manifold

---

[19] Notice that a scaling of the time t, according to the ratio of the delays, produces almost congruent curves.

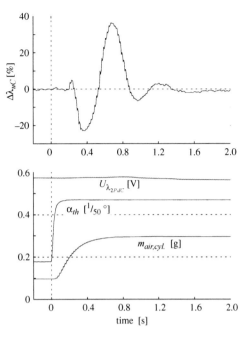

Fig. 4.33. Response of the complete control system to a step in the throttle command corresponding to a change from OP 1 to OP 2. The variable $U_{\lambda_{2PdC}}$ denotes the voltage of the switch-type air/fuel ratio sensor downstream of the catalytic converter.

dynamics. Moreover, to amplify the errors in the air/fuel ratio, the feedforward control action for the fuel injection has been reduced in these experiments.

The results of this test are shown in Fig. 4.33. In spite of the large increase in air mass and in spite of the reduced action of the feedforward control system, the downstream sensor output remains almost unchanged. This result can be taken as an indication for a constant oxygen storage in the TWC in the steady-state conditions.

4.3.4 Feedback Control: Internal-Model Control

Ideally, the efficiency of modern pollution abatement systems should be maintained over the whole lifetime of its components. To that end, it is indispensable to account for the ageing of the components involved. Specifically, the TWC and the wide-range λ sensor may undergo considerable changes in their dynamical properties during their lifetime. In section 2.8 it is shown that for a tight control of the oxygen level in the TWC with consideration of ageing effects, a more detailed, control-oriented model than the one presented in 4.23 may be necessary. In the preceding section the TWC model was included in the design process of the air/fuel ratio controller. However, with a more detailed model this neither seems to be beneficial nor tractable in practice. As depicted in Fig. 4.34, the controller for the oxygen level, based on the detailed model and the air/fuel ratio controller, are therefore structurally arranged in a cascaded control scheme, where the air/fuel ratio controller is in the inner and the oxygen level controller is in the outer loop, respectively. In this setup

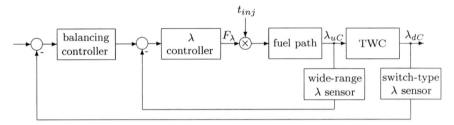

Fig. 4.34. Schematic air/fuel ratio control system.

the air/fuel ratio controller can be designed independently of the controller for the oxygen level, and the output of the level controller acts as the reference signal for the air/fuel ratio control system in the inner loop. As a consequence, the new air/fuel ratio controller needs to be optimized with respect to both disturbance rejection *and* reference tracking.

To account for variations in the wide-range sensor dynamics due to ageing of the sensor cell and clogging of the vent holes, an adaptive extension of the air/fuel ratio controller will be inevitable to meet future emission regulations. It is therefore important to choose a control structure that offers the premises for such extensions.

For processes where an input-output model is available, specifically for stable SISO processes, the internal model control (IMC) structure depicted in Fig. 4.35 offers an intuitive control design procedure that is compatible with the above-mentioned requirements of the air/fuel ratio control system. The separation of a controller into an internal model $\hat{P}(s)$ and an internal

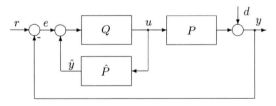

Fig. 4.35. IMC system.

controller $Q(s)$ also allows a natural handling of processes incorporating a time delay. The transfer function of the control system in Fig. 4.35 can be written as

$$T(s) = \frac{P(s)Q(s)}{1 + (P(s) - \hat{P}(s))Q(s)} \tag{4.34}$$

In the case of a perfect model, i.e., $\hat{P}(s) = P(s)$ the transfer function $T(s)$ becomes

$$T(s) = \hat{P}(s)Q(s) \tag{4.35}$$

thus, for open-loop stable plants the control system in Fig. 4.35 is stable for all stable transfer functions $Q(s)$. Furthermore, in the case of a perfect model, the signal $\hat{y}$ cancels the feedback signal y, i.e., the feedback path only is "activated" in the presence of external disturbances and modeling errors, whereas otherwise the internal controller $Q(s)$ acts as a pure feedforward controller. This property reflects the very nature of the interaction of feedforward and feedback control.

The design of the internal controller $Q(s)$ is usually done in two steps:[20] in a first step a nominal controller $\tilde{Q}(s)$ is designed that minimizes the control error e for a representative disturbance d

$$\tilde{Q} = \arg \min_{Q} \|e\|_2 \tag{4.36}$$

subject to the constraints of stability and causality of $\tilde{Q}$. With (4.35) the problem (4.36) can be rewritten as

$$\tilde{Q} = \arg \min_{Q} \|(\hat{P}Q - 1) \cdot D\|_2 \tag{4.37}$$

where D is the Laplace transform of the signal d. Obviously, for minimum-phase systems the solution of (4.37) is[21]

$$\tilde{Q} = \hat{P}^{-1} \tag{4.38}$$

In the second step an IMC filter F is added to the nominal controller $\tilde{Q}$

$$Q = \tilde{Q} \cdot F \tag{4.39}$$

to account for modeling errors and to obtain a realizable transfer function Q. Given the relative uncertainty bound W_2 of the model $\hat{P}$

$$\left| \frac{P(j\omega) - \hat{P}(j\omega)}{\hat{P}(j\omega)} \right| \leq W_2(\omega), \quad \forall \omega, \tag{4.40}$$

the robust Nyquist theorem states that the closed control loop is asymptotically stable if the inequality

---

[20] For detailed information about the design of internal model controllers the reader is referred to [145].

[21] For non-minimum-phase systems and systems with a time delay the optimal solution of (4.37) is also dependent on the shape of the disturbance d.

$$|T(j\omega)W_2(\omega)| < 1 \tag{4.41}$$

holds. With (4.35) and (4.39) the inequality (4.41) can be written as

$$|\hat{P}(j\omega)\tilde{Q}(j\omega)F(j\omega)W_2(\omega)| < 1 \tag{4.42}$$

and thus the IMC filter F can be used to shape the complementary sensitivity T to obtain a robustly stable control system. This step is often referred to as "detuning" of the nominal controller $\tilde{Q}$.

Given the model of the fuel path in (4.21) and the assumption that the main disturbances acting on the control system are step-shaped[22], the nominal air/fuel ratio controller $\tilde{Q}$ is obtained by inverting the model of the fuel path without the delay time

$$\tilde{Q}(s) = \frac{\tau s + 1}{-1}. \tag{4.43}$$

The delay time δ of the fuel path provides an upper bound for the achievable bandwidth of the control system. To account for this upper bound, the IMC filter F is chosen as

$$F(s) = \frac{1}{\frac{\delta}{\sigma}s + 1}, \tag{4.44}$$

where σ is a tuning parameter for the bandwidth of the control system. The resulting internal controller can then be written as

$$Q(s) = -\frac{\tau s + 1}{\frac{\delta}{\sigma}s + 1}. \tag{4.45}$$

With (4.35) the nominal complementary sensitivity $T(s)$ can now be written as

$$T(s) = \frac{1}{\frac{\delta}{\sigma}s + 1}e^{-s\delta}. \tag{4.46}$$

The transfer function of the nominal open loop gain $L(s)$ can be written as

$$L(s) = \frac{e^{-\delta s}}{\frac{\delta}{\sigma}s + 1 - e^{-s\delta}}. \tag{4.47}$$

Given the facts that $L(s)$ is independent of the time constant τ and that the delay time δ only appears as a factor for the Laplace variable s, the shape of the Nyquist curves shown in Fig. 4.36 is dependent only on the parameter σ and not on the parameters τ and δ. This fact can be interpreted as a constant robustness of the control system for all operating points. As shown in Fig. 4.36, large values of the bandwidth parameter σ lead to bubbles in the

---

[22] This is a valid assumption since the feedback controller is primarily designed to compensate for the static errors of the feedforward controller, whereas the bandwidth of transient errors due to the feedforward control action is much higher than the bandwidth of the feedback control system.

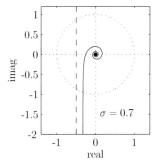 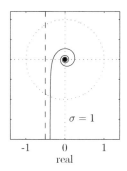 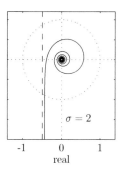

Fig. 4.36. Nyquist diagram of the nominal open air/fuel ratio IMC loop, for different values of σ. The dashed line indicates the Nyquist curve for $\sigma \to \infty$. The shape of the Nyquist curve is independent of the plant parameters.

right half plane of the Nyquist diagram. This is due to the increased derivative control action that is necessary to achieve the demanded bandwidth. Aside from the common issues with regard to amplification of measurement noise, the bubbles can lead to undesired control actions at high frequencies. More specifically, in air/fuel ratio control too high a value of σ may lead to control action due to an imbalanced air/fuel ratio in the single cylinders. This is undesirable since the model in (4.21) does not account for these effects and thus the imbalance of the cylinders may even increase due to the control action. Various experiments have shown that a value of $\sigma = 0.7$ results in a good trade-off between performance and robustness. However, when a higher bandwidth is required, the magnitude of the loop gain at high frequencies may be reduced by adding a further lowpass filter to the IMC filter.

Implementation of the IMC

Implementation of the Delay Time

The delay time in the internal model of an IMC is a critical element for its implementation in software. It is yet unclear how to efficiently implement the delay time block in the general case with fast variations in the ratio of the delay- and the sampling time. However, this issue can often be circumvented by a proper choice of the basis for the sampling time and by a physically motivated implementation of the delay time. Specifically, for the air/fuel ratio controller the sampling time t_s is chosen to be equal to the segment time t_{seg} of the engine. Figure 4.37 indicates that in the main operating range the variations of the ratio between the delay time δ of the fuel path and the segment time t_{seg} of the engine are small. When the control system is sampled at the segment frequency, the delay time can be implemented as a memory block with a constant part and variable part, where the constant part accounts for the delay time due to the engine cycle and the variable part accounts for the

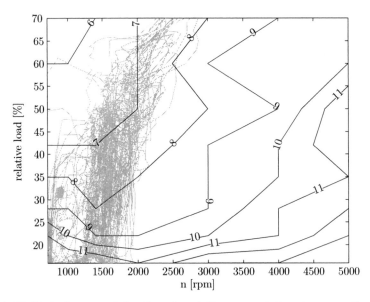

Fig. 4.37. Ratio between delay time δ and the segment time t_{seg} over the whole operating range of an engine with 5 cylinders. The gray line represents the trajectory during the FTP-75 cycle.

gas transport in the exhaust manifold. In Fig. 4.38 a possible implementation strategy of the delay time is depicted schematically. The estimated value of the delay time δ usually is obtained from an engine map as a function of the relative load and the engine speed.[23] However, the physical delay due to the gas transport in the exhaust manifold is a function of the thermodynamic properties of the exhaust gas, i.e., the estimated and the physical transport delays are "separated" by about one engine cycle. As shown in Fig. 4.38, to overcome this separation the variable part of the memory stack is placed in front of the constant part and the input to the memory is moved in accordance with the variable delay time. When the memory input is moved to the right, i.e., the variable delay time is reduced, all values in the cells on the left-hand side of the memory input are dropped and the value in the active input cell is overwritten by the new value of $\hat{y}_\tau$. This can be interpreted physically as the exhaust gas that passes the λ sensor between two sampling events. When the memory input is moved to the left, i.e., the variable delay time is increased, empty memory cells are generated. A physically correct filling of the empty cells would require an inter-sampling of the values of the model output $\hat{y}_\tau$, e.g. if the memory size is increased by one cell, the (new) first cell is filled

---

[23] In this context the engine speed is not relevant since it is shown in subsection 2.5.2 that the dynamics of the engine speed are much slower than those of the intake manifold.

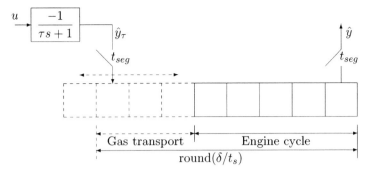

Fig. 4.38. Implementation of the delay time.

with the actual value of $\hat{y}_\tau$ at time t_k and the value of $\hat{y}_\tau$ at time t_{k-1} of the second cell is shifted to the third cell at the same time, i.e., the second cell becomes empty and should thus be filled with the value of $\hat{y}_\tau$ at time $t_{k-0.5}$. As a consequence of this process, assuming that the memory size is maximally increased by one per sample, a correct treatment of an increase in the memory size would require that the signal $\hat{y}_\tau$ is computed at twice the actual sampling frequency. This assumption is justified by the numbers shown in Table 4.2 which show the distribution of the occurrence of changes in the ratio of the delay time and the segment time during an FTP cycle.

Table 4.2. Distribution of changes in memory size per sample (segment time) during the first 1200 s of the FTP-75 cycle

Δ memory size	-2	-1	0	1
occurrence	0.02%	2.42%	96.28%	1.28%

When the given resources prohibit the use of a memory stack, the delay time could also be implemented as a finite dimensional approximation, e.g. a Padé element as shown in (4.22) and (B.18). However, in this case the approximation would contribute to the model error, which may have to be considered in the design of the IMC filter.

Static Errors

Static deviations from the stoichiometric air/fuel ratio have a detrimental effect on the performance of the TWC. It is therefore important for the controller to have a high gain at low frequencies. If the IMC has been designed according to the procedure outlined in the previous section, i.e., the static gains of the internal controller Q and the internal model $\hat{P}$ are reciprocal, it holds that

$$\lim_{\omega \to 0} \left| \frac{Q(j\omega)}{1 - \hat{P}(j\omega)Q(j\omega)} \right| = \frac{1/k}{1 - k \cdot 1/k} = \infty. \tag{4.48}$$

This results in a perfect cancellation of static disturbances. However, an ECU often allows computations at a relatively low precision only. This may lead to different static gains in the feedforward and in the feedback path, which results in a finite static gain of the controller

$$\lim_{\omega \to 0} \left| \frac{Q(j\omega)}{1 - \hat{P}(j\omega)Q(j\omega)} \right| = \frac{1/k}{1 - (k + \partial k) \cdot 1/k} = \frac{1}{\partial k}, \tag{4.49}$$

and thus a static control error of $\partial k \cdot 100\%$. It is thus important that the computations are optimized with respect to the precision of the static gains of the internal model $\hat{P}$ and the internal controller Q. A different approach to reduce static errors due to quantization is to implement the IMC in the Smith predictor structure as depicted in Fig. 4.40. The internal controller C_I of the Smith predictor can be written as

$$C_I = \frac{Q}{1 - Q\hat{P}}. \tag{4.50}$$

For the air/fuel ratio controller designed with (4.45), the internal controller C_I can be derived as

$$C_I = \frac{Q}{1 - F} = -\frac{\sigma\tau}{\delta}\left(1 + \frac{1}{\tau s}\right) \tag{4.51}$$

which is a PI controller. The computations that are performed in the two internal model feedback paths of the Smith predictor shown in Fig. 4.40 only differ by a shift register due to the delay time. Thus, under static conditions, the error e_δ is zero, irrespective of the computational precision, whereas the Smith predictor is reduced to the internal controller C_I which contains a pure integrator that can be optimized for the static precision of the control system. Figure 4.39 depicts the static error of the control system with an IMC structure and a Smith predictor structure, respectively, when a fixed-point arithmetic with a word length of 16 and a precision of 12 bits is used. The static error is evaluated for 1000 random samples of the two parameters τ and δ. The samples of both parameters are uniformly distributed, ranging from 0.07 s to 0.32 s, with the sampling time chosen to be 10 ms. The Smith predictor results in a considerably smaller static error due to its structural properties mentioned. However, the precision of both controller structures can be improved by simple measures such as an individual adjustment of the fixed-point resolution for all variables of the controller and a systematic reordering of the calculation steps. In the Smith predictor structure only the computation of the integrator in the internal controller C_I needs to be optimized for static precision, whereas in the IMC structure the computation in both the internal controller Q and the internal model $\hat{P}$ needs to be optimized.

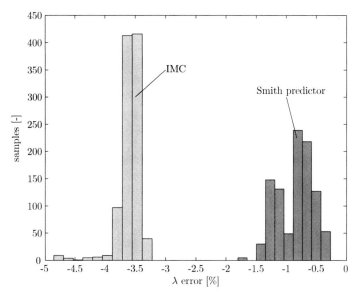

Fig. 4.39. Comparison of the static control error of the two control structures for random samples of the plant parameters τ and δ.

Integrator Windup

In classical PID control loops, actuator saturation can lead to a windup of the open integrator in the controller. This problem is typically tackled by extending the control structure with some sort of heuristic integrator windup protection (see A.7.2 for some general remarks). In the IMC structure, usually no apparent open integrator is present. However, as shown in the preceding section, the integrator action evolves from the interplay of the internal model and the internal controller. This implies that a controller windup can also be observed with an IMC system when no proper measures are taken in the case of actuator saturation. Fortunately, the IMC structure allows a systematic

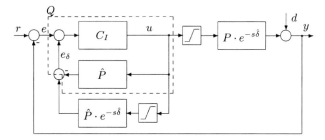

Fig. 4.40. Structure of the Smith predictor and the IMC (dashed) with anti windup measure.

treatment of actuator saturation and other input nonlinearities. To that end, a model of the nonlinearity must be included at the input of the internal model. The same strategy equivalently translates to the Smith predictor structure as depicted in Fig 4.40, where a model of the saturation is added at the input (and only there) of the internal model containing the delay. Note that, in contrast to standard PI controllers, the integrator in the internal controller C_I does not need any special treatment in the case of actuator saturations. Figure 4.41 shows the simulated system response of the air/fuel ratio IMC with actuator saturation for a disturbance pulse with an amplitude that is larger than the saturation limit. As expected, the control system with the anti windup measure has a smaller transient after the saturation period.

Experiments

Figure 4.42 depicts the response of the IMC system for steps in the nominal injection time at two different operating points. At both operating points the λ signal comprises variations of the air/fuel ratio at a relatively low frequency. These variations mainly occur at operating points where the throttle is almost closed and thus the air flow past the throttle is very sensitive with respect to the throttle position. However, these air/fuel ratio variations usually constitute no problem for the pollution abatement system and thus, need not be

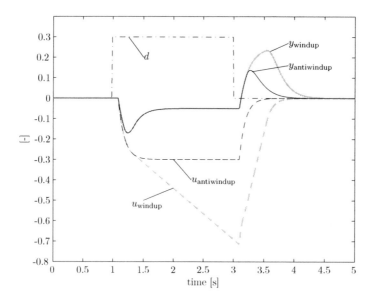

Fig. 4.41. Simulated system response for a disturbance pulse at the input of a plant with actuator saturation of ± 0.25. The black curves are with and the gray ones without anti windup measure according to Fig. 4.40. Parameters: $\tau = 0.15$ s, $\delta = 0.15$ s, $\sigma = 1$.

damped by the controller. The λ signals in Fig. 4.42 also contain fast variations due to the air/fuel ratio deviations of the individual cylinders. These frequencies must not be damped by the controller, as these effects are not modeled by 4.21 and thus, cylinder unbalancing could even be increased by the control action.

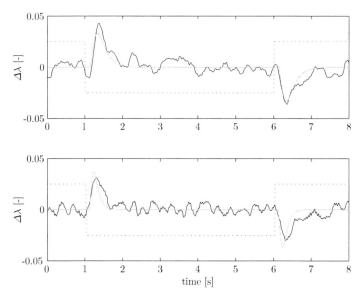

Fig. 4.42. Simulations (gray) and measurements (black) of the response of the control system with disturbances (dotted) at the input of the plant. Operation points: 1500 rpm and relative load of 30%, corresponding to a time constant of $\tau = 0.14$ s and a delay time of $\delta = 0.13$ s (upper plot), and 2250 rpm and relative load of 40% corresponding to a time constant of $\tau = 0.08$ s and a delay time of $\delta = 0.07$ s (lower plot), both with $\sigma = 0.7$.

4.3.5 Multivariable Control of Air/Fuel Ratio and Engine Speed

The time delays in the various signal paths are largest when the engine is idling. Therefore, instead of having two parallel control loops that can disturb each other, in this situation it is advantageous to combine the air/fuel ratio and the idle-speed control problems into one single MIMO control system. The classical SISO idle speed control problem is analyzed in more detail in Appendix B. In the case studied here, taken from [175], the correct handling of the multivariable nature of the combined control problem is emphasized.

The block diagram of the control system that is analyzed in this section is shown in Fig. 4.43. The reference values for the air/fuel ratio as well as

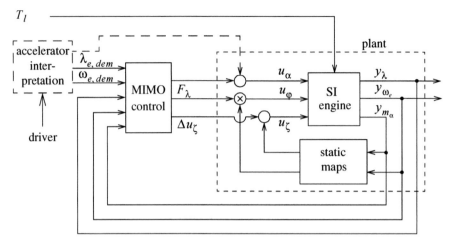

Fig. 4.43. Block diagram of the combined air/fuel ratio and idle speed control system.

for the engine speed are obtained from the accelerator pedal signal by an appropriate signal interpretation module. The sensor signals that are used by the control system are the measured air/fuel ratio, the engine speed, and the air mass flow into the intake manifold. Although there is no reference value for the air mass flow signal, including it in the loop improves the quality of the observer that is a part of the MIMO control system. Because of large time delays, the spark advance u_ζ must be used in addition to the throttle and the fuel command, as a third control channel. A basic control system that takes care of the steady-state control actions for ignition and fuel is implemented in a first step (block "static maps" in Fig. 4.43). That step, which optimizes the fuel consumption and minimizes the engine-out pollutant emissions, is not discussed here.

A naturally-aspirated and port-injected SI engine without external EGR is considered. The engine from which the experimental data shown below has been obtained has four cylinders and a displaced volume of 1.8 liter. It is mounted on a dynamic test bench that can apply load steps on the engine with a bandwidth of approximately 5 Hz. The focus of the following analysis is on the dynamic behavior of the engine system. Changes in the load torque T_l are assumed to be the main disturbances. A simple model of the corresponding engine dynamics, which will be used for the subsequent design of the control system, is shown in Fig. 4.44. Compared to the block diagram shown in Fig. 2.5, it explicitly includes the four dominant time delays and the static control maps mentioned above.

The static engine control algorithms run on a separate process controller. Only static maps for the fueling command and the spark advance are applied. The fueling command relies on the measurements of the air mass flow into

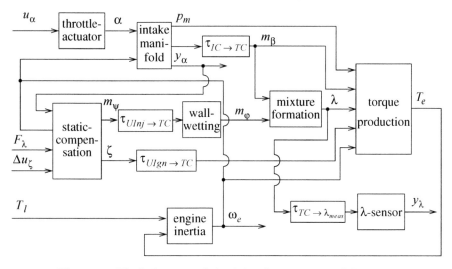

Fig. 4.44. Block diagram of the "plant" component of Fig. 4.43.

the intake manifold. Thus, the controller to be designed also must take into account errors made by this feedforward component of the fuel control system. In steady-state operation, the throttle valve correction signal must compensate all load disturbances, whereas the fueling correction signal eliminates errors in the air/fuel ratio. The spark advance correction may only be used in transients and must go back to zero under steady-state operating conditions.

For small variations of the input signal, the dynamics of the electronic throttle valve are assumed to be well approximated by a first-order lag element. For large steps, the current saturation of the DC motor becomes active, making the air flow actuator strongly nonlinear. The wall-wetting dynamics are assumed to be of first order, as described in Sect. 2.4.2. The air/fuel ratio is measured with a wide range sensor. The sensor voltage is linearized and the sensor dynamics are approximated by a first-order lag element with a time constant of 20 ms. The remaining two integrators describe the emptying and filling dynamic of the intake manifold and the rotational inertia of the engine. Figure 4.45 shows the block diagram of the resulting linear engine model.

The control system to be designed must fulfill the following requirements:

- No steady-state error in engine speed and air/fuel ratio are permitted for all load disturbances in the range $[0, 20]$ Nm.
- The controller must be robust to modeling errors, i.e., the minimum return difference of the open loop may not be smaller than 0.5.
- The spark advance control channel may only be used during the transients.
- The controller must be operated in the crank-angle domain and discretized on a segment-to-segment basis.

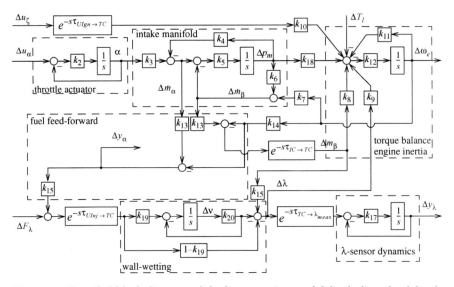

Fig. 4.45. Detailed block diagram of the linear engine model (including the delays).

Table 4.3 provides overview of the explicit numerical values of all constants used in Fig. 4.45. These parameters are valid for a four-cylinder 1.8-liter engine at idle conditions.

Table 4.3. Parameters of the linearized engine model shown in Fig. 4.45 at idle conditions.

param.	value	param.	value
k_2	40 [rad/s]	k_{13}	0.00116 [1/rpm]
k_3	43.6 [g/min/°]	k_{14}	0.000402 [g/min/rpm/rpm]
k_4	77.5 [g/min/bar]	k_{15}	2.90 [rev/g]
k_5	0.00386 [bar·min/s/g]	k_{17}	5.7 [rad/s]
k_6	686 [g/min/bar]	k_{18}	7.76 [Nm/bar]
k_7	0.422 [g/rev]	k_{19}	0.3 [-]
k_8	114 [Nm·rev/g]	k_{20}	15 [rad/s]
k_9	-16.3 [Nm]	$\tau_{Uinj \rightarrow TC}$	0.14 [s]
k_{10}	0.0984 [Nm/°CA]	$\tau_{Uign \rightarrow TC}$	0.07 [s]
k_{11}	0.00358 [Nm/rpm]	$\tau_{IC \rightarrow TC}$	0.07 [s]
k_{12}	63.7 [rpm/Nm/s]	$\tau_{TC \rightarrow \lambda_{meas}}$	0.385 [s]

The control system is designed using a model-based approach by applying the H_∞ frequency-domain optimization method. The block diagram of the linearized engine model used in that synthesis procedure is shown in Fig. 4.45. The resulting observer-based controller has 13 state variables. Five state variables are necessary to observe the relevant level variables in the engine system. Two additional state variables are used to force the steady-state errors of the air/fuel ratio λ and of the engine speed ω_e to zero. The remaining six state variables are used to approximate several delays present in the system. Using this rather large number of state variables allows to establish the achievable performance limits to be established. Of course, in a real application, an order reduction step would have to follow in order to obtain a realizable control system.

A comparison between simulation and measurement is shown in Fig. 4.46 and Fig. 4.47. Figure 4.46 shows a load step. The offset in spark advance is limited to $+5°$ crank angle. An appropriate anti-windup scheme is used.

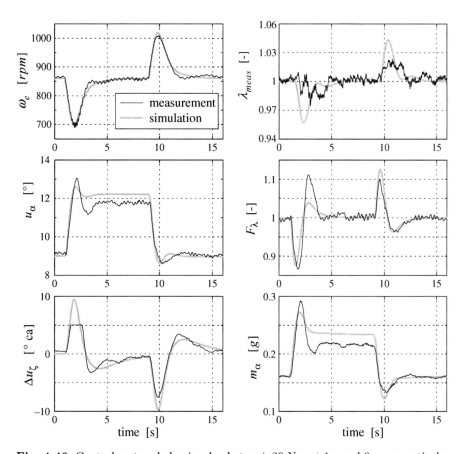

Fig. 4.46. Control system behavior, load step $\pm$ 20 Nm at 1 s and 9 s, respectively.

The tracking behavior is depicted in Fig. 4.47. Since the throttle must be actuated for steps in the demanded engine speed, the air/fuel path is excited as well. Deviations of up to 7% are observed, nevertheless they are reduced within one second to values below 3%. On the other hand, excitation of the air/fuel ratio demanded result in engine speed deviations of only 20 rpm, which are hardly noticeable.

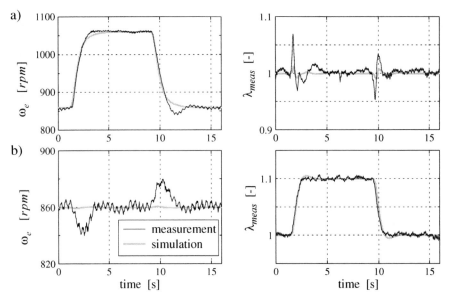

Fig. 4.47. Control system behavior, steps of the demanded values at 1 s and 9 s; respectively; a) engine speed steps (± 200 rpm), b) λ steps ($\pm 10\%$).

Finally, in Fig. 4.48, the performance of the MIMO controller designed above is compared to the performance of the series-production type controller. This controller uses two parallel SISO control loops for the air/fuel ratio and the engine speed, respectively. Since the speed set point of the series-production type controller is 770 rpm, this value was chosen in all experiments. Moreover, since the series-production type controller was not able to handle a disturbance of 20 Nm, the load step was reduced to 10 Nm in these particular experiments.

4.4 Control of an SCR System

The purpose of a selective catalytic reduction system was discussed in section 2.9. The model developed in that section forms the basis for the control system synthesis discussed below. In the case studied here, based on [178], the tailpipe NO_x emissions of a heavy-duty Diesel engine (that was certified for the Euro II limits) have been reduced to meet the Euro V emission limits. To achieve this reduction, a minimal NO_x reduction rate of 50% is necessary. Since tailpipe emissions of NH_3 are highly undesirable, a maximum mean slip of only 10 ppm of ammonia is tolerated.

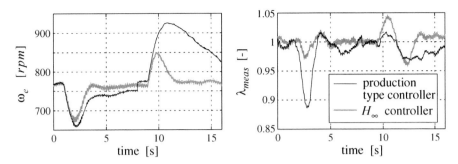

Fig. 4.48. Measured control system performance, load steps at idle speed: ± 10 Nm at 1 s and 9 s, respectively.

The engine that is considered in this section is a 10-liter supercharged heavy-duty, direct-injection, Diesel engine with intercooler. Its rated power is 275 kW. The maximum engine speed is 2150 rpm.

The amount of NO_x in the exhaust gas and the amount of ammonia injected vary strongly as functions of the engine operating point. Changing environmental conditions and the ageing components of the engine system play an important role as well. The dependency of the engine-out emissions upon the engine's operating point can be considered by implementing corresponding maps and by adding a low-pass filter to account for the thermal dynamics. All other effects (plugging of the injection system, change in ambient conditions, ageing) and the unavoidable modeling errors render a feedback control system necessary. Such an approach is only possible if information is available regarding the NO_x concentration in the exhaust upstream and downstream of the catalytic converter. The available solid-state NO_x sensors (Siemens-NGK "Smart NOx Sensor," SNS) show significant cross-sensitivities to ammonia. Therefore, close to the optimal dosage, where it is not clear whether excessive

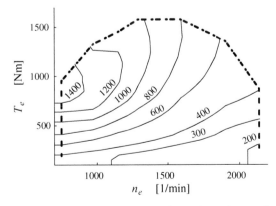

Fig. 4.49. Engine-out NO_x emissions in ppm of a 10-liter, heavy-duty Diesel engine.

amounts of NO_x or NH_3 are present, the sensor signal cannot be interpreted in a straightforward way.

Figure 4.49 shows the map that is used for the calculation of the steady-state engine-out NO_x emissions. The output of the steady-state map is filtered with a lag-lead element with a time constant of 20 s and a feed-through of 0.75. Figure 4.50 shows the measured and estimated engine-out NO_x emissions for part of an ESC test.[24]

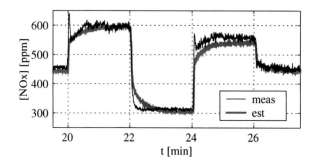

Fig. 4.50. Measured and estimated engine-out NO_x emissions for part of an ESC test. Starting at 400 Nm and 1590 rpm, a step to 1365 Nm and 1900 rpm is performed, then a step to 350 Nm, followed by a step to 1020 Nm, and finally a step to 695 Nm, all at 1900 rpm.

Figure 4.51 shows the relevant variables for a sweep in ammonia dosage. The cross-sensitivity to the ammonia concentration for higher-than-stoichiometric ammonia dosages is obvious. The "true" NO concentration is measured with a standard emission testing device, whereas the ammonia concentration is measured with a dedicated instrument.

Due to ammonia storage effects in the catalytic converter, the SNS sensor signal is substantially less noisy when the value of α/α_{OD}[25] is larger than one. Using this property, an additional periodic modulation of the dosage signal can be used to separate the $\alpha/\alpha_{OD} < 1$ portion from the $\alpha/\alpha_{OD} > 1$ part. In fact, the amplitude reduction caused by the catalytic converter is used to decide in which regime the system is working and, thus, to calculate the output indicated by SNS,pp that represents an estimation of the NO_x concentration. Notice that in the case of excess ammonia this value can be negative.

This interpretation of the NO_x sensor signal allows for the implementation of a true SCR feedback control system. Of course, this component has to be complemented by an appropriate feedforward control scheme in order to

---

[24] The ESC test cycle is the equivalent of an MVEG-95 cycle for heavy-duty Diesel engines.

[25] The variable α_{OD} is the optimal dosage ammonia feed ratio, defined as the ratio $\frac{\dot{n}_{NH_3,u}}{\dot{n}_{NO_x,u}}$ where an ammonia slip of 10 ppm is observed.

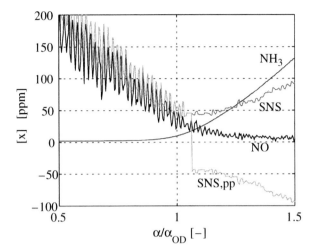

Fig. 4.51. Output signals of a solid state NO_x sensor (SNS), true NO signal (NO), and NH_3 concentration in the exhaust gas for a sweep in ammonia dosage. The post-processed sensor signal (SNS,pp) is also shown.

achieve the best possible bandwidths. Figure 4.52 shows the block diagram of the combined control scheme.

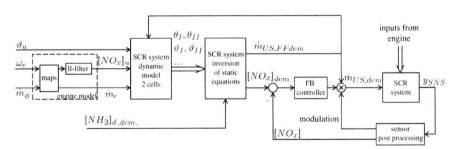

Fig. 4.52. Block diagram of the combined feedforward/feedback SCR control system.

The main elements of the proposed SCR control system are:

- Engine model: The signals available in the standard ECU (engine speed and load, temperature and pressure in the intercooler, temperature upstream and downstream of the catalytic converter, etc.) are used to calculate an estimate of the engine-out concentration of NO_x and the engine-out mass flow. As indicated, this can be done by using appropriate maps. The steady-state NO_x engine-out emissions are then filtered with a first-order lag-lead element with a time constant of 20 s in order to account for the thermal dynamics of the system.

- Dynamic model of the SCR system: The output signals of the engine model and the measured exhaust gas temperature upstream of the converter system are the inputs to an observer for the SCR system. The observer uses the simplified two-cell model described in Sect. 2.9. All states of the model are calculated, taking into account the actual injected amount of urea solution.

- Feedforward control: By inverting (2.248) and (2.249), the demanded value for the amount of urea solution injected and the SCR system-out (tailpipe) emissions of NO_x can be calculated. In this step, the demanded value for the tailpipe ammonia slip is chosen to be 10 ppm.

- Feedback control: The demanded amount of injected urea solution is multiplied by the output of the feedback controller. The control error is the difference between the calculated amount of tailpipe emissions of NO_x and the interpreted sensor output. The feedback element is a simple PI controller that is designed using classical frequency-domain methods, as explained in Appendix A. This PI control system may be chosen to be rather "slow" because it only has to cope with the disturbances originating from the environment and from ageing components.

- Signal processing: In order to separate the NO_x information from the NH_3 information in the sensor output signal, the product of the outputs of the feedback and feedforward control systems is modulated with a periodic signal with an amplitude of approximately 10%. This filtered output forms the reference signal for the urea dosage unit. A local dosage control system guarantees a good agreement of the mass flow of urea injected with this reference value.

Figure 4.53 shows the performance of the SCR control system for an ESC test cycle. The experimental verification of this control strategy showed an average NO_x reduction of almost 90%, while the ammonia slip was kept at an average value of 10 ppm. Note that some ammonia slip is present even if the exact stoichiometric urea dosage is applied all of the time. Therefore, lower ammonia slips can be achieved only by accepting a deterioration of the efficiency of the NO_x reduction. Moreover, the measured value of 10 ppm lies below the limit which is considered to be harmful and is below the human perception limit as well.

A further improvement of the performance of the control system can be achieved by accepting ammonia slips that are proportional to the tailpipe NO_x emissions. As before, equations (2.248) and (2.249) can be used for the calculation of the demanded amount of urea solution and the estimated tailpipe NO_x emissions. The resulting control system structure remains essentially the same, but when necessary, higher ammonia slips are accepted for the sake of an improved NO_x reduction.

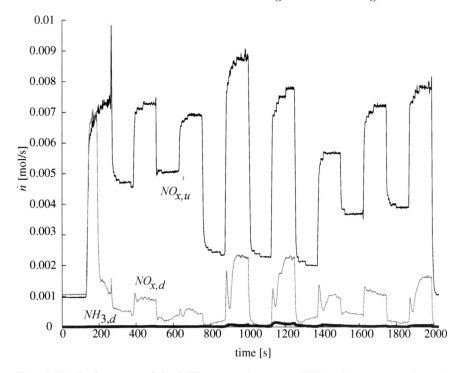

Fig. 4.53. Performance of the SCR system during an ESC cycle on a supercharged heavy-duty engine with $V_d = 10l$. $NO_{x,u}$ and $NO_{x,d}$ are the upstream and downstream molar flows of NO_x, respectively, and $NH_{3,d}$ represents the downstream molar flow of ammonia.

4.5 Engine Thermomanagement

4.5.1 Introduction

In automotive applications, approximately one third of the fuel energy delivered to the engine must be removed by the cooling system [97]. In high-speed and high-load operating points this amounts to significant amounts of heat flux. Specifically, at high ambient temperatures and low vehicle velocities, large radiators, additional fans, and large thermostats must be used to guarantee that the cooling system works properly and that the engine does not overheat. Accordingly, when the engine runs in part-load conditions, which is the case most of its lifetime, the cooling effect is too large. This leads to lower fuel economy due to lower engine temperatures and therefore higher friction losses. In addition, considerable losses are caused by the mechanically driven water pump and by the inefficient thermostat used to control the coolant temperature.

While the drawbacks of a mechanically coupled water pump are obvious, the drawbacks of the conventional thermostat must be explained in more detail. As shown in Fig. 2.44, the thermostat is a valve whose position determines the flow distribution between the radiator and its bypass duct. Under stationary conditions, the position of the thermostat is defined by the coolant temperature. The active element of the thermostat (a wax element) has dynamics that cannot be neglected. The time constant of this element amounts to approximately 30 s. For cold starts at high loads, the engine temperature must be kept from overshooting to avoid damage to the cylinder head gaskets and other components. Therefore, the thermostat's static characteristic must allow a certain amount of coolant to flow through the radiator, even at relatively low temperatures (70°C). On the other hand, at 90°C, the thermostat must be fully open to enable maximum cooling. These requirements lead to a nonlinear static gain function as shown in Fig. 4.54. The thermostat acts as a proportional controller. Since the plant itself does not have any integrating behavior, this simple control system cannot avoid errors in steady-state conditions. In particular, at higher engine speeds and at lower loads, this temperature deviation from the demanded value leads to higher friction losses in the engine and, thus, higher fuel consumption.

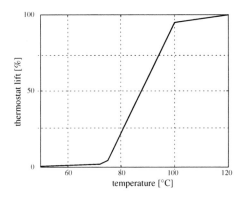

Fig. 4.54. Lift characteristic of a conventional thermostatic valve.

4.5.2 Control Problem Formulation

Modern engine cooling systems include electrically driven water pumps and electronically controllable bypass valves. The higher cost of these components must be compensated for by improvements in the overall fuel efficiency of the system. This requires improved control strategies which are able to operate the engine at part load at higher temperatures, yet still guarantee proper system cooling at high loads. A detailed mathematical model of such a cooling system was developed in Sect. 2.6.5. Here, the design of the corresponding control system is discussed.

The new cooling system offers two independent control inputs. The additional degree of freedom can be used to simultaneously control the temperatures of the coolant entering and leaving the engine. Similarly to the conventional approach, the temperature ϑ_{eo} of the coolant leaving the engine is set to a demanded value of $\vartheta_{eod} \approx 90\,°\mathrm{C}$. The temperature ϑ_{ei} of the coolant entering the engine is controlled using the actual value of ϑ_{eo}. In order to avoid detrimental mechanical stress effects, the temperature difference across the engine may not be larger than approximately $15\,°\mathrm{C}$.

Notice that in this situation the system to be controlled has a multiple-input, multiple-output (MIMO) structure. Compared to a SISO problem, the control of such MIMO systems generally is substantially more difficult. Therefore, whenever possible, the problem must be decoupled into several SISO loops using the knowledge of the physical properties of the plant. In the case discussed here, such a simplification is possible with the separation of four different cases in which four different control strategies can be applied. Table 4.4 summarizes these four cases that are treated separately below.

Table 4.4. Summary of the decoupling strategy used in the engine-cooling control system.

	$y_2 = \vartheta_{ei}$	$y_1 = \vartheta_{eo}$	$u_1 = \dot{m}_{by}/\dot{m}_c$	$u_2 = \dot{m}_c$
case 1	$< \vartheta_{eod}$	$< \vartheta_{eod}$	100%	minimal
case 2	$> \vartheta_{eid}$	$> \vartheta_{eod}$	$\Delta\vartheta_{eo} \cdot G_{c1}(s)$	minimal
case 3	$< \vartheta_{eid}$	$> \vartheta_{eod}$	$\Delta\vartheta_{eo} \cdot G_{c1}(s)$	$\Delta\vartheta_{ei} \cdot G_{c2}(s)$
case 4	$> \vartheta_{eid}$	$> \vartheta_{eod}$	0%	$\Delta\vartheta_{eo} \cdot G_{c3}(s)$

Case 1 applies in the situation in which the engine is cold. The temperatures ϑ_{ei} of the coolant entering and ϑ_{eo} of the coolant leaving are lower than their demanded values $\vartheta_{...d}$.[26] Therefore, the bypass valve remains open, and the coolant flow through the water pump is kept at its minimum value.[27] In this situation, the temperatures entering and leaving the engine rise almost simultaneously.

As soon as ϑ_{eo} reaches its demanded value (case 2), the bypass valve must start opening, otherwise the engine might overheat. The temperature difference over the engine starts to rise, until, with the coolant flow through the engine still at its minimum, the heat cannot be removed anymore without violating the maximum temperature difference condition.

---

[26] Notice that the reference values $\vartheta_{...d}$ can be functions of the engine operating point. For instance, higher reference temperatures may be chosen at low loads such that fuel consumption is reduced while at full load, in order to prevent knocking, lower temperature set points must be used.

[27] The coolant flow may not be completely shut off. This could cause damage at local "hot spots." Moreover, a minimum coolant flow is necessary to guarantee the system to be observable.

Therefore, the coolant mass flow must now be increased (case 3). Only in this situation a true MIMO control problem is encountered. Fortunately, the cross-coupling between the two control channels is not strong, so the two SISO control loops yield satisfactory results. In the first loop, the coolant mass flow defined by the water pump speed is used to control the temperature difference across the engine. In the second loop, the bypass valve position is used to control the engine-out temperature.

As soon as the bypass valve reaches its fully open position (case 4), the engine-out temperature is controlled by changing only the coolant mass flow.

The design of the three SISO controllers $G_{c...}(s)$ can be accomplished using any of the classical methods presented in Appendix A. A PI structure offers sufficient degrees of freedom in most cases. The plant must be linearized around typical operating points (cold start, engine fully warmed up, full load, etc.), and the resulting PI controllers are parameterized using several possible operating modes. Figure 4.55 shows, as an example, the resulting open-loop frequency response for case 2. Because of the strongly varying plant parameters, it is important to achieve sufficient robustness. The loop gain depicted in Fig. 4.55 shows that, in this example, this objective has been achieved.

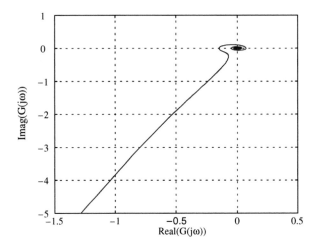

Fig. 4.55. Nyquist diagram of the open-loop transfer function with the plant channel from the input u_1 to the output ϑ_{eo} and a PI controller.

4.5.3 Feedforward Control System

Due to the large delays present in this system, the compromise between short system response times and robustness of the feedback loop cannot be satisfied by a feedback-only control system. In fact, large overshoots of the engine temperature in cold start transients and oscillations at steady-state can be observed only if a feedback control system is used.

A feedforward control system can help to improve the system response in both problem areas. The model of the external cooling circuit that has been derived in Sect. 2.6.5 forms the basis of the synthesis of this feedforward control system. The starting point is the equation

$$\vartheta_{ei}(t) = \vartheta_{mix}(t - \tau_{4 \to 6}) \tag{4.52}$$

where

$$\vartheta_{mix}(t) = u_1(t) \cdot \vartheta_{eo}(t - \tau_{1 \to 2} - \tau_{3 \to 4}) + (1 - u_1(t)) \cdot \vartheta_{ro}(t - \tau_{5 \to 4}) \tag{4.53}$$

Using (4.53) the control input $u_1(t)$ can be calculated as follows

$$u_1(t) = \frac{\dot{m}_{by}(t)}{\dot{m}_c(t)} = \frac{\vartheta_{mix}(t) - \vartheta_{ro}(t - \tau_{5 \to 4})}{\vartheta_{eo}(t - \tau_{1 \to 2} - \tau_{3 \to 4}) - \vartheta_{ro}(t - \tau_{5 \to 4})} \tag{4.54}$$

Inserting the demanded values for the mixing temperature, the engine-out temperature, and the measured values for the radiator temperature, this equation can be used for the design of the feedforward controller. Since the mixing point and the engine-in point of the coolant are usually quite close to each other, the demanded value for the engine-in temperature can be used for the mixing temperature as well. The demanded temperature for the coolant entering the engine may be calculated from the differential equation of the engine-out temperature (ϑ_{eo}) which has been derived in Sect. 2.6.5

$$\frac{d}{dt}\vartheta_{eo}(t) = \frac{\dot{Q}_{w,c}(t) - \dot{Q}_{c,eb}(t) + c_c \cdot \dot{m}_c \cdot [\vartheta_{ei}(t) - \vartheta_{eo}(t)]}{c_c \cdot m_c} \tag{4.55}$$

and

$$\dot{Q}_{w,c}(t) = \alpha_c \cdot A_c \cdot \left[\vartheta_w(t) - \frac{\vartheta_{eo}(t) + \vartheta_{ei}(t)}{2}\right] \tag{4.56}$$

$$\dot{Q}_{c,eb}(t) = \alpha_{eb} \cdot A_{eb} \cdot \left[\frac{\vartheta_{eo}(t) + \vartheta_{ei}(t)}{2} - \vartheta_{eb}(t)\right] \tag{4.57}$$

Using these equations, the engine-in temperature can be calculated as a function of engine-out temperature, its derivative, the engine-block temperature, and the engine cylinder-wall temperature

$$\vartheta_{ei} = f\left(\vartheta_{eo}, \frac{d\vartheta_{eo}}{dt}, \vartheta_w, \vartheta_{eb}\right) \tag{4.58}$$

The problems associated with the feedback control system arise mainly at cold start, especially in the transition phase from case 1 to case 2. For this situation the temperature of the radiator can be assumed to be constant and equal to the ambient temperature. Therefore, the expression (4.54) can be used to calculate the bypass valve position as a function of engine-in and engine-out temperatures. To calculate the demanded value for the engine-in temperature, ϑ_{eo} in (4.58) is replaced by a *constant* reference value for the engine-out temperature, ϑ_{eod}, with its derivative being equal to zero.

$$\vartheta_{eo} = \vartheta_{eod} \tag{4.59}$$

$$\frac{d}{dt}\vartheta_{eo}(t) = 0 \tag{4.60}$$

Since the temperature ϑ_w of the cylinder wall and ϑ_{eb} of the engine block cannot be measured, they must be observed. A model-based open-loop estimator, whose structure is shown in Fig. 4.56, is used for the estimation of the corresponding values $\hat{\vartheta}_w$ and $\hat{\vartheta}_{eb}$, respectively. Once the desired temperature ϑ_{eod} of the coolant leaving the engine is known, the corresponding coolant temperature at the engine input can be expressed as a function of the known variables

$$\vartheta_{eid} = f\left(\vartheta_{eod}, 0, \hat{\vartheta}_w, \hat{\vartheta}_{eb}\right) \tag{4.61}$$

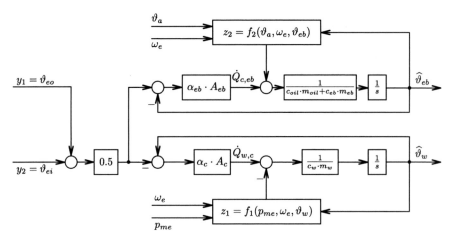

Fig. 4.56. Block diagram of the open-loop observer for engine block and engine wall temperatures.

According to Fig. 2.47 and Fig. 2.48, the disturbance $z_1 = \dot{Q}_{g,w}$ can be calculated as a function of the operating point (described by ω_e and p_{me}) and of the cylinder wall temperature ϑ_w. The disturbance $z_2 = \dot{Q}_{if} - \dot{Q}_{eb.a}$ is modeled according to Sect. 2.5.1 using the following equation

$$z_2 = \dot{Q}_{if} - \dot{Q}_{eb.a}$$

$$= p_{me0f} \cdot V_d \cdot \frac{\omega_e}{4\pi} - A_{eb} \cdot \alpha_{eb,a} \cdot (\vartheta_{eb} - \vartheta_a) \tag{4.62}$$

$$\tag{4.63}$$

where p_{me0f} is defined in (2.112).

The block diagram of the resulting control system is depicted in Fig. 4.57.

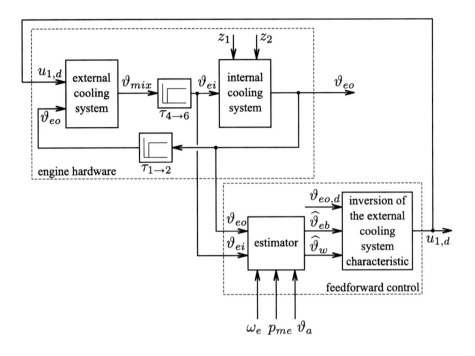

Fig. 4.57. Feedforward control system for the bypass valve channel.

4.5.4 Experimental Results

Figure 4.57 shows the block diagram of the control system for the case in which only feedforward action is applied. Note that this system is not a true feedforward controller because it uses measured values of the engine-in and engine-out temperatures to drive the open-loop observer. Additionally, since this feedforward control system is primarily designed for the warming phase, a minimum coolant flow is assumed. This control system was implemented in a rapid-prototyping hardware, and experiments were run on the engine system described in [45]. The simulated and the measured behavior of the feedforward system are shown in Fig. 4.58. In the simulation, ϑ_{eo} remains exactly at the demanded value of $\vartheta_{eod} = 90\,°C$. The temperature ϑ_{ei} of the coolant entering the engine rises simultaneously with ϑ_{eod} until $\vartheta_{eod} = 90\,°C$ is reached. Then the temperature ϑ_{ei} of the coolant drops again in order to remove the appropriate amount of heat from the engine and to keep ϑ_{eo} constant.

As shown in the lower two plots of Fig. 4.58, neither the engine-in nor the engine-out temperature remain at their desired values in these experiments. This had to be expected because the engine model cannot describe precisely the real system behavior and because the feedforward control action does not include any error-correcting mechanisms. In addition to the temperature of the coolant entering and leaving the engine, the opening of the bypass valve is shown in Fig. 4.58. In the warming phase, the bypass valve is fully open and the coolant flow completely bypasses the radiator. As soon as the engine-out temperature reaches its demanded value, the bypass valve is partially closed causing the excess heat to be discarded in the radiator.

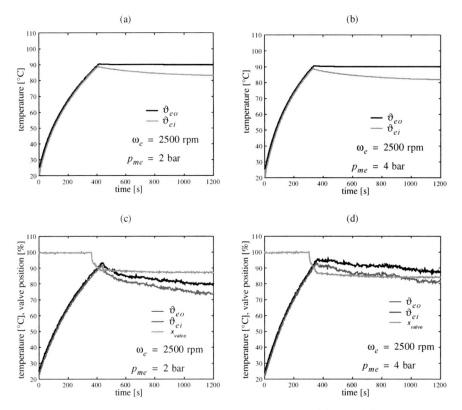

Fig. 4.58. Coolant temperatures at cold start. Graphs (a) and (b): results from simulations, $p_{me} = 2bar$ and $p_{me} = 4bar$, respectively. Graphs (c) and (d): results from measurements at the same operating points.

The feedforward controller must be complemented by a feedback component which is to account for modeling inaccuracies and unmeasurable disturbances. The block diagram of the combined feedforward and feedback controller is shown in Fig. 4.59. As discussed above, the feedback control systems are realized as PI controllers. The specific input and output structures are listed in Table 4.4. Compared to the feedback-only loop, whose frequency response is shown in Fig. 4.55, the gain of the feedback component in this control system can be substantially reduced because of the beneficial action of the feedforward controller.

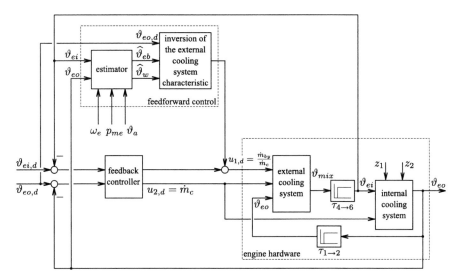

Fig. 4.59. Combined feedforward and feedback control system.

The experimental comparison of the effects of the feedback-only and the combined feedforward-feedback strategy is shown in Fig. 4.60. The feedback-

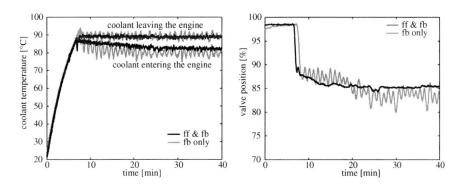

Fig. 4.60. Coolant temperature (left) and valve position (right) for the feedback-only and combined feedforward/feedback control structures.

only control system is able to bring the engine-out temperature to its desired value. However, this approach produces a noticeable overshoot when the desired temperature is first reached and, due to the large time delays present in the plant, causes a non-vanishing limit cycle behavior in steady-state conditions. The combined feedforward and feedback control system avoids both unwanted effects and, moreover, reduces the settling time.

Of course the constant operating point experiments are not very informative and do not show any real benefit in terms of reduced fuel consumption as yet. To assess the benefits of the proposed control system, experiments were conducted that compared the regular cooling system performance with the performance of the proposed novel control system. The fuel consumption benefits achievable for several constant operating points are shown in Fig. 4.61.

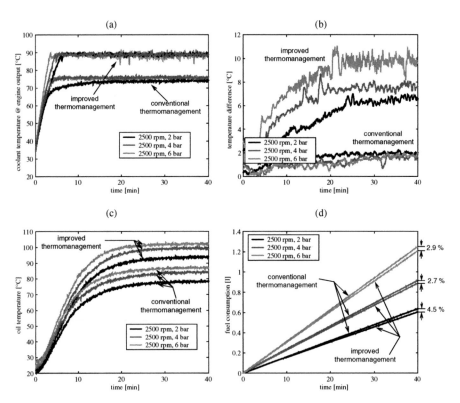

Fig. 4.61. Cold-start measurements for various steady-state engine operating points.

Because of the higher engine temperatures, a reduction in fuel consumption of up to 4.5% was realized in that specific case. The highest relative reduction in fuel consumption is achieved for the lowest load because of the

largest increase in coolant temperature. Plot b) shows the larger temperature difference over the engine with the new control system. This is a consequence of the small coolant flow used. For higher engine loads, the temperature difference increases. Also note that, compared to the standard case
in the advanced engine-temperature control approach, the oil temperature is
significantly higher.

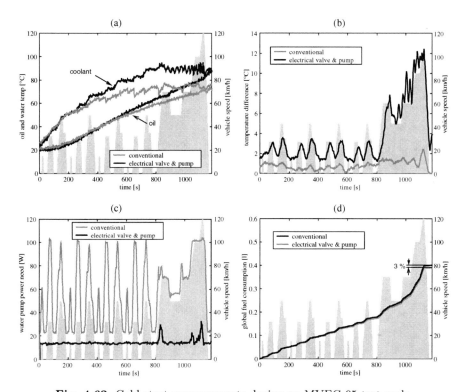

Fig. 4.62. Cold-start measurements during an MVEG-95 test cycle.

The fuel-economy benefits realizable in transient conditions were assessed
using the European test cycle. The results of these experiments are shown
in Fig. 4.62. With respect to the control scheme defined in Table 4.4, it was
found that case 3, where the power of the coolant pump had to be increased
in order to reduce the temperature over the engine, was only used for a short
time at the highest acceleration in the last phase of the MVEG-95 test cycle.[28]

Two final remarks remain to be made:

---

[28] Because it is calculated as the difference of two relatively large signals, the temperature difference over the engine is a very noisy signal. Therefore, an appropriate
signal filter is necessary.

- Because the mechanically driven coolant pump is replaced by one that is electrically driven raises the question of whether this results in higher energy consumption. Since the electrically driven pump works at approximately one-sixth of the nominal mass flow of the mechanical one, this is not the case.
- The proposed approach not only increases fuel efficiency, but at the same time improves comfort and safety. In fact, quicker engine heating and a higher engine temperature result in a more comfortable cabin climate and faster windshield defrosting in winter.

A

Basics of Modeling and Control-Systems Theory

This appendix collects some of the most important definitions and results of the theory of system modeling and control systems analysis and design. To avoid using any complex mathematical formulations, all systems are assumed to have only one input and one output.

This appendix is not intended to serve as a self-contained course on control systems analysis and design (this would require much more space). Rather, it is intended to synchronize notation and to recall some points that are used repeatedly throughout the main text. Readers who have never had the opportunity to study these subjects are referred to the many standard text books available, some of which are listed at the end of this chapter.

A.1 Modeling of Dynamic Systems

General Remarks

This section recalls briefly the most important ideas in system modeling for control. In general, two classes of models can be identified:

1. models based purely on (many) experiments; and
2. models based on physical first principles and (a few) parameter identification experiments.

In this text, only the second class of models is discussed. General guidelines to formulate such a control-oriented model encompass the following six steps at least:

1. Identify the system boundaries (what inputs, what outputs, etc.).
2. Identify the relevant *reservoirs* (mass, energy, information, ...) and the corresponding *level variables*.
3. Formulate the differential equations for all relevant reservoirs as shown in eq. (A.1).

$$\frac{d}{dt}(\text{reservoir content}) = \sum \text{inflows} - \sum \text{outflows} \qquad (A.1)$$

4. Formulate the (usually nonlinear) algebraic relations that express the *flows* between the reservoirs as functions of the level variables.
5. Identify the unknown system parameters using experiments.
6. Validate the model with experiments other than those used to identify the system parameters.

In step two, it is important to distinguish between slow (curve c) in Fig. A.1), relevant (curve b) and fast (curve a) dynamic effects (very large, relevant and very small reservoirs). The relevant time scales are defined by the main variable b) that is to be controlled. Type a) variables are simplified as being algebraic functions of the other variables and the inputs, while type c) variables are assumed to be constants.

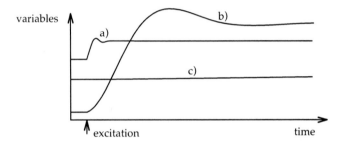

Fig. A.1. Classification of variables into "algebraic" (a), "dynamic" (b) and "static" (c).

The first four steps are shown here in further detail by deriving a model of a cylindric water-tank system as depicted in Fig. A.2:

1. The input is $\dot{m}_{in}(t)$, i.e., the inflowing water mass flow.
   The output is $h(t)$, i.e., the measured height of the water level in the tank.
2. The only relevant reservoir is the mass $m(t)$ of water in the tank.
   The dynamic effects in the measuring device are very fast and can therefore be neglected.
   The water temperature (and therefore its density) is assumed to change very slowly such that it may be assumed to be constant.
3. The resulting differential equation is shown in (A.2).

$$\frac{d}{dt}m(t) = \dot{m}_{in}(t) - \dot{m}_{out}(t) \qquad (A.2)$$

4. The input mass flow is assumed to be the control signal and can therefore be set arbitrarily by the controller to be designed.
   The outgoing mass flow can be expressed using Bernoulli's law, shown in

(A.3), assuming incompressible and frictionless flow conditions, resulting in (A.4).

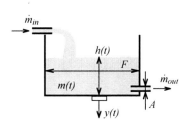

Fig. A.2. Water tank system, $m(t)$ mass of water in tank, and corresponding height $h(t)$, F tank-floor area, A = area of outflow orifice.

$$\dot{m}_{out}(t) = A\rho v(t), \quad v(t) = \sqrt{2\Delta p(t)/\rho}, \quad \Delta p(t) = \rho g h(t) \tag{A.3}$$

$$\frac{d}{dt}m(t) = \rho F \frac{d}{dt}h(t) = \dot{m}_{in}(t) - A\rho\sqrt{2gh(t)} \tag{A.4}$$

The models obtained thus are (usually) nonlinear and have non-normalized variables. Unfortunately, there is, in general, no "neat" controller design method for nonlinear systems, and non-normalized variables can lead to numerical problems and are hard to compare with one another.

For these reasons, the model (A.4) must first be normalized using constant nominal values ("set points"), i.e.,

$$h(t) = h_n x(t), \qquad \dot{m}_{in}(t) = \dot{m}_{in,n} u(t) \tag{A.5}$$

such that the new variables $x(t)$ and $u(t)$ will be dimensionless and around 1 in magnitude. Using the definitions (A.5), (A.4) can be rewritten as

$$\rho F h_n \frac{d}{dt}x(t) = \dot{m}_{in,n}u(t) - A\rho\sqrt{2gh_n}\sqrt{x(t)} \tag{A.6}$$

Notice that if the set point is chosen to be an equilibrium point (which is usually the case), the nominal amounts of in-flowing and out-flowing water must be equal, i.e.,

$$\dot{m}_{in,n} = A\rho\sqrt{2gh_n} \tag{A.7}$$

In order to linearize the system, only small deviations from its set points are analyzed, such that the following new variables

$$x(t) = 1 + \delta x(t) \text{ with } |\delta x| \ll 1, \qquad u(t) = 1 + \delta u(t) \text{ with } |\delta u| \ll 1 \tag{A.8}$$

can be introduced. Expanding (A.6) into a Taylor series and neglecting all the terms of second and higher order yields a linear differential equation

$$\rho F h_n \frac{d}{dt}\delta x(t) \approx \dot{m}_{in,n}(1 + \delta u(t)) - A\rho\sqrt{2gh_n}\left(1 + \frac{1}{2}\delta x(t)\right) \qquad (A.9)$$

which, using (A.7), can be simplified to

$$\rho F h_n \frac{d}{dt}\delta x(t) = \dot{m}_{in,n}\delta u(t) - A\rho\sqrt{2gh_n}\frac{1}{2}\delta x(t) \qquad (A.10)$$

or, by rearranging terms and again using (A.7)

$$\tau \frac{d}{dt}\delta x(t) = 2\delta u(t) - \delta x(t) \qquad (A.11)$$

where the *time constant* τ is defined by

$$\tau = \frac{F}{A}\sqrt{\frac{2h_n}{g}} \qquad (A.12)$$

This equation shows that the system becomes slower when either the ratio F/A or the set point h_n is increased. The *gain* of the system is 2 in its normalized form and equal to $2h_n/\dot{m}_{in,n}$ in its non-normalized representation.

Example Cruise-Control Problem

The objective of a series of examples is to design a robust cruise-control system,[1] i.e., a control loop that keeps the vehicle speed at its set point despite unobservable disturbances (grading, wind, etc.) and which is robust with respect to changing system parameters (speed set point, vehicle mass, gear ratio, etc.) and to neglected dynamic effects (engine torque delays, sensors, etc.). In this section a mathematical model of the dynamic behavior of a simplified vehicle (see Fig. A.3) is derived.

Nonlinear System Description

First, the nonlinear and nonnormalized system equations are derived. The following assumptions are made in this step:

1. The clutch is engaged such that the gear ratio γ is piecewise constant.
2. No drivetrain elasticities and wheel slip effects need be considered, i.e., the following relations are valid: $v(t) = r_w \cdot \omega_w(t)$ and $\omega_w(t) = \gamma \cdot \omega_e(t)$.
3. The vehicle must overcome rolling friction (the force acting on the vehicle is $F_r = c_r \cdot m \cdot g$) and aerodynamic drag ($F_a(t) = 1/2 \cdot \rho \cdot c_w \cdot A \cdot v^2(t)$).

---

[1] Of course, due to space limitations many important aspects of a real cruise-control system cannot not be addressed in this example.

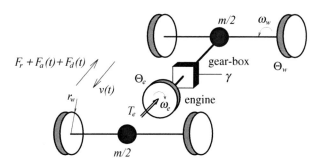

Fig. A.3. Simplified vehicle structure, vehicle mass m (wheel masses included) concentrated in the two axles.

4. All other forces are lumped into an unknown disturbance $F_d(t)$.
5. The mechanical energy of a moving part is $1/2 \cdot m \cdot v^2$ (pure translation) and $1/2 \cdot \Theta \cdot \omega^2$ (pure rotation).

The procedure outlined in Sect. A.1 is followed now.

Step 1: The input to the system is the engine torque T_e, which can be assumed to be arbitrarily controllable (the time delay caused by the engine dynamics is very small compared to the typical time constants of the vehicle speed and is therefore assumed to be an algebraic variable, see Fig. A.1). The system output is the vehicle speed $v(t)$.

Step 2: The relevant reservoirs are the kinetic energies stored in the vehicle's translational and rotational degrees of freedom, i.e.,

$$E_{tot} = \frac{1}{2}mv^2(t) + 4\frac{1}{2}\Theta_w\omega_w^2(t) + \frac{1}{2}\Theta_e\omega_e^2(t) \tag{A.13}$$

Using the assumptions listed above, (A.13) can be simplified as follows

$$E_{tot} = \frac{1}{2}\left[m + \frac{4\Theta_w}{r_w^2} + \frac{\Theta_e}{\gamma^2 r_w^2}\right]v^2(t) \tag{A.14}$$

Obviously, the level variable is the vehicle speed $v(t)$.

Step 3: The flows acting on the system are the mechanical powers affecting the system, i.e.,

$$P_+(t) = T_e(t)\omega_e(t), \quad \text{and} \quad P_-(t) = (F_r + F_a(t) + F_d(t))v(t) \tag{A.15}$$

The energy conservation principle yields the equation

$$\frac{d}{dt}E_{tot}(t) = P_+(t) - P_-(t) \tag{A.16}$$

Inserting (A.14) and (A.15) yields

$$\frac{1}{2}\left[m + \frac{4\Theta_w}{r_w^2} + \frac{\Theta_e}{\gamma^2 r_w^2}\right] 2v(t)\frac{d}{dt}v(t) = T_e(t)v(t)\frac{1}{r_w\gamma} - (F_r + F_a(t) + F_d(t))v(t)$$
(A.17)

Step 4: This step simply consists of rearranging (A.17) to obtain

$$M(\gamma,m)\frac{d}{dt}v(t) = \frac{1}{r_w\gamma}T_e(t) - \left(c_rmg + \frac{1}{2}\rho c_w Av^2(t) + F_d(t)\right)$$
(A.18)

where

$$M(\gamma,m) = \left[m + \frac{4\Theta_w}{r_w^2} + \frac{\Theta_e}{\gamma^2 r_w^2}\right]$$
(A.19)

is the total inertia of the system.

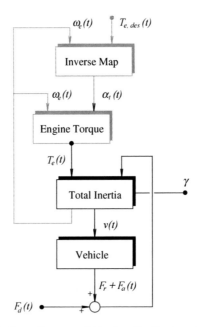

Fig. A.4. Cause-and-effect diagram of the longitudinal system dynamics. Explanations of the gray-shaded blocks will follow.

Normalization and Linearization

The next steps are the choice of a meaningful set point and the normalization and linearization of the equations obtained. The set point is $\{v_n, T_{e,n}, 0\}$ and the normalized system variables are

$$v(t) = v_n x(t), \qquad T_e(t) = T_{e,n} u(t), \qquad F_d(t) = \rho c_w A v_n^2 d(t) \qquad \text{(A.20)}$$

Note that since the disturbance set point is zero, the disturbance force is normalized by twice the nominal aerodynamic force (normalization with rolling friction is also possible). The normalized version of (A.18) thus has the form

$$M(\gamma, m) v_n \frac{d}{dt} x(t) = \frac{T_{e,n}}{r_w \gamma} u(t) - \left(c_r m g + \frac{1}{2} \rho c_w A v_n^2 x^2(t) + \rho c_w A v_n^2 d(t) \right) \qquad \text{(A.21)}$$

The set point $\{v_n, T_{e,n}, 0\}$ is again assumed to be an equilibrium point such that

$$\frac{T_{e,n}}{r_w \gamma} = c_r m g + \frac{1}{2} \rho c_w A v_n^2 \qquad \text{(A.22)}$$

Linearizing the system using the small deviations

$$x(t) = 1 + \delta x(t), \quad |\delta x| \ll 1 \qquad \text{(A.23)}$$

$$u(t) = 1 + \delta u(t), \quad |\delta u| \ll 1 \qquad \text{(A.24)}$$

$$d(t) = \delta d(t), \quad |\delta d| \ll 1 \qquad \text{(A.25)}$$

yields

$$M(\gamma, m) v_n \frac{d}{dt} \delta x = \frac{T_{e,n}}{r_w \gamma} (1 + \delta u) - c_r m g - \frac{1}{2} \rho c_w A v_n^2 \left[1 + 2\delta x + \delta x^2 + 2\delta d \right] \qquad \text{(A.26)}$$

(for reasons of space, the time dependencies have been omitted). Neglecting the second-order terms and inserting (A.22) yields the desired linear differential equation

$$M(\gamma, m) v_n \frac{d}{dt} \delta x(t) \approx \frac{T_{e,n}}{r_w \gamma} \delta u(t) - \rho c_w A v_n^2 (\delta x(t) + \delta d(t)) \qquad \text{(A.27)}$$

This equation can be put into its normal form by dividing all terms by $\rho c_w A v_n^2$ and using equations (A.22) and (A.19)

$$\tau(\gamma, m, v_n) \frac{d}{dt} \delta x(t) = k(m, v_n) \delta u(t) - \delta x(t) - \delta d(t) \qquad \text{(A.28)}$$

where

$$\tau(\gamma, m, v_n) = \frac{1}{\rho c_w A v_n} \left(m + \frac{4\Theta_w}{r_w^2} + \frac{\Theta_e}{\gamma^2 r_w^2} \right) \qquad \text{(A.29)}$$

and

$$k(m, v_n) = \frac{1}{2} + \frac{c_r m g}{\rho c_w A v_n^2} \qquad \text{(A.30)}$$

Both τ and k are functions of the system parameters mass m and gear-ratio γ (the other parameters will vary much less) and of the vehicle speed set point. This is emphasized by using the explicit notation $...(\gamma, m, v_n)$.

System Parameter Variations

Of course, the physical system parameters are not known precisely and vary within some reasonable limits during normal operation of the vehicle. Table A.1 lists examples of such limits.

Table A.1. Examples of system parameters.

$r_w = 0.25\ m$	$\Theta_w = 1.0\ kg\,m^2$	$\Theta_e = 0.25\ kg\,m^2$
$\rho = 1.2\ kg/m^3$	$c_r = 0.012\ -$	$c_w A = 0.75\ m^2$
$m = 1500\ kg$	$m = 1600\ kg$	$m = 2000\ kg$
$\gamma_3 = 0.19\ -$	$\gamma_4 = 0.25\ -$	$\gamma_5 = 0.37\ -$
$v_n^* = 20\ m/s$	$v_n^* = 30\ m/s$	$v_n^* = 40\ m/s$

Masses correspond to empty vehicle, one passenger and fully loaded, gear ratios correspond to third, fourth and fifth gears (cruise control is assumed to be disabled in first and second gear).

The parameters τ and k of the control-oriented model (A.28) depend upon all of these physical parameters in a complex way. Therefore, it would be interesting to know the limits

$$\tau_{min} \leq \tau \leq \tau_{max}, \qquad k_{min} \leq k \leq k_{max} \qquad (A.31)$$

within which they vary. This problem is best solved by writing a small program that does a reasonable "gridding" in the parameter space $\{m, \gamma, v_n\}$ and that finds (estimates) of the model parameter intervals. Note that this approach, in general, is *not* guaranteed to yield exact results (readers interested in that aspect are referred to [1]).

Remark: The nominal vehicle speed v_n substantially influences the system's dynamic behavior. In order to avoid a poor closed-loop system behavior, the controllers are usually scheduled with regard to nominal vehicle speed (gain scheduling will be discussed in Sect. A.7.1). For the sake of simplicity, only three scheduling speeds v_n^* are assumed in the following. Each value will be active while the vehicle is in a 10 m/s interval, e.g., for $v \in [25, 35]$ m/s a v_n^* of 30 m/s will be chosen (cruise control is assumed to be disabled for speeds below 15 m/s, which corresponds to approximately 50 km/h or 30 mph, and top speed of the vehicle is assumed to be at 45 m/s, which corresponds to 160 km/h or 100 mph). Table A.2 shows the intervals of the system parameters τ and k obtained for these three nominal speeds.

Table A.2. Extreme values of the control model parameters for the scheduling speeds v_n^*.

$v_n^* = 20\ m/s$	$\tau_{min} = 68.7\ s$	$\tau_{max} = 155.\ s$	$k_{min} = 0.813$	$k_{max} = 1.663$
$v_n^* = 30\ m/s$	$\tau_{min} = 49.0\ s$	$\tau_{max} = 92.8\ s$	$k_{min} = 0.660$	$k_{max} = 0.919$
$v_n^* = 40\ m/s$	$\tau_{min} = 38.2\ s$	$\tau_{max} = 66.3\ s$	$k_{min} = 0.596$	$k_{max} = 0.714$

Some Comments

The maximum variation intervals shown in Table A.2 are not independent. As Fig. A.5 shows, control-oriented parameters like τ and k generally depend in a nonlinear way upon the physical parameters (mass, gear ratio, nominal speed, etc.). Assuming that τ and k are independent, i.e., assuming that *all* values inside the rectangle shown in Fig. A.5 must be expected, yields sufficient, but not necessary, conditions for stability and performance levels. Accordingly, such designs will be "conservative," i.e., they cannot realize some of the available potential.

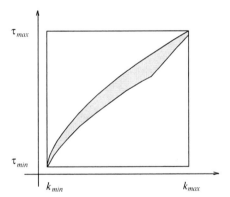

Fig. A.5. Set of possible control-oriented parameters τ and k for varying physical parameters m, γ and v_n^*.

Another comment concerns the chosen input signal. Of course, the assumption that the engine torque is the *input* to the system is not physically correct:

1. The real system input is the throttle angle α_t, as shown in Fig. A.4.
2. The *static* engine torque depends on the engine speed and on the throttle angle in a nonlinear way $T_e = f(\omega_e, \alpha_t)$, as shown in Fig.A.6.

3. The *dynamic* engine torque reacts with some delay to changes in the throttle position, mainly due to the intake manifold dynamics and the induction to power stroke delays (see Section 2.1).

However, the nonlinearity can be compensated by a static inverse map, as shown with gray lines in Fig. A.4, such that $T_e(t) \approx T_{e,des}(t)$ and the dynamic effects, as mentioned, are relatively small and can be dealt with by designing sufficient robustness into the system.

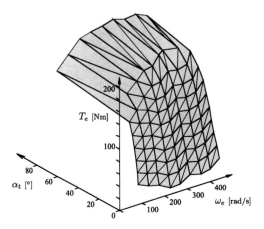

Fig. A.6. Static relation between engine speed ω_e, throttle valve opening angle α_t and engine torque T_e (2-liter SI engine).

A.2 System Description and System Properties

General Remarks

This section summarizes several ways to describe and analyze dynamic SISO systems. One important description for dynamic systems was used in the last section. In fact, ordinary differential equations

$$\frac{d}{dt}x(t) = f(x(t), u(t), t), \quad x(t) \in \Re^n, \quad u(t) \in \Re, \quad x(0) = x_0 \qquad (A.32)$$

are the typically used first form to describe the dynamic behavior of a system. Once the level variable $x(t)$ is known, the output $y(t)$ depends algebraically on this variable and on the input signal $u(t)$

$$y(t) = g(x(t), u(t)), \quad y(t) \in \Re \qquad (A.33)$$

Normalizing and linearizing (A.32) and (A.33) yields the following linear system equations

$$\frac{d}{dt}\delta x(t) = A \cdot \delta x(t) + b \cdot \delta u(t)$$
$$\delta y(t) \quad = c \cdot \delta x(t) + d \cdot \delta u(t)$$

$$(A.34)$$

The system matrices $\{A, b, c, d\}$ are defined by

$$A = \left.\frac{\partial f(x)}{\partial x}\right|_{x_0,u_0} \in \Re^{n \times n}, \quad b = \left.\frac{\partial f(x)}{\partial u}\right|_{x_0,u_0} \in \Re^{n \times 1}$$

$$c = \left.\frac{\partial g(x)}{\partial x}\right|_{x_0,u_0} \in \Re^{1 \times n}, \quad d = \left.\frac{\partial g(x)}{\partial u}\right|_{x_0,u_0} \in \Re^{1 \times 1}$$

$$(A.35)$$

where x_0 is the system's equilibrium point associated to the input u_0 at which the system is linearized. Usually, the prefix δ is omitted when using this "state-space form."

The solution for (A.32) can be found for the given initial conditions x_0 and for the input signal u using *numerical* routines running on digital computers. This process, denoted by system simulation, has become a routine operation since easily programmable and reliable software tools exist today.

While simulations are very useful to *analyze* the system's behavior, they cannot be used for controller *synthesis*. The basic problem is that, in general, closed-form solutions for (A.32), or even for its linearized version (A.34), are not easy to find.

However, at least for linear or linearized systems (A.34), such solutions can be found in the frequency domain using *Laplace transformation* techniques. The main idea of Laplace transformations is to replace the independent (real) variable t by another (complex) variable s such that differentiation (with respect to time, $\frac{d}{dt}$) is transformed to a multiplication with s in the frequency domain. Three simple rules are important:

- Variables $x(t)$ are transformed to variables $X(s)$.
- Linear multiplications are invariant under Laplace transformation, i.e., $k \cdot x(t)$ is transformed to $k \cdot X(s)$.
- Derivatives of variables are transformed according to the rule $\frac{d}{dt}x(t) \rightarrow sX(s) - x(0)$.

Applying these rules to the water-tank equation

$$\tau \frac{d}{dt}\delta x(t) = 2\delta u(t) - \delta x(t)$$

$$(A.36)$$

yields

$$\tau\left(s\delta X(s) - \delta x(0)\right) = 2\delta U(s) - \delta X(s)$$

$$(A.37)$$

which is an algebraic equation that can now be solved easily

$$\delta X(s) = \frac{2}{\tau s + 1} \cdot \delta U(s) + \frac{\tau}{\tau s + 1} \cdot \delta x(0)$$

$$(A.38)$$

The initial condition $\delta x(0)$ is often zero. Specifically for feedback control design, the important component is the *input-output* behavior characterized by the plant's *transfer function*

$$P(s) = \frac{\delta X(s)}{\delta U(s)} = \frac{2}{\tau s + 1} \tag{A.39}$$

This representation, which in general has the form

$$P(s) = \frac{b_m s^m + b_{m-1} s^{m-1} + \ldots + b_1 s + b_0}{s^n + a_{n-1} s^{n-1} + a_{n-2} s^{n-2} + \ldots + a_1 s + a_0}, \tag{A.40}$$

is useful for system analysis and synthesis. The zeros of the denominator polynomial (the transfer function's poles) are the eigenvalues of the original linear system, which describe the qualitative behavior of the system (unstable/stable, damping, etc.). The zeros of the numerator polynomial (the transfer functions zeros) also play an important role (e.g., they limit the achievable performance of any feedback controller).

In synthesis, another property of transfer functions is especially useful, *viz.* the series interconnection of two dynamic systems results in a simple multiplication of their transfer functions, e.g., for

$$\tau_1 \frac{d}{dt} \delta x(t) = k_1 \delta u(t) - \delta x(t), \quad \tau_2 \frac{d}{dt} \delta z(t) = k_2 \delta x(t) - \delta z(t) \tag{A.41}$$

the frequency domain representation is

$$P(s) = \frac{\delta Z(s)}{\delta U(s)} = \frac{k_1}{\tau_1 s + 1} \cdot \frac{k_2}{\tau_2 s + 1} = \frac{k_1 k_2}{\tau_1 \tau_2 s^2 + (\tau_1 + \tau_2)s + 1} \tag{A.42}$$

Transfer functions are completely equivalent to differential equations (no information is lost going from one domain to the other), i.e., they describe the full dynamic behavior of the system. If one is interested only in the system's response to harmonic excitations, and, moreover, if only the steady-state component of the response is considered, then the simpler (but very useful) description as a *frequency response* is sufficient.

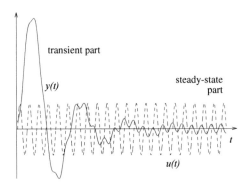

Fig. A.7. Response $y(t)$ (solid) of an asymptotically stable system to a sinusoidal input $u(t)$ (dashed).

In steady-state, i.e., after the transients due to initial conditions have vanished, the system's response is also a sinusoidal curve having exactly the same frequency as the excitation (see Fig. A.7). However, the *amplitude* of the output signal is different and its *phase* usually lags behind the input signal.

$$u(t) = cos(\omega t) \quad \rightarrow \quad y(t) = A(\omega) \cdot cos(\omega t + \varphi(\omega)) \tag{A.43}$$

Since these two quantities depend on the excitation frequency, it is useful to display this dependency in amplitude/frequency and phase/frequency diagrams which are called *Bode diagrams*. (Fig. A.8 shows the Bode diagram of the water-tank system.)

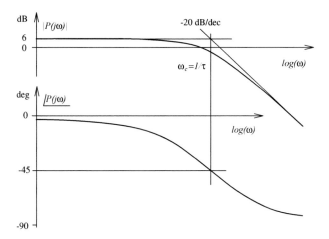

Fig. A.8. Bode diagram of the water-tank system, where ω_c is the *corner frequency* defined by $\omega_c = 1/\tau$.

To facilitate series interconnections of systems, Bode diagrams are usually plotted with a logarithmic scale in the amplitude and a linear scale in the phase, such that series connections like (A.41) can be visualized by simply adding up the amplitudes and phases of the two Bode diagrams. Often the phase is given in degrees and the amplitude in dB defined as

$$A_{dB} = 20 \cdot log_{10}(A) \tag{A.44}$$

Although very useful for quantitative reasonings, Bode diagrams have the drawback that the phase and the magnitude information is separated. This problem is avoided in *Nyquist diagrams* which combine this information by using a frequency-implicit representation (see Fig. A.9).

Notice that the frequency response is obtained by substituting the complex variable s in the transfer function $P(s)$ by the purely imaginary variable $j\omega$ ($j = \sqrt{-1}$). For instance, assume that the transfer function of a system is given by

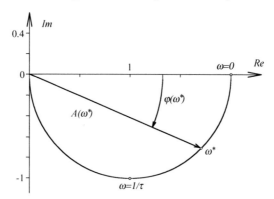

Fig. A.9. Nyquist diagram of the water-tank system.

$$P(s) = \frac{bs + 1}{\gamma s^2 + \beta s + 1} \tag{A.45}$$

then its frequency response is simply

$$P(j\omega) = \frac{bj\omega + 1}{\gamma(j\omega)^2 + \beta j\omega + 1} = \frac{1 + j\omega b}{(1 - \gamma\omega^2) + j\omega\beta} \tag{A.46}$$

The maximum value of the magnitude of $P(j\omega)$ over all frequencies ω is an important parameter. This variable is therefore denoted by a special symbol

$$||P(s)||_\infty = \max_{\omega \in \Re}\{|P(j\omega)|\} \tag{A.47}$$

If $P(s)$ is a transfer function that describes the influence of a disturbance d on the system output y, then minimizing $||P(s)||_\infty$ corresponds to minimizing the maximum influence over all frequencies that d can have on y in steady-state.

The three mentioned system representations, i.e., results of simulations, Bode diagrams and Nyquist diagrams, are the "language" with which all the following controller design and analysis ideas will be explained.

The last point discussed in this section is the question of system stability. The stability of a linear and time-invariant system can be assessed by computing (or measuring) its response to impulse inputs. In this case, the system output $y(t)$ can have three forms:

- $\lim_{t\to\infty} |y(t)| = 0$ in which case the system is *asymptotically stable*
- $\lim_{t\to\infty} |y(t)| = c < \infty$ in which case the system is *stable* (but not *asymptotically* stable)
- $\lim_{t\to\infty} |y(t)| = \infty$ in which case the system is *unstable*.

It can be shown that a system with transfer function $P(s)$ is asymptotically stable if and only if all of the poles of $P(s)$ have negative real parts.

One of the most useful stability analysis tools is the Nyquist criterion, which will be presented below for the case where both the plant and the

controller are asymptotically stable. The Nyquist criterion is able to predict
the stability of a closed-loop system based on the *frequency response* of the
open-loop system. In other words, the steady-state behavior of the open loop
is sufficient to predict the transient behavior of the closed loop. This rather
surprising fact is a very powerful tool for system analysis and synthesis, as
will be shown in the following sections.

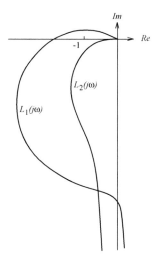

Fig. A.10. Open-loop systems which are unstable $(L_1(j\omega))$ and stable $(L_2(j\omega))$
when the loop is closed with unity feedback.

Nyquist Stability Criterion (simplified) Assume that the controller
$C(s)$ and the plant $P(s)$ have no poles in the complex right-half plane. Then
the closed-loop system $T(s) = L(s)/(1 + L(s))$ will be asymptotically stable
if and only if the frequency response $L(j\omega)$ of the open-loop transfer function
$L(s)$ does not encircle the critical point $z = -1$.

Examples of a stable, $L_2(s)$, and an unstable, $L_1(s)$, closed-loop system
are shown in Fig. A.10.

Example Cruise-Control Problem

As an example, the Bode and the Nyquist diagrams of the cruise-control
system for all possible values of the parameters and the scheduling speed
$v_n^* = 30 \ m/s$ are computed here. Fig. A.11 shows the resulting Bode sets for
"all" possible parameter values listed in Table A.1. The analogous Nyquist
set is shown in Fig. A.13. Of course, this plot has been derived by "gridding"
the (three-dimensional) uncertain parameter space and therefore there is no
guarantee that the exact limits are found in this way. Notice that – in general
– it is not a vertex in the parameter space (i.e., a combination of extreme

parameter values) that produces the boundary of the Bode sets and that the nominal parameters are not the mean values of the interval. For more details see [1].

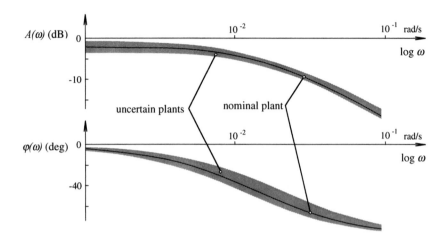

Fig. A.11. Bode sets corresponding to all possible parameter values as indicated in Table A.1 and for the scheduling speed $v_n^* = 30 \; m/s$ (speed interval $25 \; m/s \leq v_n < 35 \; m/s$).

Notice that when using linear stability theory the implicit assumption is that a nonlinear and parameter-varying system can be described by a set of linear time-invariant equations. This assumption is by no means obvious and can be justified only if the set points and the parameter changes are substantially slower than the relevant dynamic effects that are to be controlled.

For the cruise-control problem, this is indeed the case, i.e., the vehicle mass remains constant during a transient (i.e., a grading step) and the vehicle speed set points and the gear ratios remain piecewise constant for a finite interval.

A.3 Model Uncertainty

General Remarks

In the last section it was shown that uncertain systems can be described as families of linear systems centered around a nominal model. However, the resulting sets generally are much too complicated to be analyzed directly. Therefore, a description of the model uncertainty must be found that is both simple enough to work with and sufficiently complex to permit useful conclusions about the system behavior when feedback controllers are designed.

There are many approaches that satisfy these requirements. In this text only the most simple method will be presented. Simplicity will always induce conservativism, i.e. since the uncertainty descriptions will exaggerate the actual uncertainty, the resulting controllers will be able to cope with larger sets of uncertain systems than the one at hand, at the price of not fully exploiting the possibilities offered by the uncertain system.

The starting point of all uncertainty descriptions is always a nominal system that must be centered in the family such as to minimize conservativism. Here, this is equivalent to the nominal plant being in the center of the Nyquist set of the uncertain system (black curve in Fig. A.13). The uncertain system (darker grey set in Fig. A.13) is now "overbounded" by the lighter grey set, which has to be as simple as possible in its description.

An obvious idea is to parameterize the uncertainty by the set

$$\Pi = \{P(s)(1 + W(s)\Delta(s)) \mid \ |\Delta(s)| \leq 1, \mathrm{Arg}(\Delta(s)) \in \{-\pi, \pi\}\} \quad (A.48)$$

where:

- $P(s)$ is the nominal plant (exactly known, centered in the uncertain set, and used below to design the nominal controller).
- $W(s)$ represents the a-priori knowledge about the uncertainty; essentially it is a frequency-dependent upper bound on the modulus of the uncertainty.
- $\Delta(s)$ is a unit disc centered at the origin.

The set Π is best visualized in a Nyquist plot as shown in Fig. A.12. For each fixed frequency ω^*, the set $\Pi(\omega^*)$ is a *disc* with the radius $|P(j\omega^*)W(j\omega^*)|$ which is centered at $P(j\omega^*)$. By choosing an appropriate weighting $W(s)$, the actual uncertain set is covered, whereby $W(s)$ has to be chosen such that as little overlap as possible occurs. Note that only the magnitude of $W(s)$ is important. Its phase information is neutralized by $\Delta(s)$.

The uncertainty parametrization shown here is one of the simplest available. Other parametrizations are known, most being more useful but also more complex. The most important limitation of the multiplicative parametrization is the restriction that the number of unstable poles may not change for the entire uncertain set.

Fig. A.12 also shows the fictitious true plant $P_t(s)$ that below will be assumed to exist. The design rules discussed in the next section will guarantee some desired properties for all possible true plants as long as they are completely inside the set Π. Of course, such a true plant is only an idealization (linear, time invariant) and therefore the "guarantees" obtained should be used with the necessary caution.

Example Cruise-Control Problem

In this example, the two quantities $P(s)$ and $W(s)$, as introduced above, will be defined for the cruise-control example. The nominal system $P(s)$ is derived

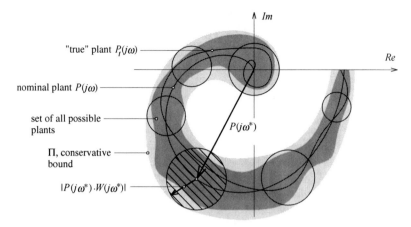

Fig. A.12. Geometric interpretation of the set Π.

first. This problem is solved iteratively by finding a well-centered nominal system. Sometimes physical intuition is helpful, but usually it is not the case that the average physical parameter values define a good nominal system. Note that there might be many parameter combinations that produce good nominal systems. In this case, the following parameters were chosen: $m = 1850\ kg$, $\gamma = 0.28$ – and $v_n = 29\ m/s$. The resulting Nyquist curve is shown in Fig. A.13.

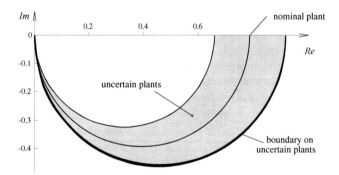

Fig. A.13. Nyquist sets of the cruise-control example, nominal, parametric uncertain (gray) and conservative boundary set (black).

The second step is to find a tight uncertainty bound $W(s)$. Again, this step involves several iterations in order to find a satisfactory solution. As a starting point, the gain of the weighting $W(s)$, i.e., its value at $s = 0$ must be found. If the nominal system is well centered at $s = 0$, the value of $W(0)$ is minimized and corresponds to half the extension of the uncertain set at

$s = 0$. As can be seen in Fig. A.13, a gain of approximately $k = 0.165$ is a good starting point.

This information provides the first choice possible for $W(s) = k$. This is, of course, the simplest possible uncertainty. Since the weighting $W(s)$ is constant for all frequencies, it overestimates the uncertainties at higher frequencies and thus the resulting control system is far from fully exploiting the potential of the real system.

To reduce that effect, an all-pass element

$$W(s) = 0.165 \cdot \frac{20s + 1}{60s + 1} \tag{A.49}$$

can be added such that the weight $W(s)$ better fits the (real) plant uncertainty. The result of this choice is shown in Fig. A.13 as well. All of the following considerations will use this uncertainty bound $W(s)$.

Remark: Often a system is not modeled on the basis of physical first principles, but its input-output behavior is measured using devices such as frequency analyzers. If these measurements are repeated several times (e.g., for different operating points), a set of Nyquist plots is obtained that is very similar to those presented here. Fitting a (fictitious) nominal system and a conservative uncertainty bound $W(s)$ is then completely analogous to the approach introduced in the previous sections. All of the following arguments will be applicable to this case as well.

A.4 Control-System Design for Nominal Plants

General Remarks

Control System Structure

This section discusses the *nominal* controller design step, i.e., the procedure for finding a good controller $C(s)$ for the nominal plant $P(s)$. In this step it is very important to distinguish between feedforward and feedback controllers. Figure A.14 shows the setup that is analyzed in this section.

In addition to the input signal $u(t)$, measurable and unmeasurable disturbances influence the system output through the corresponding transfer functions. The system output is not available. Only the variable $y(t)$, which is affected by an unknown noise, is available to the control system. The four variables that are accessible to the combined controller are the observable disturbance $d_o(t)$, the measured output $y(t)$ and the reference signal $r(t)$, which are the inputs to the controller, and the input to the plant, $u(t)$.

A controller will therefore have the structure shown in Fig. A.15. The main idea in the feedforward component is to use all available knowledge, i.e., the nominal plant model $P(s)$ and the nominal disturbance model $Q_o(s)$, to design

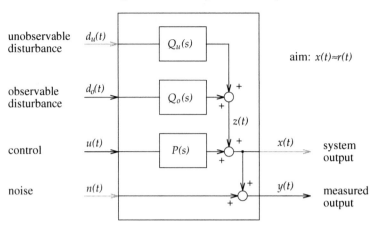

Fig. A.14. System setup for controller design.

the filters $K_o(s)$ and $F(s)$. The design of these feedforward components is relatively straightforward for plant and disturbance models that are minimum phase (no zeros in the right-half complex plane). Essentially, this step requires the plant dynamics to be "inverted," i.e., a filter must be designed such that when connected to the input of the plant, it produces a near-unity transfer function over the frequency range of interest. The feedforward filter must include some low-pass elements with high corner frequencies in order to be realizable and in order not to amplify any sensor noise.

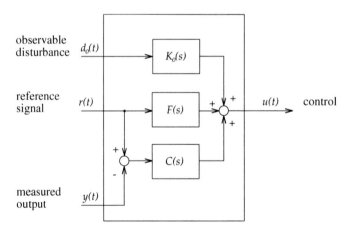

Fig. A.15. General control system structure.

For the controller design process, the following general guidelines are important:

- Use feedforward (open-loop) control whenever possible. It is fast, does not cause any additional stability problems and it does not introduce the measurement noise into the system.
- All models are uncertain and plants are always affected by not measurable disturbances. The sole purpose of feedback control is to reduce the uncertainty introduced by these unobservable disturbances and modeling errors.
- All feedback controllers must be designed such that stability and performance are guaranteed as long as the actual disturbances and modeling errors remain inside the bounds that have been determined a priori.

Design Guidelines

The design of the nominal feedback controller $C(s)$ is an interesting task for which a large amount of literature is available. In this section, only single-input-single-output (SISO) systems are discussed which, moreover, are assumed to be "nice" in the sense that they have no unstable poles and no "unstable" zeros (i.e., no zeros in the right complex plane, a system feature which is often called *minimum-phase*). A plant that does not have all three of these properties requires more caution, and the straightforward methods presented below will not work properly.

The following definitions will be useful

$$L(s) = C(s) \cdot P(s) \qquad \text{loop gain}$$

$$S(s) = 1/(1 + L(s)) \qquad \text{sensitivity} \qquad\qquad (A.50)$$

$$T(s) = L(s)/(1 + L(s)) \qquad \text{complementary sensitivity.}$$

These transfer functions describe how the exogenous inputs affect the closed-loop system output when only feedback control is applied, i.e., $F(s) = K_0(s) = 0$, $Q_u(s) = 1$, $d_o(t) = 0$ (cf. Figs. A.14 and A.15)

$$y(s) = S(s) \cdot d_u(s) + T(s) \cdot r(s) \qquad\qquad (A.51)$$

Many controller designs are founded on the following observations:

- Most plants are low-pass elements, i.e., they have a finite gain at very low frequencies and attenuate input signals of higher frequencies.
- High loop gains are useful for reducing the effects of unmeasurable disturbances and to improve the set-point tracking behavior, as expressed by eq. (A.50).
- Phase lag, model uncertainty and noise limit the frequency range where high loop gains are permitted; otherwise an unstable or badly damped closed-loop system behavior results.

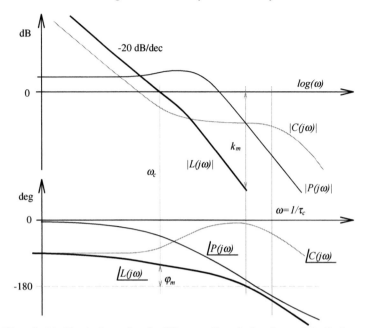

Fig. A.16. Typical result of a PI controller design for a "nice" plant.

PI Controllers

Therefore, at low frequencies the controller must have high gains (they may increase without limits with frequencies tending to zero), whereas at high frequencies, finite or even decreasing gains, guaranteeing a quick *roll-off*, are desirable. Since PI controllers satisfy these conditions and are relatively easy to implement and tune, the majority of control loops are realized with such controllers

$$C(s) = k_p \cdot \left(1 + \frac{1}{\tau_i \cdot s}\right) \cdot \frac{1}{(\tau_c \cdot s + 1)^\rho}, \quad \rho \geq 1 \qquad (A.52)$$

Figure A.16 shows a typical PI controller design in a Bode diagram. In general, the following characteristics are important:

- the crossover frequency ω_c is that frequency at which the loop gain magnitude $|L(j\omega)|$ crosses the 0 dB line;
- the phase margin φ_m is the distance of the loop gain phase at ω_c to the -180 deg line; and
- the gain margin k_m is the distance of the loop gain magnitude to the 0 dB line at that frequency at which the phase of the loop gain is -180 deg.

The design of a PI controller consists of finding controller parameters $\{k_p, \tau_i, \tau_c\}$ such that the desired crossover frequency and phase and gain margins are attained. Useful rules of thumb are:

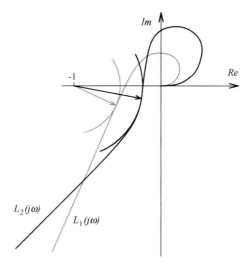

Fig. A.17. Loop shaping to increase minimum return difference (from $L_1(j\omega)$ to $L_2(j\omega)$).

- $\omega_c \approx 2.3/t_{des}$ where t_{des} is the desired settling time of the closed-loop system after a step disturbance;
- $\varphi_m > 70\ldots80\ deg$ is required if no overshoot to disturbance steps is acceptable, a smaller phase margin can be acceptable, but at least 45 deg should be guaranteed for the nominal plant; and
- $k_m > 4\ dB$ is required which corresponds to a loop-gain error tolerance of approximately 50% .

The best approach to assessing the robustness of a specific design is to look at the Nyquist plot of the loop gain and to try to maximize the minimum distance d_{min} (A.53) to the critical point

$$d_{min} = \min_{\omega}\{|1 + L(j\omega)|\} \tag{A.53}$$

while realizing the desired crossover frequency ω_c. Figure A.17 shows an example of two Nyquist curves obtained with two sets of parameters $\{k_p,\ \tau_i,\ \tau_c\}$ and the corresponding d_{min}. Section A.5 introduces lead/lag elements (A.58) that can be used to improve a PI control system design in specific frequency bands.

Experimental Tuning Methods

If no mathematical model is available, the parameters of a PI controller (A.52) can be determined using one the following two experimental methods. Note that these approaches work well in *some* cases; however, there is no guarantee that a satisfactory design is obtained in *every* case.

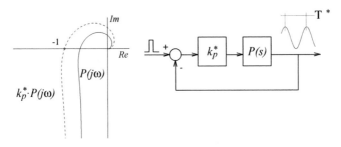

Fig. A.18. Illustration of the Ziegler-Nichols PI tuning method.

The setup used in the first method is illustrated in Fig. A.18. The integrator time constant τ_i of the controller (A.52) is set to infinity and the gain k_p is increased until the loop oscillates, i.e., until the poles of the closed-loop system are on the stability boundary. The corresponding critical gain k_p^* and critical oscillation period τ^* are used to derive the parameters of the PI controller as indicated in Table A.3.

Table A.3. Experimental design of a PI controller.

method	k_p	τ_i
Ziegler-Nichols	$0.45 \cdot k_p^*$	$0.85 \cdot \tau^*$
Chien-Hrones-Reswick, 0% overshoot	$0.6/\alpha$	$4.0 \cdot \delta$
Chien-Hrones-Reswick, 20% overshoot	$0.7/\alpha$	$2.3 \cdot \delta$

The second approach is illustrated in Fig. A.19. Here, a step response is the starting point for the estimation of the parameters of the PI controller. First, the time t^* must be found at which the second derivative $\partial^2 y/\partial t^2$ equals zero. The intersection of the tangent at $y(t)$ in that point with the axes yields the two necessary parameters: a delay time δ and the starting point of the affine tangent α. Again, once these two parameters are known, the PI controller gains can be found using the rules indicated in Table A.3.

The Ziegler-Nichols approach usually yields the more precise results. Unfortunately, its application is sometimes not possible or too time consuming. The Chien-Hrones-Reswick method is easy to apply, but estimating the parameters is not always easy, especially for noisy output signals. More information on these and substantially more powerful design methods for PI control systems can be found in [12].

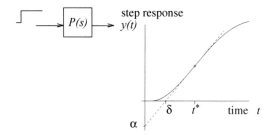

Fig. A.19. Illustration of the Chien-Hrones-Reswick PI tuning method.

Example Cruise-Control Problem

In this section of the cruise-control example, a nominal controller and a feed-forward filter will be designed. The linearized and normalized plant and the disturbance model of the cruise-control system has the following form

$$\delta y(s) = \frac{k(m, v_n)}{\tau(\gamma, m, v_n)s + 1} \cdot \delta u(s) + \frac{-1}{\tau(\gamma, m, v_n)s + 1} \cdot \delta d(s) \qquad \text{(A.54)}$$

and therefore

$$P(s) = \frac{k(m, v_n)}{\tau(\gamma, m, v_n)s + 1}, \quad \text{and} \quad Q_u(s) = \frac{-1}{\tau(\gamma, m, v_n)s + 1} \qquad \text{(A.55)}$$

Following the rules of thumb listed above, a crossover frequency of $0.4 \ rad/s$ is aimed at. This corresponds to a settling time of approximately 6 s, which is much slower than the neglected torque dynamics, but sufficiently fast to satisfy customer demands. Overshoot should be avoided (at least in the nominal case) such that a phase margin of at least 70 deg must be achieved. Gain margin is not a problem in the cruise-control case since the dynamics of the system are essentially first-order only.

After several iterations, the values $k_p = 33$ and $\tau_i = 10$ s are found. The additional low-pass filter is chosen at $\tau_c = \tau_i/20 = 0.5$ s such that it is robust with respect to any neglected dynamics (torque dynamics, drive train ringing, etc.). As Fig. A.20 shows, the closed-loop system is stable since the Nyquist criterion is satisfied. This figure also shows that the return difference (A.53) has a minimal magnitude of approximately 0.87.

The disturbance response is the relevant time-based information since set-point tracking can be influenced by the design of the feedforward filter (see below). Figure A.21 shows the closed-loop system response for a disturbance step $\delta d = 0 \rightarrow 1$, which for the chosen operating point corresponds to a step in demanded power of $17kW \rightarrow 40 \ kW$, or a disturbance force of $0 \rightarrow 756 \ N$. The settling time, i.e., the time needed for the engine torque to reach its new steady-state value, as expected, is around 5 s. Also, the speed trajectory satisfies the "no overshoot" condition (torque slightly overshoots around $t = 20$ s which, of course, could be perceived by the driver as a small jerk).

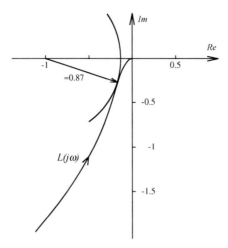

Fig. A.20. Nyquist diagram of a design of a PI controller for the cruise-control problem; also indicated is the minimum value of the *return difference*.

Figure A.21 shows the simulation results obtained when applying the PI controller to both the original nonlinear plant (A.18) and its linearization (A.28). Although the magnitude of the disturbance step is quite large, the two trajectories of the original nonlinear plant and of the linearized plant are indistinguishable. This is a consequence of the fact that vehicle speed does not change much during this transient (less than 5%). No engine torque limits have been considered yet such that for larger disturbance steps the system's trajectories would not change substantially. In Sect. A.7, however, it will be

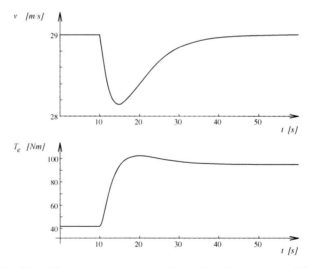

Fig. A.21. Closed-loop system response for a disturbance step of $\delta d = 0 \rightarrow 1$.

shown that (unavoidable) engine torque limitations will have very important consequences.

As mentioned, the overall control system performance can be improved substantially by adding feedforward action. In the cruise-control example, such a feedforward filter $F(s)$ will have the form

$$F(s) = \frac{\tau \cdot s + 1}{k \cdot (\tau_F \cdot s + 1)^\rho}, \quad \rho \geq 1 \tag{A.56}$$

The available knowledge about the system is "inverted" and is used to derive the driving signal, i.e., the desired engine torque. Since for low-pass systems a full inversion is not possible, additional low-pass filters with time constants τ_F are used that must be chosen as a compromise between fast system behavior and low control efforts and noise sensitivity.

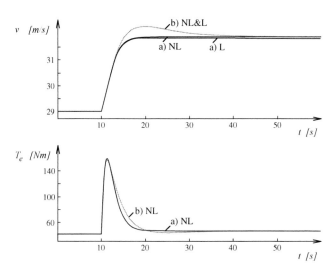

Fig. A.22. Speed and engine torque response to a set-point step for a) feedforward action only and b) feedback action only (NL: nonlinear plant model, L: linear plant model).

For the cruise-control example, a time constant $\tau_F = \tau/60$ turns out to be a good choice. Figure A.22 shows the set-point tracking behavior of the system in the case a) with feedforward only $(C(s) = 0)$ and case b) with feedback only $(F(s) = 0)$. In the case a), the response of the system to a step change in desired speed is clearly better without requiring any higher control efforts. Notice however that in the case a) (no feedback control), the behavior of the linear and nonlinear plants differs substantially (an unavoidable consequence of the missing feedback correction action). In the case b), a clear tendency to speed overshoot is observed. If this overshoot is eliminated by reducing the controller gains, the disturbance rejection deteriorates, which, of course, is not desirable. A combination of the two control blocks, i.e., a feedforward and a feedback part, can achieve good disturbance rejection *and* good tracking behavior simultaneously.

A.5 Control System Design for Uncertain Plants

General Remarks

In this section, the controller designed for the nominal system is tested with respect to its stability properties for the uncertain system. The main objective will be to "guarantee" stability of the closed-loop system for all possible true plants in the set defined by (A.48). The problem of how to improve a given design with respect to its stability properties is discussed below. However, the *robust performance* problem, i.e., how to guarantee that even in the uncertain case a design performs according to the chosen specifications, is not treated here. The interested reader will find this information and many other important points in [53].

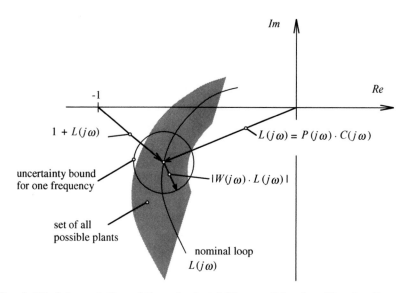

Fig. A.23. Interpretation of the robust stability condition in a Nyquist diagram.

The key idea of *robust stability* is to use the conservative bound $W(s)$ described in Sect. A.3 to test the uncertain system for closed-loop stability. If the following assumptions are satisfied:

- a nominal plant $P(s)$, a conservative bound on its uncertainty $W(s)$, and a fixed controller $C(s)$ are known;

- the nominal closed-loop system is asymptotically stable; and
- the following inequality is satisfied for all frequencies ω

$$|1 + C(j\omega) \cdot P(j\omega)| \geq |C(j\omega) \cdot P(j\omega) \cdot W(j\omega)| \quad , \tag{A.57}$$

then the controller $C(s)$ guarantees the closed-loop system to be asymptotically stable for all true plants $P_t(s)$ in the set Π.

This robust stability theorem is very useful since it involves only one test on three known transfer functions. Its main limitation is the assumption that a true plant exists. Figure A.23 shows a graphic interpretation of the condition (A.57).

If an insufficient minimum robust return difference is obtained, an iteration in the controller design is required. First attempts will be to optimize the controller parameters (k_p, τ_i and τ_c in the PI case). If this does not produce the desired results (insufficient stability margins or too slow a nominal system behavior), then the order of the controller must be increased. Often this is done by adding *lead/lag* elements which shape the open loop in specific frequency intervals such that the design specifications are reached.

These lead or lag elements are defined by

$$C_{ll} = \frac{\tau \cdot s + 1}{\alpha \cdot \tau \cdot s + 1} \tag{A.58}$$

where $\tilde{\omega} = 1/(\tau \sqrt{\alpha})$ is the corner frequency (the frequency where the loop-shaping action is centered) and α is the element's gain which determines the character of the element ($\alpha < 1$ for a lead element with differentiating behavior, $\alpha > 1$ for a lag with integrating behavior). As Fig. A.24 shows, a lead element can be used to shift the loop phase towards positive values (e.g., to increase the return difference in some frequency band). The drawback of that is an increased loop gain at higher frequencies. The lag element is used to decrease the loop gain (e.g., to improve the roll-off behavior), but it increases the phase lag in a certain frequency band. Because of these partially contradictory effects, several design iterations generally are required before a satisfactory result is obtained. More systematic procedures exist, but they require more evolved mathematical concepts. The interested reader is referred to [53].

Of course, the ultimate goal of all control system designs is not to merely achieve robust stability, but robust *performance*, i.e., the closed-loop system is expected to perform as specified as long as the true plant is included in the set (A.48). To achieve this goal, an additional sufficient stability margin must be maintained by the controller $C(s)$.

Example Cruise-Control Problem

Figure A.25 shows the same quantities as those depicted in the schematic representation of Fig. A.23 for the cruise-control example. Clearly, the controller

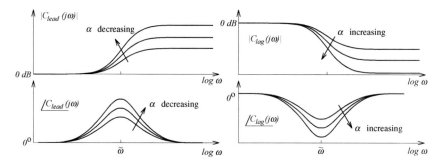

Fig. A.24. Bode diagrams of lead and lag elements.

$C(s)$ designed in the last section satisfies the robust stability condition (A.57). The minimum distance of the critical point to the uncertainty bounding set can be interpreted as a *robust return difference*, which has to be larger than zero in order to guarantee robust stability. For the cruise-control example, a minimal robust return difference of 0.84 is obtained with the PI controller designed in the last section. Since the minimal *nominal* return difference was 0.87, shown in Fig. A.20, the deterioration caused by the uncertainty has been limited by the chosen controller to just a small percentage.

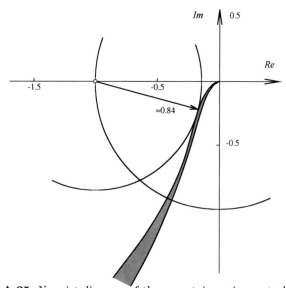

Fig. A.25. Nyquist diagram of the uncertain cruise control loop.

As mentioned above, the controller designed in Sect. A.4 meets the robust stability requirement (A.57) with a large additional stability margin. This reserve in stability indicates that the *performance* of the closed-loop system remains close to the nominal performance as long as the plant remains within the set Π of (A.48). This assertion is supported by the results of a series of numerical simulations of the linear controller applied to the nonlinear plant model. As Fig. A.26 shows, the disturbance rejection properties of the controller $C(s)$ designed in Sect. A.4 remain satisfactory for the full range of expected nominal vehicle speeds and parameter variations.

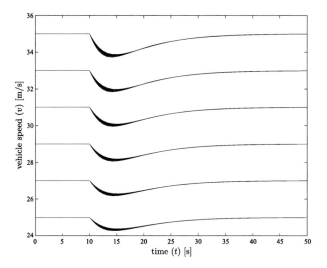

Fig. A.26. Disturbance rejection behavior for a scheduling speed $v_n^* = 30 \; m/s$, parameter intervals as shown in Table A.2.

Note that these simulations use the nonlinear and non-normalized plant model. The interface between this plant and the controller that was designed for the normalized and linearized plant is shown in Fig. A.27. This structure is similar to the one encountered in the later control system realization. Some of the most important additions will be discussed in the next section.

A.6 Controller Discretization

Structure of a Discrete-Time Control System

Almost all control systems in passenger cars are implemented using digital computers. These electronic devices can only handle information using a *sampled-data* approach. The schematic layout of such a control system is shown in Fig. A.28.

The analog-to-digital converter (AD) samples the continuous-time signal $(e(t))$ and transforms the sampled value to a binary electronic signal $(e^*(k))$

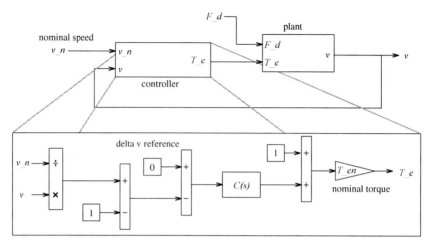

Fig. A.27. Interface structure of the closed-loop system.

that can be processed in the microcontroller. Usually, the amplitude quantization involved in that step can be neglected. Correctly handling *overflow problems*, which are due to the fixed-point arithmetic of low-cost microprocessors, is often more important.

The discrete-time controller $C(z)$, whose design will be discussed below, computes the control signal $u^*(k)$. This signal is transformed by the digital-to-analog (DA) converter to a continuous time signal $u(t)$. The DA converter always includes a hold element. In most cases, this is a zero-order hold (ZOH) element defined by

$$u(t) = u^*(k), \quad \forall\, t \in [k \cdot \tau_s, (k+1) \cdot \tau_s) \tag{A.59}$$

where τ_s is the sampling time.

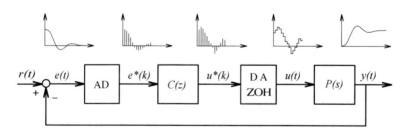

Fig. A.28. Discrete-time control system structure.

As the commutation diagram in Fig. A.29 shows, there are two approaches for the design of discrete-time control systems. The direct approach, denoted by "A" in Fig. A.29, consists of first discretizing the plant and then designing a discrete-time controller using the discrete-time plant model $P(z)$. This approach is mandatory for cases where the sampling time τ_s is large compared to the plant's time constants. The problem of discretizing a plant will be discussed below, including some basic remarks on the $\mathcal{Z}$ transformation that is the discrete-time counterpart of the Laplace transformation.

In the alternative approach, denoted by "B" in Fig. A.29, a continuous-time controller $C(s)$ is designed in a first step. This system is then transformed to an equivalent discrete-time transfer function $\widetilde{C}(z)$ that emulates, as well as possible, the closed-loop behavior of $C(s)$ in the sampled-data control loop shown in Fig. A.28. This approach provides no guarantees with respect to stability, robustness, etc. However, for relatively small sampling times τ_s, it works well in most cases.

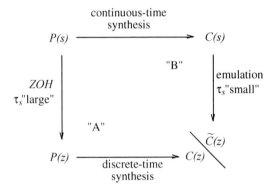

Fig. A.29. The possible design procedures to obtain a discrete-time control system for a continuous-time plant.

Plant Discretization

The starting point of the direct approach "A" is the discretization of the continuous-time plant

$$\tfrac{d}{dt}x(t) = A \cdot x(t) + b \cdot u(t), \quad x \in \Re^n, \; u \in \Re$$
$$y(t) = c \cdot x(t) + d \cdot u(t), \quad y \in \Re \tag{A.60}$$

using the fact that the input $u(t)$ is piecewise constant (A.59). The result of this discretization is the discrete-time plant

$$x((k+1) \cdot \tau_s) = F \cdot x(k \cdot \tau_s) + g \cdot u(k \cdot \tau_s)$$
$$y(k \cdot \tau_s) = c \cdot x(k \cdot \tau_s) + d \cdot u(k \cdot \tau_s) \tag{A.61}$$

where

$$F = e^{A \cdot T_s}, \quad g = \int_0^{T_s} e^{A\sigma}\, d\sigma\, b = A^{-1} \cdot \left[e^{A \cdot T_s} - I \right] \cdot b \qquad (A.62)$$

The second expression for g is, of course, only applicable for plants with no poles in the origin ($\det(A) \neq 0$). The term $e^{A \cdot T_s}$ denotes the matrix exponential. Several reliable numerical software packages are available for its computation.

For first-order systems, scalar exponential functions are sufficient, i.e., such a first-order system can be represented in any of the following four forms (for the sake of brevity, the independent variable $k \cdot T_s$ is replaced by k):

1. continuous-time, time domain ($\{a, b, c, d\}$ real scalars)

$$\frac{d}{dt} x(t) = a \cdot x(t) + b \cdot u(t)$$
$$y(t) = c \cdot x(t) + d \cdot u(t) \qquad (A.63)$$

2. continuous-time, frequency domain

$$Y(s) = \frac{d \cdot s + [c \cdot b - a \cdot d]}{s - a} \cdot U(s) \qquad (A.64)$$

3. discrete-time, time domain ($a \neq 0$)

$$x(k+1) = e^{a \cdot T_s} \cdot x(k) + \frac{(e^{a \cdot T_s} - 1) \cdot b}{a} \cdot u(k)$$
$$y(k) = c \cdot x(k) + d \cdot u(k) \qquad (A.65)$$

4. discrete-time, frequency domain

$$Y(z) = \frac{d \cdot z + [c \cdot (e^{a \cdot T_s} - 1) \cdot b/a - e^{a \cdot T_s} \cdot d]}{z - e^{a \cdot T_s}} \cdot U(z) \qquad (A.66)$$

In the fourth representation, the $\mathcal{Z}$ transformation of the discrete-time signal $x(k)$ has been used[2]

$$X(z) = \mathcal{Z}\{x(k)\} = \sum_{k=0}^{k=\infty} x(k) \cdot z^{-k} \qquad (A.67)$$

The $\mathcal{Z}$ transformation is the counterpart of the Laplace transformation and may be used to solve a difference equations using the rules

$$z \cdot X(z) - z \cdot x_0 \leftrightarrow x(k+1), \quad z^{-1} \cdot X(z) \leftrightarrow x(k-1) \qquad (A.68)$$

The first relation can be used to solve the difference equation (A.61) in the frequency domain

---

[2] The signal $x(k)$ is assumed to be equal to zero for all $k < 0$.

$$z \cdot X(z) = F \cdot X(z) + g \cdot U(z) + z \cdot x(0) \tag{A.69}$$

$$Y(z) = c \cdot X(z) + d \cdot U(z) \tag{A.70}$$

where the variables $X(z)$ and $U(z)$ represent the $\mathcal{Z}$ transformation of their corresponding counterparts $x(k)$ and $u(k)$. Assuming a zero initial condition, i.e., $x(0) = 0$, the *transfer function*

$$P(z) = \frac{Y(z)}{U(z)} = c \cdot [z \cdot I - F]^{-1} \cdot g + d \tag{A.71}$$

can be derived. This transfer function plays the same role as its continuous-time counterpart $P(s)$, i.e., it describes how any input u is transformed to the system output y.

Once $P(z)$ is known a controller, $C(z)$ can be designed with any of the many discrete-time design methods available. The general idea behind all these methods is the same as in the continuous-time case; viz. to guarantee a desired closed-loop system performance *and* a sufficient robustness to modeling errors. The interested reader is referred to the classical texts on this subject, e.g., [66].

Note that, compared to the corresponding continuous-time formulation, some systems can be described much better using a discrete-time approach. One interesting example of such an element is a pure time delay that is represented, in the different domains, by the following four forms:

$$y(t) = u(t - \tau_s) \tag{A.72}$$

$$Y(s) = e^{-\tau_s \cdot s} \cdot U(s) \tag{A.73}$$

$$y(k) = u(k - 1) \tag{A.74}$$

$$Y(z) = z^{-1} \cdot U(z) \tag{A.75}$$

The two frequency domain representations can be used to derive the important correspondence

$$z = e^{\tau_s \cdot s} \tag{A.76}$$

Emulation Methods

In practice, another approach is often chosen that consists of designing the controllers as shown in Sect. A.4 and to transform these systems into their discrete-time equivalents using an approximation (path "B" in the commutation diagram A.29). One of the most popular is the *Tustin transformation* defined by

$$s \approx \frac{2}{\tau_s} \cdot \frac{z - 1}{z + 1} \tag{A.77}$$

where τ_s is the sampling time. Using this relation, a continuous-time linear controller $C(s)$ can be transformed into its approximate discrete-time equivalent $C(z)$ by substituting the variable s by the variable z.

These rules allow any given controller $C(z)$ to be transformed into a series of commands which are executable on a digital computer. This is best explained with the simple example of a low-pass controller. First, the system is Tustin-transformed

$$C(s) = \frac{c}{\tau \cdot s + 1} \quad \rightarrow \quad C(z) \approx \frac{c \cdot (z+1)}{(2\frac{\tau}{\tau_s} + 1) \cdot z + (1 - 2\frac{\tau}{\tau_s})} = \frac{U(z)}{E(z)} \quad \text{(A.78)}$$

This can be written as

$$U(z) \cdot \left[\left(2\frac{\tau}{\tau_s} + 1 \right) \cdot z + \left(1 - 2\frac{\tau}{\tau_s} \right) \right] = E(z) \cdot c \cdot (z+1) \quad \text{(A.79)}$$

or

$$U(z) \cdot \left[\left(2\frac{\tau}{\tau_s} + 1 \right) + \left(1 - 2\frac{\tau}{\tau_s} \right) \cdot z^{-1} \right] = E(z) \cdot c \cdot (1 + z^{-1}) \quad \text{(A.80)}$$

The corresponding time-domain equation is obtained using the transformation rules (A.68)

$$u_k \cdot \left(2\frac{\tau}{\tau_s} + 1 \right) + u_{k-1} \cdot \left(1 - 2\frac{\tau}{\tau_s} \right) = c \cdot (e_k + e_{k-1}) \quad \text{(A.81)}$$

Solving this equation for the unknown variable u_k, a recursive formulation of the controller is found

$$u_k = \left(2\frac{\tau}{\tau_s} + 1 \right)^{-1} \left[c \cdot (e_k + e_{k-1}) - u_{k-1} \cdot \left(1 - 2\frac{\tau}{\tau_s} \right) \right] \quad \text{(A.82)}$$

This expression can then be transformed to a computer-executable program.

Notice that this approach is *not* guaranteed to yield satisfactory results. Specifically, if the sampling time τ_s is relatively large compared to the time constants of the system, a badly damped or even unstable closed-loop system can result (see the example below).

Frequency Domain Representation of Discrete-Time Systems

The frequency response $P_c(j\omega)$ of a continuous-time system $P_c(s)$ has been shown to contain much valuable information on the dynamic properties of that system. The graphic representations of $P_c(j\omega)$ in Nyquist or Bode diagrams are standard tools in the analysis and design of control systems. It is therefore interesting to discuss what form a discrete-time frequency response will have. Unfortunately, the sampling and holding process has several unexpected effects on the frequency response such that a complete discussion is beyond the scope of this appendix. Only the most basic effects will be discussed below using intuitive rather than mathematically precise arguments.

As indicated in Fig. A.30, a sampled-data control loop can always be broken at two different points, yielding two representations of the same system. Breaking the loop at the point 1 yields a *continuous-time* loop transfer function

$$L_c(s) = \frac{Y(s)}{E(s)} \quad , \tag{A.83}$$

while breaking the loop at the point 2 yields a *discrete-time* loop transfer function

$$L_d(z) = \frac{V(z)}{U(z)} \tag{A.84}$$

The transfer function $L_c(s)$ describes the behavior of the system for *each* time t, while $L_d(z)$ only gives information on the system behavior at the discrete points in time $t = k \cdot \tau_s$. Not surprisingly, it is more difficult to correctly formulate the mathematical form of $L_c(s)$ than that of $L_d(z)$.

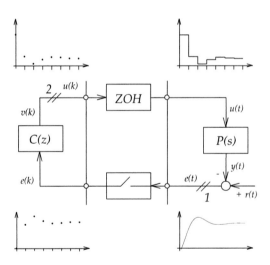

Fig. A.30. Sample-data system.

Continuous-Time Interpretation

In the continuous-time formulation, i.e., using $L_c(s)$, the sampling process is modeled by an impulse sampling (see [66]). This mechanism introduces spectral aliasing.[3] Moreover, the zero-order hold element, which can be modeled using the transfer function

---

[3] To avoid spurious effects caused by aliasing, sampled-data systems usually include continuous-time anti-aliasing filters at the inputs of the control system.

$$ZOH(s) = \frac{1 - e^{-T_s \cdot s}}{T_s \cdot s} \quad , \qquad (A.85)$$

introduces zeros at frequencies equal to integer multiples of the sampling frequency. The Bode diagram of the frequency response $ZOH(j\omega)$ of this element is shown in Fig. A.31.

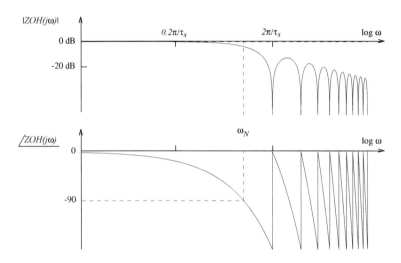

Fig. A.31. Bode diagram of the ZOH system; $\omega_N = \pi/T_s$ is the Nyquist frequency.

Discrete-Time Interpretation

In this setting, the discrete-time transfer function $L_d(z)$ is obtained using one of the approaches shown above. Equation (A.76) shows that the imaginary axis $j\omega$, which is used in frequency responses as an independent variable, is transformed to the unit circle $e^{j\omega}$ when the system representation is changed from continuous-time to discrete-time:

$$s \to j\omega \quad \Rightarrow \quad z = e^{T_s \cdot s} \to e^{j\omega T_s} \qquad (A.86)$$

Accordingly, the Bode and Nyquist diagrams of a discrete-time system $L_d(z)$ are obtained by plotting $L_d(e^{j\omega T_s})$. For frequencies higher than the Nyquist frequency, the function $e^{j\omega T_s}$ is not one-to-one. Therefore, the frequency range in these plots must be limited to $\omega_N = \pi/T_s$, where ω_N is usually referred to as the Nyquist frequency.

Figure A.32 shows the Bode diagrams of the continuous-time system

$$L_c(s) = \frac{1}{\tau \cdot s + 1} \qquad (A.87)$$

and of the two discrete-time systems obtained for a sampling time of τ_s

$$L_{d,ZOH}(z) = \frac{1 - e^{-\tau_s/\tau}}{z - e^{-\tau_s/\tau}} \tag{A.88}$$

$$L_{c,Tustin}(z) = \frac{z + 1}{(2 \cdot \frac{\tau}{\tau_s} + 1) \cdot z + (1 - 2 \cdot \frac{\tau}{\tau_s})} \tag{A.89}$$

with the ZOH direct method and the Tustin emulation approach.

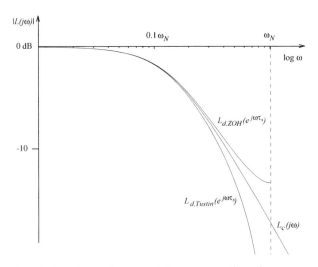

Fig. A.32. Bode diagram of the example (A.87); $\tau = 7\ s$.

Example Cruise-Control Problem

In a series-production vehicle, the PI controller designed in Sect. A.5 will be realized as a discrete-time control system. Accordingly, the continuos-time controller

$$C(s) = k_p \cdot \left(1 + \frac{1}{\tau_i \cdot s}\right) \tag{A.90}$$

must be transformed into an equivalent discrete-time controller $C(z)$ using either an emulation method or designing it based on the discretized plant model $P(z)$. Here, the first approach is chosen.

This controller $C(z)$ is inserted in a structure that is similar to the one shown in Fig. A.27, with the lower component replaced by the system shown in Fig. A.33.

In this example, the discrete-time controller $C(z)$ is found using a Tustin emulation (A.77) of the continuous-time controller $C(s)$ (A.90)

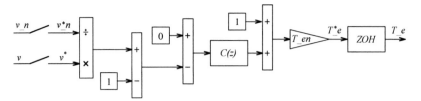

Fig. A.33. Discrete-time control system that replaces the controller in Fig. A.27.

$$C(z) = k_p \cdot \frac{z \cdot \left(\frac{T_s}{2\tau_i} + 1\right) + \left(\frac{T_s}{2\tau_i} - 1\right)}{z - 1} \tag{A.91}$$

The results shown in Fig. A.34 are obtained by inserting the numerical values for the parameters $\{k_p, \tau_i\}$ of the PI controller found in Sect. A.4 and simulating the closed-loop system for different sampling times τ_s.

The response obtained with a very fast sampling $\tau_s = 0.1$ s is almost the same as that obtained with the original continuous-time controller. Increasing the sampling time to $\tau_s = 1, \ldots, 4$ s yields differing, but still acceptable closed-loop system responses.

For sampling times around $\tau_s \approx 4.7$ s, the closed-loop system starts to oscillate and eventually reaches its stability limit at $\tau_s \approx 5.63$.

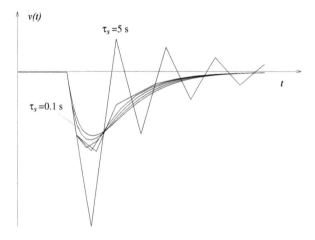

Fig. A.34. Disturbance rejection behavior of the closed-loop cruise-control system. Controller (A.91) realized with the sampling times $\tau_s = \{0.1, 1, 2, 3, 4, 5\}$ s.

A.7 Controller Realization

The realization of a control system is, of course, a vast subject that includes many problem areas. This section briefly discusses only two important theoretical aspects of controller realization, namely:

- controller gain scheduling, i.e., the adaptation of the controller gains to changing operating conditions; and
- controller reset-windup, i.e., the problems that can arise when plant input saturation and integral controller action are combined.

A.7.1 Gain Scheduling

Gain scheduling is a straightforward method to adapt the controller parameters to changing operating conditions. The key idea is to parametrize the controller with respect to a *slowly* changing operating point around which the plant is linearized and normalized. This scheduling point is then assumed to be representative for a certain set of operating points in its neighborhood. The compromise to be made is between a finer "gridding" (large design and implementation efforts, but better controller performance) and a coarser "gridding" (smaller design efforts, but more conservative and therefore slower controllers).

Figure A.35 shows a block representation of such a controller gain scheduling scheme. The controller parameters (gains, time constants, etc.) are denoted by p and the scheduling variable by o. The block "linearization and normalization" is the interface between the nonlinear and the linear components, as shown in the inset in Fig. A.27.

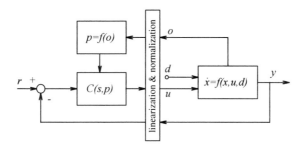

Fig. A.35. Gain scheduling controller structure.

In the cruise-control example, the scheduling variable is the desired vehicle speed (gear ratio and mass have been shown to have less influence on plant behavior). This variable changes only when the driver changes the reference values, but it remains constant during a finite interval.

A major challenge in designing a gain scheduling system is the bumpless and smooth transition between one controller setting and the next. Interpolation of controller values, ramped transitions and several other "tricks" have to be used here. Also, some type of hysteresis must be implemented in order to avoid a limit-cycle behavior at the boundaries between two scheduling zones.

Note that there is no simple way to guarantee that a gain scheduling system is asymptotically stable, even if this is true for all of the closed loops with a fixed controller. It is therefore important to monitor the system behavior in order to detect any instabilities induced by the scheduling process.

A.7.2 Anti-Reset Windup

A very important nonlinearity, which in most cases is present at the input of the plant, is a control signal saturation. For instance, in the cruise-control problem, the engine torque is always limited by $T_{e,min}(\omega_e) \leq T_e \leq T_{e,max}(\omega_e)$. This saturation can cause problems if the controller has any components which are not asymptotically stable. Note that this is always the case if the controller includes integral action (which is stable, but not asymptotically stable). This combination can lead to an unexpected overshoot for large disturbances or set-point changes, even when the linear (small-signal) behavior is perfectly damped.

In order to suppress such phenomena (often called *reset windup*), a special controller structure must be implemented. Figure A.36 shows a possible structure for discrete-time systems (similar structures exist for continuous-time controllers).

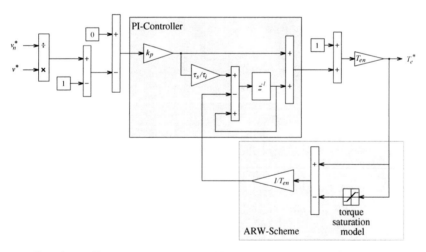

Fig. A.36. Discrete-time PI controller with anti-reset windup circuit.

A.8 Further Reading

A good first book for those who wish to enter the field of control systems analysis and design is [65]. An extension of that first text that contains a classical treatment of digital (discrete-time) control systems is [66]. The text [53] extends the classic concepts to the more modern robust control perspective. A good discussion of the design of PID controllers is found in [12]. The subject of anti-reset windup is discussed in detail in [73]. A good overview on gain-scheduling techniques can be found in [184].

B

Case Study: Idle Speed Control

The control loop discussed in this case study is the well-known idle-speed control system for which a rich literature is available (see [166] for an early work or [100] for an overview paper). The objective of this appendix is not to show all steps of the design and realization of such a control system. For instance the control problem here will be formulated as an SISO problem (see [155] and Sect. 4.3.5 for a true MIMO formulation) and only feedback action will be investigated (see [35] for a design that includes both feedback and feedforward parts). Instead, the focus will be on showing how some of the basic building blocks introduced in the main text can be combined to form a mathematical model and how the parameters of this model can be identified using measurements. Then, using this mathematical model, a simplified control system is designed and tested using the methods introduced in Appendix A.

Specifically, the following four subproblems will be discussed in this case study:

1. *Synthesis of a mathematical model of the dynamic behavior of the relevant engine parts using some of the building blocks introduced in Chapter 2.*
2. *Parameter identification and model validation using data obtained on a dynamic test bench with a 2.8-liter SI engine.*
3. *Model linearization and delay approximation to obtain a finite-dimensional linear system description.*
4. *Synthesis of a simplified idle speed control system (air path actuation only) using the mathematical model derived in step 3. Implementation of this controller on a digital signal processor and test of the closed-loop system on an engine test bench.*

B.1 Modeling of the Idle Speed System

B.1.1 Introduction

Modern SI engines have rather low idle speeds of around 700 rpm in order to minimize fuel consumption and pollutant emission. The drawback is that sudden changes in engine torque (electric loads on the alternator, AC compressor, etc.) may stall the engine. A fast and robust idle-speed control-system (ISCS) is therefore mandatory.

The basic structure of a modern ISCS for SI engines is shown in Fig. B.1. The engine speed is the main input signal, other engine variables (oil temperature, intake-air pressure and temperature, etc.) are used to compute the reference speed at which the engine should run.

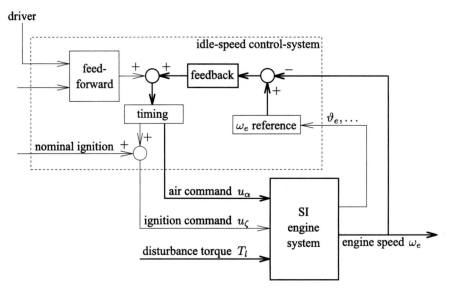

Fig. B.1. Complete ISCS structure; parts discussed in this case study are drawn with thick black lines.

As Fig. B.1 shows, the disturbance torque is compensated in a coordinated way using both the air command and the spark command. This approach permits a very fast reaction to unmeasurable load disturbances. In fact, as shown in Sect. 3.2.1, the ignition input is able to change the engine torque almost instantaneously, albeit at the price of a reduced engine efficiency and higher pollutant-emission levels. Therefore the "timing" block in the ISCS utilizes the spark channel in a first phase in which the air command has not yet produced the desired torque change. As soon as that command starts to produce the desired action, the ignition command is phased back to its nominal value.

In conventional engine systems with mechanical throttle valves the air command actuates a parallel idle speed control bypass valve. In engine systems with electronically controlled throttle valves the idle speed controller usually directly commands that element. Fig. 4.1 shows the layout of such a modern engine control system.

The design of a complete ISCS is well beyond the scope of this case study. The following simplifications are therefore made:

- The engine speed reference value is kept constant (warmed engine, etc.).
- Only the air command path is considered, i.e., only those parts in Fig. B.1 that are drawn with thick lines.
- The injection control system is assumed to be able to keep the air/fuel ratio at its stoichiometric value of $\lambda = 1$.
- The ignition angle is not changed from its value as defined by the nominal ignition map (see Fig. 4.6). It turns out that this is a rather critical simplification for the performance of the system.
- The disturbance torque is assumed not to be measurable. No feedforward control is therefore possible and only the feedback part will be considered.

The main signal measured will be the engine speed. Four other signals (intake manifold pressure, temperature, air mass-flow and load torque) are available for the parameter identification and model validation parts.

The engine used in this case study is a 2.8-liter V6 SI engine with conventional port injection. The ignition and the injection are actuated using the basic maps of the ECU. Only the air loop will be redesigned and implemented in a digital signal processor real-time system. The output of that control system is the throttle valve position command signal u_α.

B.1.2 System Structure

The system to be modeled is shown in Fig. B.2. It is a simplified version of the general SI engine system introduced in Sect. 2.2.1. According to the assumptions listed in the previous section, the fueling and the ignition parts may be omitted. The measured variables and the two input signals are explicitly shown in Fig. B.2.

As introduced in Chapter 2, the engine is modeled as a mean-value volumetric pump with the intake manifold and the intake air throttle upstream of the intake manifold. The torque produced by the engine is linked to the amount of air aspirated by the engine. It acts positively on the constant engine inertia. The engine is assumed to be decoupled from the rest of the power train (this is the worst-case situation because then the inertia is minimal) and the load torque is presumed to act on the engine flywheel directly.

One problem is the generation of known load disturbances for test purposes. The dynamometer cannot be used since its inertia would change the dynamic behavior of the system considerably. A viable alternative is to utilize the engine's alternator as a variable load source by connecting it to an

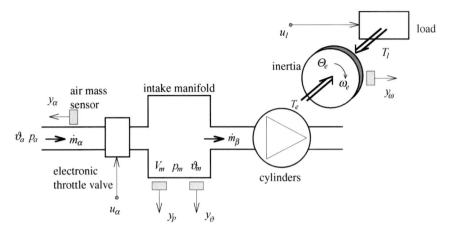

Fig. B.2. Structure of the mean-value engine model used by the ISCS.

electronically controllable load. In this case a load of 0 to 1 kW is used, which yields a load disturbance of up to 15 Nm (more on that will be said in Sect. B.2). The electronic load and alternator subsystem has its own dynamics that will have to be modeled.

The corresponding cause and effect diagram of the combined ISCS is shown in Fig. B.3. As in the main text, while the shaded boxes indicate dynamic modules, the non-shaded boxes represent algebraic modules. The control input is u_α, i.e., the command signal for the electronic throttle valve. The disturbance input u_l, which in the series application is assumed not to be measurable, is the only exogenous disturbance signal. In order to test the performance of the ISCS, u_l has to be designed in this case study.

The measured system variables are:

- y_α: the air mass flowing through the throttle into the intake manifold;
- y_ω: the engine speed;
- y_p: the intake manifold pressure;
- y_ϑ: the intake manifold temperature;
- y_U: the voltage at the alternator; and
- y_I: the current flowing through the alternator.

The intake manifold temperature, $y_\vartheta(t)$, is measured. Its dynamics thus need not be modeled.

B.1.3 Description of Subsystems

Most of the subsystems indicated by boxes in Fig. B.3 have been discussed in Chapter 2. Here those results are adapted to the ISCS problem and a formal description of each block is given to facilitate the subsequent parameter identification and control-system synthesis.

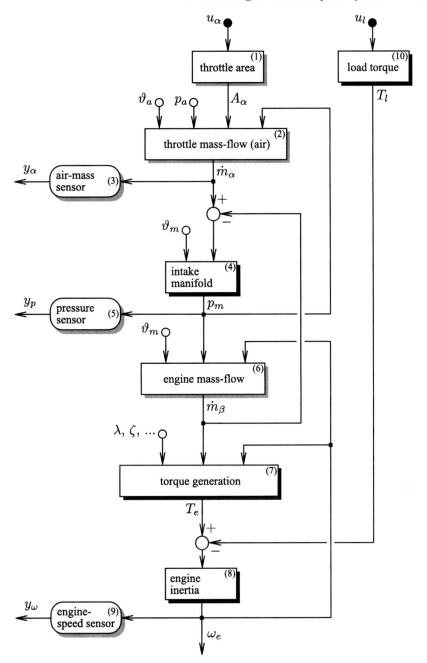

Fig. B.3. Cause and effect diagram of the simplified ISCS engine system.

Throttle Area (1)

Purpose: Models the open area as a function of the input signal $u_\alpha(t)$.

Signals: Input: $u_\alpha(t)$ $[0\ldots1]$ control signal
                 Output: A_α $[m^2]$ open area

Assumption: The dynamics of the actuator may be neglected.

Equations: $\alpha_{th} = \alpha_{th,0} + \left(\frac{\pi}{2} - \alpha_{th,0}\right) \cdot u_\alpha$

$A_\alpha(\alpha_{th}) = \frac{\pi d_{th}^2}{4} \cdot \left(1 - \frac{\cos(\alpha_{th})}{\cos(\alpha_{th,0})}\right) + A_{th,leak}$

where $\alpha_{th,0}$: angle when throttle fully closed,
where $A_{th,leak}$: throttle opening area when $\alpha_{th} = \alpha_{th,0}$

Throttle Mass-Flow (2)

Purpose: Models the air mass flow through throttle as well as the pressure drop.

Signals: Inputs: A_α $[m^2]$ open area
                           p_a $[Pa]$ upstream (ambient) pressure
                           ϑ_a $[K]$ upstream (ambient) temperature
                           $p_m(t)$ $[Pa]$ downstream (manifold) pressure
                 Output: $\dot{m}_\alpha(t)$ $[kg/s]$ air mass-flow

Assumption: Air is a perfect gas, throttle is isenthalpic (see Section 2.3.2).

Equation: $\dot{m}_\alpha(t) = \begin{cases} A_\alpha(t)\frac{p_a}{\sqrt{R\vartheta_a}}\frac{1}{\sqrt{2}} & \text{if } \frac{p_m(t)}{p_a} < 0.5 \\[3mm] A_\alpha(t)\frac{p_a}{\sqrt{R\vartheta_a}}\sqrt{2\frac{p_m(t)}{p_a}\left[1 - \frac{p_m(t)}{p_a}\right]} & \text{else} \end{cases}$

Air-Mass Sensor (3)

Purpose: Produces signal proportional to the air mass flow.

Signals: Input: $\dot{m}_\alpha(t)$ $[kg/s]$ inflowing air mass
                 Output: $y_\alpha(t)$ $[V]$ sensor signal

Assumption: Sensor is a static linear system with known characteristics.

Equation: $y_\alpha(t) = k_a \cdot \dot{m}_\alpha(t)$

Intake Manifold (4)

Purpose: Describes emptying and filling dynamics of the manifold.

Signals: Inputs: $\Delta \dot{m}$ $[kg/s]$ difference of mass flows $\dot{m}_\alpha(t) - \dot{m}_\beta(t)$
                         ϑ_m $[K]$ air temperature in manifold
                Output: $p_m(t)$ $[Pa]$ manifold pressure

Assumption: Lumped-parameter model, isothermal conditions.

Equation: $\frac{d}{dt} p_m(t) = \frac{R \cdot \vartheta_m}{V_m} \cdot [\dot{m}_\alpha(t) - \dot{m}_\beta(t)]$

Pressure Sensor (5)

Purpose: Models the behavior of the intake manifold pressure sensor.

Signals: Input: $p_m(t)$ $[Pa]$ manifold pressure
                Output: $y_p(t)$ $[V]$ sensor output signal

Assumption: Sensor is a static linear system.

Equation: $y_p(t) = k_p \cdot p_m(t)$

Engine Mass-Flow (6)

Purpose: Models the mass flow the engine inducts, which is equal to the air mass flow that exits the intake manifold.

Signals: Inputs: $p_m(t)$ $[Pa]$ manifold pressure
                         ω_e $[rad/s]$ engine speed
                         ϑ_m $[K]$ air temperature in manifold
                         p_e $[Pa]$ exhaust-manifold pressure
                Output: $\dot{m}_\beta(t)$ $[kg/s]$ air mass-flow exiting manifold

Assumptions: No pressure drop in the intake runners.

              The temperature drop from the evaporation of the fuel is canceled out by the heating effect of the hot intake duct walls. The temperature in the cylinder is thus the same as in the intake manifold.

              The air/fuel ratio is constant and equal to stoichiometry.

              Volumetric efficiency depends on manifold pressure as shown in Section 2.3.3.

              The exhaust manifold pressure is constant.

Equations: Mixture entering cylinders:
              $$\dot{m}_e(t) = \frac{p_m(t)}{R\vartheta_m(t)} \cdot \lambda_l(\omega_e(t), p_m(t)) \cdot V_d \cdot \frac{\omega_e(t)}{4\pi}$$

              Air exiting intake manifold:
              $$\dot{m}_\beta(t) = \frac{\dot{m}_e(t)}{1 + 1/(\lambda\,\sigma_0)} \quad \text{with } \sigma_0 \approx 14.7 \text{ (stoichiometric AFR)}$$

              Volumetric efficiency:
              $$\lambda_l(\omega_e, p_m) = \lambda_{lw}(\omega_e) \cdot \eta_{lp}(p_m)$$
              $$\lambda_{lw}(\omega_e) = \gamma_0 + \gamma_1 \cdot \omega_e + \gamma_2 \cdot \omega_e^2$$
              $$\lambda_{lp}(p_m) = \frac{V_c + V_d}{V_d} - \frac{V_c}{V_d} \cdot \left(\frac{p_e}{p_m(t)}\right)^{1/\kappa}$$

Torque Generation (7)

Purpose: Estimation of torque produced by the engine.

Signals: Input: $\dot{m}_\beta(t)$ $[kg/s]$ air mass-flow entering cylinders
                    Output: $T_e(t)$ $[Nm]$ engine torque

Assumptions: The engine may be simplified as a Willans machine, see
               (2.100).
               The delay $\tau_{inj \to TC}$ is calculated according to the definition
               shown below.
               As the engine system is equipped with a fast, research- ori-
               ented control unit, the calculation time $\tau_{calculation}$ and the
               sampling time t_{sample} can be neglected.

Equations:

$$T_e(t) = p_{me} \frac{V_d}{4\pi}$$

$$= (e \cdot p_{m\varphi} - p_{me0f} - p_{me0g}) \cdot \frac{V_d}{4\pi}$$

$$e \cdot p_{m\varphi} = (\eta_0 + \eta_1 \cdot \omega_e) \cdot \frac{H_l \cdot m_\varphi}{V_d}$$

$$= (\eta_0 + \eta_1 \cdot \omega_e) \cdot H_l \cdot \frac{\dot{m}_\varphi \cdot 4\pi}{\omega_e \cdot V_d}$$

$$= (\eta_0 + \eta_1 \cdot \omega_e) \cdot H_l \cdot \frac{\dot{m}_\beta(t - \tau_{inj \to TC}) \cdot 4\pi}{\lambda \cdot \sigma_0 \cdot \omega_e(t - \tau_{inj \to TC}) \cdot V_d}$$

$$p_{me0f} = (\beta_0 + \beta_2 \cdot \omega_e^2) \cdot \frac{4\pi}{V_d}$$

$$p_{me0g} = p_a - p_m$$

$$\tau_{inj \to TC} \approx \tau_{U inj \to TC}$$

$$= \tau_{inj} + \tau_{IVO \to IC} + \tau_{IC \to TC}$$

Remark: As the delay occurs within the feedback loop, it has a strong
                    influence on the dynamic behavior of the idle speed control
                    system.

Engine Inertia (8)

Purpose: Models engine speed dynamics.

Signals: Inputs: $T_e(t)$ $[Nm]$ engine torque
                          $T_l(t)$ $[Nm]$ load torque
                 Output: $\omega_e(t)$ $[rad/s]$ engine speed

Assumption: The engine has a constant inertia and all internal friction is
                 included in T_e.

Equation: $\frac{d}{dt}\omega_e(t) = \frac{1}{\Theta_e}\left[T_e(t) - T_l(t)\right]$

Engine-Speed Sensor (9)

Purpose: Models the behavior of the engine speed sensor.

Signals: Input: $\omega_e(t)$ $[rad/s]$ engine speed
                 Output: $y_\omega(t)$ $[V]$ sensor signal

Assumption: The sensor is linear and very fast.

Equation: $y_\omega(t) = k_\omega \cdot \omega_e(t)$

Load Torque (10)

Purpose: Models the dynamic behavior of the load torque unit.

Signals: Input: $u_l(t)$ $[V]$ desired load-torque signal
                 Output: $T_l(t)$ $[Nm]$ actual load torque

Assumption: The dynamics of the alternator and electronic load may be
                 lumped into one first-order system.

Equation: $\frac{d}{dt}T_l(t) = \frac{1}{\tau_l} \cdot \left[-T_l(t) + k_l \cdot u_l(t)\right]$

   All of these blocks form a mathematical description that approximates the
dynamic behavior of the idle speed system. The engine speed is limited to the
range of interest (say 70 to 120 rad/s). With the exception of the equations
of block [7], all system descriptions are ordinary differential equations that
could be combined into one large vector-valued ODE. However, this is not
recommended for such complex systems. Here the alternative approach of
encoding these equations using a CACSD software tool will be chosen. This
representation is then used for all subsequent analysis and synthesis steps.

B.2 Parameter Identification and Model Validation

These two steps are closely linked. Some measurements are utilized to estimate the as yet unknown parameters of the model, while others are to assess the quality of the model in predicting the system behavior. In this section, the static and the dynamic behavior of the plant is analyzed in two different steps. This helps to cope with the large number of unknown parameters.

B.2.1 Static Behavior

The plant has two inputs (u_α and u_l) and three[1] outputs (y_α, y_p, and y_ω). The static relations between these quantities are investigated first.

The engine operating conditions are chosen such that the corresponding input signals represent the widest possible range of meaningful values.

For the identification of the model, the load torque has to be measured. This is not a trivial task, especially because a dynamometer cannot be used since it would increase the rotational inertia. Instead, the alternator is used to produce the desired load torque. This variable is inferred from two easily measurable quantities; electrical current and voltage.

$$T_{alt} = \frac{P_{alt}}{\omega_e} = \frac{U_{alt}\, I_{alt}}{\omega_e\, \eta_{alt}} = T_l \tag{B.1}$$

The voltage U_{alt}, the current I_{alt} and the alternator speed can be measured. The efficiency η_{alt} is estimated using the efficiency map shown in Fig. B.4. As this figure shows, an assumed average efficiency of 0.6 will produce errors of less than 10% for all operating points of interest here (low speeds, zero to full torque).

The steady-state behavior of the plant is defined by setting all time derivatives to be equal to zero, i.e., the following two non-trivial conditions have to be satisfied

$$\dot{m}_\alpha(t) = \dot{m}_\beta(t) \tag{B.2}$$

$$T_e(t) = T_l(t) \tag{B.3}$$

Some of the system parameters introduced in the last section are not relevant for the steady-state behavior of the system and cannot be derived using the last equations. Moreover, a closed-form solution of the coupled nonlinear static equations following from (B.2) and (B.3) is not possible such that numerical methods will have to be applied.

The set of unknown parameters influencing the steady-state behavior is

$$\hat{\Pi}_s = \{\alpha_{th,0}, A_{th,leak}, p_a, \vartheta_a, R, \vartheta_m, k_p, V_d, V_c, \\ \gamma_0, \gamma_1, \gamma_2, \eta_0, \eta_1, \beta_0, \beta_2, k_\omega, k_l\} \tag{B.4}$$

---

[1] Since the temperature ϑ_m is measured, no model is needed to predict the behavior of this variable.

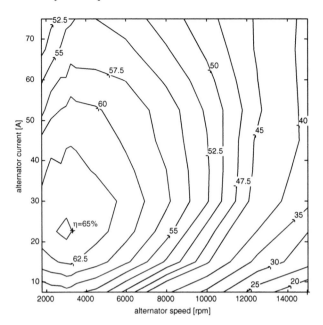

Fig. B.4. Alternator efficiency map. Alternator speed is twice the engine speed.

Some of these parameters can be determined easily (from direct measurements, physical handbooks or manufacturer data sheets) such that only a subset will have to be estimated using the measured data. In principle, this could be done using nonlinear least-squares methods, with the error to be minimized defined by

$$E = \sum_{i=1}^{N} \left[k_1 \left(p_{m,i} - \tilde{p}_{m,i} \right)^2 + k_2 \left(\dot{m}_{\alpha,i} - \tilde{\dot{m}}_{\alpha,i} \right)^2 + k_3 \left(\omega_{e,i} - \tilde{\omega}_{e,i} \right)^2 \right] \quad \text{(B.5)}$$

Here the variables $\{p_{m,i}, \dot{m}_{\alpha,i}, \omega_{e,i}\}$ are the *measured* and $\{\tilde{p}_{m,i}, \tilde{\dot{m}}_{\alpha,i}, \tilde{\omega}_{e,i}\}$ are the *predicted* system outputs, whereas $i = 1, \dots, N$ are the individual measurements taken. The constants k_i are weighting factors that are chosen to normalize the three different parts in the sum.

However, such a direct approach is not very likely to succeed, especially in the noisy environment around SI engines. Usually, a modular approach is more promising, i.e., the first subsystems in the causality chain are identified first using intermediate system outputs. Moreover, whenever possible, the problems have to be reformulated such that equations result that are linear in the parameters (lip), in which case a linear LS estimation becomes possible. This approach is used below starting with the throttle valve as a first element.

Identification of the Throttle Behavior

Since the manifold pressure p_m is smaller than 0.5 bar in all operating conditions relevant for the idle-speed control problem, this simplifies the identification problem because in this case the nonlinear function Ψ is constant.

Second, since the diameter of the throttle d_{th} is introduced in the equations in a nonlinear way, the following substitution has to be used in order to formulate a lip problem

$$\delta = \frac{\pi \cdot d_{th}^2}{4} \qquad (B.6)$$

The following definitions simplify the subsequent derivations

$$\tilde{y}_\alpha = \begin{bmatrix} \dot{\tilde{m}}_{\alpha,1} \\ \dot{\tilde{m}}_{\alpha,2} \\ \cdots \\ \dot{\tilde{m}}_{\alpha,N} \end{bmatrix} \in \Re^{N \times 1}, \; M_\alpha = k_\alpha \cdot \begin{bmatrix} 1 & 1 - \frac{\cos \alpha_{th}}{\cos \alpha_{th0,1}} \\ 1 & 1 - \frac{\cos \alpha_{th}}{\cos \alpha_{th0,2}} \\ \cdots & \cdots \\ 1 & 1 - \frac{\cos \alpha_{th}}{\cos \alpha_{th0,N}} \end{bmatrix} \in \Re^{N \times 2} \; (B.7)$$

where $k_\alpha = \frac{p_a}{\sqrt{R\vartheta_a}} \frac{1}{\sqrt{2}}$.

The system behavior as expressed in the equations of blocks (1) and (2) can now be compactly written as follows

$$\tilde{y}_\alpha = M_\alpha \cdot [A_{th,leak}, \delta]^T \qquad (B.8)$$

The solution $[A_{th,leak}, \delta]^T$ of this equation which minimizes the error

$$E_{LS} = \sum_{i=1}^{N} \left(\dot{m}_{\alpha,i} - \dot{\tilde{m}}_{\alpha,i} \right)^2 \qquad (B.9)$$

between the measured air mass flow $\dot{m}_{\alpha,i}$ and the predicted air mass flow $\dot{\tilde{m}}_{\alpha,i}$ will then be

$$[A_{th,leak}, \delta]^T = \left[M_\alpha^T \cdot M_\alpha \right]^{-1} \cdot M_\alpha^T \cdot y_\alpha \qquad (B.10)$$

where y_α is the measurement vector that collects all measured values of $\dot{m}_{\alpha,i}$.

Once δ is identified with (B.10), the substitution (B.6) can be inverted and the diameter of the throttle thus becomes

$$d_{th} = 2 \cdot \sqrt{\frac{\delta}{\pi}} \qquad (B.11)$$

The results of this identification can be found in Fig. B.5.

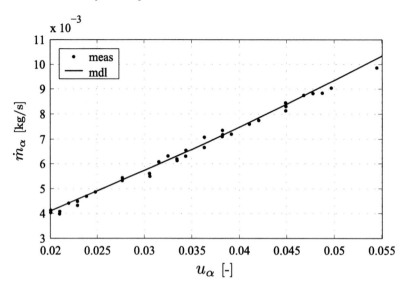

Fig. B.5. Comparison of measured and modeled values of the mass flow $\dot{m}_\alpha$ after the identification of the throttle behavior.

Identification of the Volumetric Efficiency

The coefficients γ_i of the speed-dependent part of the volumetric efficiency are found using the same approach (notice that in steady-state operation the equation $\dot{m}_{\beta,i} = \dot{m}_{\alpha,i}$ is fulfilled). The starting point is given by the equations of block [6]. If the air mass flow $\dot{m}_{\alpha,i}$, the intake manifold pressure $p_{m,i}$, and the engine speed $\omega_{e,i}$ are known (i is the counter that indicates the i-th measurement), the volumetric efficiency can be written as

$$\gamma_0 + \gamma_1 \cdot \omega_{e,i} + \gamma_2 \cdot \omega_{e,i}^2 = \psi(\omega_{e,i}, p_{m,i}, \dot{m}_{\alpha,i}) \qquad \text{(B.12)}$$

where

$$\psi(\omega_{e,i}, p_{m,i}, \dot{m}_{\alpha,i}) =$$

$$\frac{R\vartheta_m}{p_{m,i}} \cdot \left(\frac{V_c + V_d}{V_d} - \frac{V_c}{V_d} \left(\frac{p_a}{p_{m,i}} \right)^{1/\kappa} \right)^{-1} \cdot \dot{m}_{\alpha,i} \cdot \frac{1 + \sigma_0}{\sigma_0} \cdot \frac{4\pi}{\omega_{e,i} V_d} \qquad \text{(B.13)}$$

Analogously to the definitions introduced above for the throttle parts, the following new variables are defined here

$$\tilde{y}_\eta = \begin{bmatrix} \psi_1 \\ \psi_2 \\ \dots \\ \psi_N \end{bmatrix} \in \Re^{N \times 1}, \quad M_\eta = \begin{bmatrix} 1 & \omega_{e,1} & \omega_{e,1}^2 \\ 1 & \omega_{e,2} & \omega_{e,2}^2 \\ \dots & \dots \\ 1 & \omega_{e,N} & \omega_{e,N}^2 \end{bmatrix} \in \Re^{N \times 3} \qquad \text{(B.14)}$$

The solution of this equation, which minimizes the prediction error in the LS sense, is given by

$$\left[\gamma_0\ ,\ \gamma_1\ ,\ \gamma_2\right]^T = \left[M_\eta^T \cdot M_\eta\right]^{-1} \cdot M_\eta^T \cdot \tilde{y}_\eta \qquad (B.15)$$

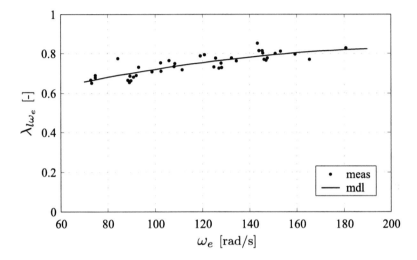

Fig. B.6. Comparison of the measured and the modeled speed-dependent factor λ_{lw_e} of the volumetric efficiency λ_l.

Identification of the Torque Generation Module

The parameters of this subsystem can be identified using the outputs of the previously identified subsystems as inputs. However, this approach yields less precise results than the one used below. In this alternative approach the throttle valve command u_α is used as input and the following error criterion is minimized

$$E = \sum_{i=1}^{N} \left[(\omega_{e,i} - \tilde{\omega}_{e,i})^2\right] \qquad (B.16)$$

The variable $\omega_{e,i}$ is the measured and the variable $\tilde{\omega}_{e,i}$ is the engine speed predicted by the model. The index $i = 1, \ldots, N$ indicates the individual measurements.

B.2.2 Dynamic Behavior

Several parameters cannot be determined using steady-state measurements. This set of parameters is

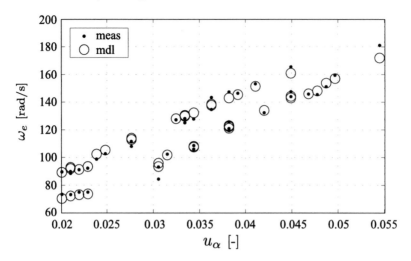

Fig. B.7. Comparison of measured and modeled correlation between the throttle signal (u_α) and the resulting engine speed (ω_e).

$$\Pi_d = \{V_m, \Theta_e, \tau_{inj \to TC}, \tau_l\} \tag{B.17}$$

The time constant of the load can be estimated very well by separate experiments or theoretical considerations. The remaining unknown "dynamic" parameters are therefore the intake manifold volume V_m, the engine inertia Θ_e and the transport delay from injection to torque center $\tau_{inj \to TC}$. These parameters are identified using measured frequency responses. For both of the following experiments, the input signal is the throttle valve command u_α. In a first experiment the engine is connected to a dynamometer and the engine speed is kept constant. The measured output signal is the manifold pressure p_m.

The only parameter having any influence on the transfer function $u_\alpha \to p_m$ is the volume of the manifold V_m. This parameter is adapted until the measured and the predicted frequency responses match well. The result shown in Fig. B.8 indicate a good agreement up to a frequency of 10 rad/s.

The delay from injection to torque center ($\tau_{inj \to TC}$) is identified according to (3.5). Figure B.9 shows the phase plotted linearly over the excitation frequency. At higher frequencies, at which the influence of the finite-dimensional dynamic subsystems are no longer relevant, the phase rolls off linearly. The gradient of this line is proportional to the unknown delay.

A second frequency analysis is carried out with the engine decoupled from the dynamometer. This allows the measurement of the transfer function from the throttle angle to the engine speed.

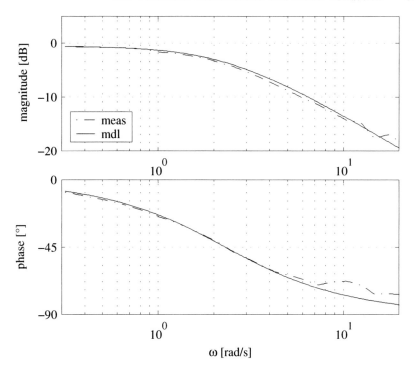

Fig. B.8. Comparison of the measured and modeled transfer function from the throttle position to the pressure in the intake manifold ($\omega_e = 105$ $[rad/s]$, $T_l = 0$ $[Nm]$).

With the transport delay identified, the phase in the Bode plot can be corrected for that effect. Figure B.10 shows the corresponding transfer function from the throttle angle to the speed of the engine.

The result of this identification is evident in Fig. B.11, where the measured dynamic response of the engine is compared to the results of the simulation.

B.2.3 Numerical Values of the Model Parameters

The results of the dynamic and the static parameter identification process is a set of model parameters. For the sake of completeness these parameters are all listed below.

Throttle Area (1)

Parameters:				
	Angle	$\alpha_{th,0}$	$[°]$	$\{7.9\}$
	Diameter	d_{th}	$[m]$	$\{58.7 \cdot 10^{-3}\}$
	Leakage area	$A_{th,leak}$	$[m^2]$	$\{5.3 \cdot 10^{-6}\}$

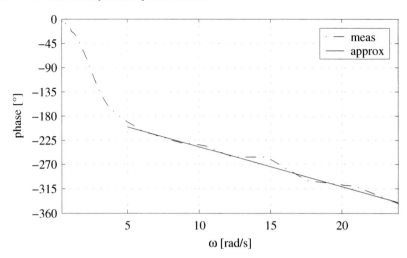

Fig. B.9. Phase response from the throttle (u_α) to the engine speed (ω_e). By plotting the phase linearly over the excitation frequency, the transport delay can be identified directly ($u_{\alpha 0} = 0.023[-]$, $T_l = 0[Nm]$)

Throttle Mass-Flow (2)

Parameters:	Gas constant air	R [J/kg K] {287}
	Ambient temperature	ϑ_a [K] {298}
	Ambient pressure	p_a [Pa] {$0.98 \cdot 10^6$}

Air-Mass Sensor (3)

| Parameter: | Static gain | k_a [V s/kg] {1} |

Intake Manifold (4)

| Parameters: | Volume intake manifold | V_m [m^3] {$5.8 \cdot 10^{-3}$} |
| | Temperature air in manifold | ϑ_m [K] {340} |

Pressure Sensor (5)

| Parameter: | Gain | k_p [V/Pa] {$0.1 \cdot 10^{-3}$} |

Engine Mass-Flow (6)

Parameters:	Coefficient 1	γ_0 [-] {0.45}
	Coefficient 2	γ_1 [s] {$3.42 \cdot 10^{-3}$}
	Coefficient 3	γ_2 [s^2] {$-7.7 \cdot 10^{-6}$}
	Displacement	V_d [m^3] {$2.77 \cdot 10^{-3}$}
	Compression volume at TDC	V_c [m^3] {$0.277 \cdot 10^{-3}$}
	Isentropic exponent air	κ [-] {1.35}
	Back pressure exhaust manifold	p_e [Pa] {$1.08 \cdot 10^5$}

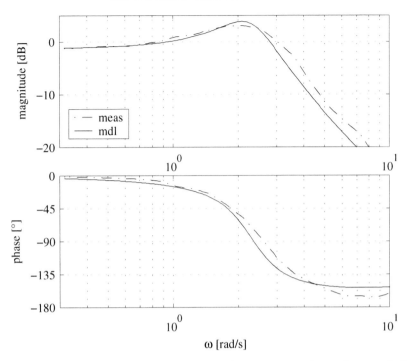

Fig. B.10. Transfer function from the throttle position to the engine speed. The phase is corrected by the amount of transport delay $\varphi = \omega \cdot \tau_{inj \rightarrow TC}$ ($u_\alpha = 0.23$ $[-]$, $T_l = 0$ $[Nm]$).

Torque Generation (7)

Parameters: Willans parameter 1: η_0 [J/kg] $\{0.16\}$
                Willans parameter 2: η_1 [J s/kg] $\{2.21 \cdot 10^{-3}\}$
                Willans parameter 3: β_0 [Nm] $\{15.6\}$
                Willans parameter 4: β_2 [Nm s^2] $\{0.175 \cdot 10^{-3}\}$
                Transport delay $\tau_{inj \rightarrow TC}|_{\omega_e = 90 rad/s}$ [s] $\{0.125\}$

Engine Inertia (8)

Parameter: Engine inertia Θ_e $[kg/m^2]$ $\{0.2\}$

Engine-Speed Sensor (9)

Parameter: Gain k_ω [V s/rad] $\{0.1\}$

Load Torque (10)

Parameters: Gain k_l [Nm/V] $\{1\}$
               Time constant τ_l [s] $\{0.02\}$

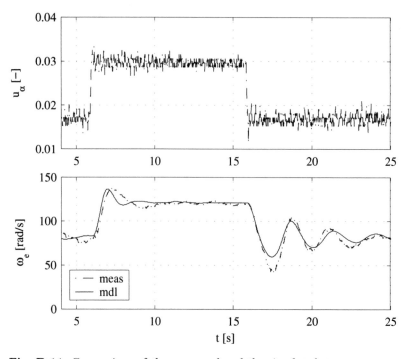

Fig. B.11. Comparison of the measured and the simulated step responses.

B.3 Description of Linear System

To design the idle speed control system a linear system representation is needed. The main problem encountered in this step is that the system derived in the last sections contains a substantial delay. In the time domain, such delays cannot be represented exactly by any finite-dimensional system description such that some sort of approximation will be needed. The key idea is to use Padé all-pass elements to approximate these delays

$$e^{-s \cdot \delta} \approx \frac{\sum_{j=0}^{n_{apx}} a_j \cdot (-s \cdot \delta)^j}{\sum_{j=0}^{n_{apx}} a_j \cdot (s \cdot \delta)^j} \qquad a_j = \frac{(2 \cdot n_{apx} - j)! \cdot n_{apx}!}{j! \cdot (n_{apx} - j)! \cdot (2 \cdot n_{apx})!} \qquad \text{(B.18)}$$

Usually, a small order n is sufficient, while for larger system orders other, numerically more reliable forms are recommended. Matlab offers appropriate software tools for those computations (see `help pade`).

Notice that Padé approximations introduce non-minimum phase system zeros which are known to limit the achievable closed-loop bandwidth. This is not a problem of the chosen delay approximation but a limitation imposed by the unavoidable IPS delay itself.

For this case study a sixth-order Padé element is chosen to approximate the IPS delay ($\tau_{inj \to TC}$) in block [7]. The variable delay is approximated by the nominal one

$$\tau_{inj \rightarrow TC, 0} = \frac{\Delta}{\omega_0} \tag{B.19}$$

where ω_0 is the reference idle speed. Figure B.12 shows the effect of these approximations by comparing the step responses of the system model with variable delay and with fixed-delay Padé approximations.

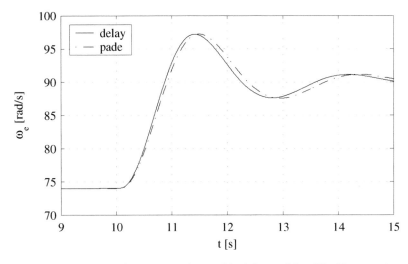

Fig. B.12. Comparison of systems with variable delay and fixed Padé approximation by their step responses.

The system equations can be linearized using a purely numerical approach by approximating the partial derivatives by finite differences. Again, Matlab offers appropriate software tools for that step (see `help linmod2`). Notice that the system can be linearized (in closed form or numerically) only if its differential equations are sufficiently "smooth" (continuous first partial derivatives in all variables) around the nominal point.

When the numerical approach is used, the differences have to be chosen small enough to avoid nonlinear effects and, at the same time, large enough to avoid numerical problems. Some iterations are therefore usually necessary to find the optimum differences.

Fig. B.13 shows the Bode diagram of the control disturbance channel. Also shown is the influence of the IPS delay as approximated by a sixth-order Padé element. Its contribution to the phase lag of the system has a substantial influence on the control system design process.

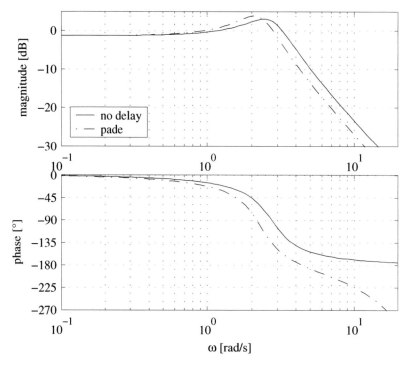

Fig. B.13. Comparison of Bode diagrams of the control channel of the linearized system (with $u_\alpha = 0.02$ [-] and $u_l = 0$ [-]) using Padé approximation (dashed) and without any delays (solid).

B.4 Control System Design and Implementation

In this section a control system is designed using the model derived above. The specifications that have to be met by this control system are listed below:

- The reference speed is constant at 90 rad/s. No substantial changes in this reference signal have to be expected.
- Load disturbances are the main phenomenon for which the controller has to be tuned.
- The steady-state error caused by a constant load disturbance has to be zero.
- The engine speed may not fall below 60 rad/s in order to prevent engine stall.
- The torque ripple induced by the reciprocating engine behavior may not excite the closed-loop system.
- The settling time to load disturbances has to be around 5 s.
- Overshoot of engine speed after a load disturbance is acceptable but should be limited to 10% of the reference value.

- A minimum return difference of 0.5 has to be achieved in order to cope with the large model uncertainties and neglected nonlinearities.

Obviously, to satisfy these specifications the controller has to include an integral part. The engine has a very low damping rate around $2.5\,rad/s$ (see Fig. B.15). This resonance turns the design of the idle speed controller into a nontrivial task.

The controller is designed using an observer-based pole-placement approach. In order to avoid a steady-state error, the state feedback controller is augmented by an integrator. Figure B.14 shows a detailed signal flow chart of the controller.

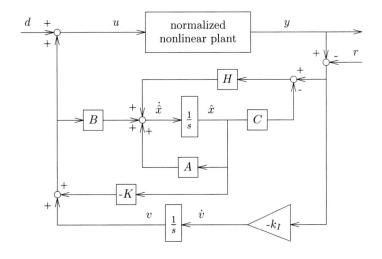

Fig. B.14. Detailed signal-flow chart of the idle-speed controller.

The differential equations of the augmented observer-based state feedback controller can be written as follows:

$$\begin{bmatrix} \dot{\hat{x}} \\ \dot{v} \end{bmatrix} = \begin{bmatrix} A - BK - HC & B \\ 0 & 0 \end{bmatrix} \begin{pmatrix} \hat{x} \\ v \end{pmatrix} + \begin{bmatrix} 0 & H \\ 0 & -k_I \end{bmatrix} \begin{pmatrix} d \\ y - r \end{pmatrix} \quad \text{(B.20)}$$

$$[u] = \begin{bmatrix} -K & 1 \end{bmatrix} \begin{pmatrix} \hat{x} \\ v \end{pmatrix} + \begin{bmatrix} 1 & 0 \end{bmatrix} \begin{pmatrix} d \\ y - r \end{pmatrix} \quad \text{(B.21)}$$

Figure B.15 shows the Bode diagrams of the plant $P(s)$, the controller $C(s)$ and the open loop $L(s) = C(s) \cdot P(s)$. The controller damps the frequencies with a critical phase shift and corrects the phase where the amplification of the gain is critical.

The gain margin of the implemented controller is 2.3 and the phase margin is 48°.

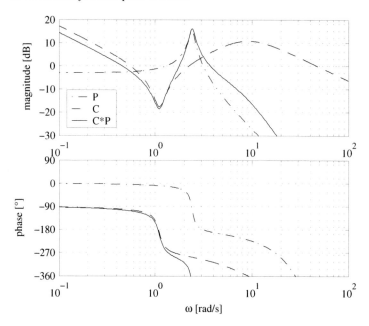

Fig. B.15. Bode plots, plant $P(s)$, controller $C(s)$ and open loop gain $L(s) = C(s) \cdot P(s)$.

The last step is to test the closed-loop system on an engine test bed. Figure B.16 shows the results. The controller fulfills all the required specifications, but cannot completely eliminate the steady-state oscillations of the system. Adding a second fast control loop that utilizes the spark advance as an additional control action would considerably reduce these oscillations.

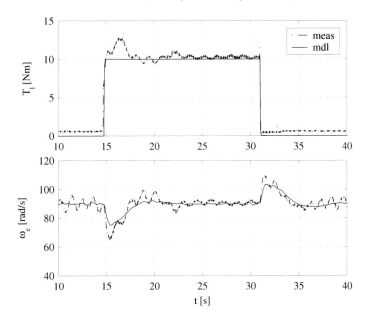

Fig. B.16. Comparison of the closed-loop behavior (solid = simulated and dash-dot = measured).

C

Combustion and Thermodynamic Cycle Calculation of ICEs

There exists a vast amount of literature on the calculation of the thermody-namic cycle of ICEs. Excellent introductions are given, for instance, in [97] and [163]. Readers interested in more details are referred to these books and the references indicated there. In the following sections only a short summary is given in order to emphasize some control-oriented aspects.

C.1 Fuels

A large variety of different fuels can be used in combustion engines. The basic properties of the fuels, such as the range of their boiling-temperatures, their ignition temperatures, etc., determine which combustion process is the most suitable for its use in an ICE. Table C.1 lists some important properties of Diesel, gasoline and methane. Crude oil is the source of many hydrocarbons. They range from natural gas (mostly methane) down to tar, exhibiting vastly different volatility values. The cracking process in the refinery transforms the crude oil mainly to Diesel and gasoline. The ratio between these two products can be chosen within certain limits, depending on the composition of the crude oil.

Table C.1. Properties of Diesel, gasoline and methane.

fuel	density at $20\,^{\circ}C$ $[\frac{kg}{dm^3}]$	lower heating value H_l $[MJ]$	H/C ratio [-]	boiling temperature $[^{\circ}C]$	ignition temperature $[^{\circ}C]$
Diesel	0.81 ... 0.86	42.5	~ 2	150 ... 360	250
gasoline	0.73 ... 0.78	42.7	~ 2	25 ... 210	400
methane	0.000835	50	4	-162	650

Since Diesel evaporates at rather high temperatures, it is very difficult to obtain an ignitable mixture of Diesel and air under ambient conditions. An injection technique producing a very fine spray or a high-temperature pre-mixing chamber is necessary. Today's injection systems use pressures of more than 2000 bar. Such extreme pressures cause the injection pulses to be very short and the fuel to be practically "atomizing" when it is injected into the cylinder. Since temperature and pressure conditions are at self-ignition level for Diesel fuel, the droplets in the outer part of the spray burn immediately. The inner part, however, still needs time to mix with the surrounding air (or, using another notion, burns as soon as it advances to the surface of the spray), see Fig. 2.61. The Diesel combustion thus is split into a premixed (almost instantaneously burning) part and a slowly burning diffusion part. Since the combustion of Diesel takes place on the surface of the spray, it is necessary to run the engine under lean conditions in order to always have enough air available. Therefore, large amounts of excess air pose no problem but is even desirable, enabling the Diesel engine to run un-throttled. This is, with respect to efficiency, an inherent advantage over the spark-ignited, fully-premixed charge engine.[1] However, excess air and locally high temperatures increase the amount of produced NO_x. Therefore, too high air-to-fuel ratios and too high local temperatures have to be avoided by exhaust-gas recirculation. Another advantage over SI engines are the higher possible compression ratios. Since the combustion is triggered by the injection, engine knock (spontaneous self-ignition of a homogeneous mixture at high temperature and pressure, see Section 4.2) is not a problem and thus the compression ratio is only restricted by mechanical limits.[2]

Gasoline evaporates rapidly even in ambient conditions and therefore is ideally suited for premixed combustion processes not necessitating a direct injection into the cylinder at high pressure. A low-pressure injection into the intake port of the engine suffices to produce a very homogenous mixture of air and fuel. Since gasoline self-ignites at higher temperatures than Diesel, an artificially produced spark is needed to trigger the combustion. Even if a strong enough spark is available, the mixture does not ignite if the air-to-fuel ratio λ exceeds rather narrow limits around stoichiometry ($\lambda = 1$): The upper limit lies at approximately $\lambda = 1.4$ (lack of fuel molecules), whereas for values below $\lambda < 0.7$ there is a lack of oxygen which results in large amounts of unburnt hydrocarbons and CO. The colder the engine, the closer to stoichiometry these limits get. A cold engine hardly runs at $\lambda > 1.1$, whereas the rich limit is less critical. Given these bounds and the fact that a stoichiometric charge is required for reasons of emissions, see Sections 2.7 and 2.8, the engine torque cannot be controlled by the amount of fuel injected but has to be regulated

---

[1] Note that with respect to peak power, this is a drawback: The oxygen available in the cylinder is not fully used.

[2] Highly turbocharged Diesel engines already reach peak cylinder-pressures of more than 200bar.

by the air mass in the cylinders. A simple and almost uniquely used way of controlling this is to lower the intake-manifold pressure by closing a throttle and thus reducing the density of the air aspirated by the cylinders. This however reduces the engine's efficiency by introducing additional pumping losses: The engine has to pump the mixture from a lower (in the intake manifold) to a higher pressure level (at least ambient) in the exhaust manifold. Gasoline engines can be operated on cylinder-average air-to-fuel ratios higher than 1.4, but only if the cylinder charge is stratified: A zone with stoichiometric mixture surrounds the spark plug, enclosed by pure air (in the ideal case).[3]

Methane accounts for the largest fraction of natural gas (NG). It is ideally suited for premixing because no evaporation has to take place. Due to its high hydrogen-to-carbon ratio, its combustion produces less CO_2 for the same energy output. Similarly to gasoline, methane requires an external ignition due to its high self-ignition temperature. A general disadvantage of gaseous fuels is the fact that when they are injected outside the cylinder the air charge is reduced: Having roughly the same specific volume, it supplants a significant amount of air, which reduces the achievable full-load torque. Due to its high self-ignition temperature, methane is very knock-resistant, which enables high compression ratios and thus partly compensates for the drawback described above. The late light-off in the catalytic converter due to the higher required temperatures, however, makes attaining a high efficiency of the after-treatment system a challenge.

C.2 Thermodynamic Cycles

Assuming perfect gas, ideal valves (fast opening and closing, immediate pressure equalization) and adiabatic conditions, ideal engine cycles can be derived. Such a cycle consists of state changes with either constant volume (isochoric), pressure (isobaric) or entropy (isentropic, $s = p \cdot v^\kappa$) only. It is thus advantageous to plot the engine cycles in a double-logarithmic pV-diagram, since by doing so any polytropic (even more general; $p \cdot v^n = const.$) process is represented by a straight line. Subsequently, the three fundamental ideal cycles, as depicted in Figure C.1, are outlined:

Otto Cycle

This cycle represents an ideal premixed combustion. As described above, a homogeneous mixture of air and fuel (gasoline or gas) burns very fast. Therefore, an isochoric (and thus instantaneous) combustion at TDC is assumed.

---

[3] Note the similarity to the Diesel combustion. In fact, there are currently many new concepts trying to combine the advantages of both combustion processes.

Diesel Cycle

The combustion process in early Diesel engines was slow and almost completely diffusion-controlled, i.e. there was almost no premixed combustion at
all[4] This can be best approximated by an isobaric state-change.

Seiliger (or Dual) Cycle

The Seiliger cycle combines Otto and Diesel cycles by attributing an isochoric
as well as an isobaric part to every combustion process. Depending on the
ratio of these two parts, it allows a rather good approximation of real gasoline
and Diesel combustion.

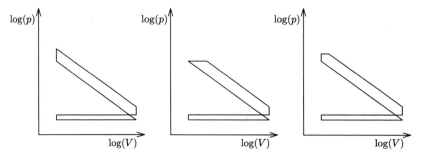

Fig. C.1. Otto, Diesel, and Seiliger cycle in double-logarithmic representation.

C.2.1 Real Engine-Cycle

If real engine cycles have to be simulated or analyzed, a more complex treatment is necessary. According to (2.3) and (2.4), the balances of inner energy
and mass yield the following two equations

$$\frac{dm_c}{d\phi} = \frac{dm_{ent\ cyl}}{d\phi} - \frac{dm_{exit\ cyl}}{d\phi} \tag{C.1}$$

$$\frac{dU_c}{d\phi} = m_\varphi H_l \frac{dx_B}{d\phi} - p_c \frac{dV_c}{d\phi} - \frac{dQ_w}{d\phi} + \frac{dH_{ent\ cyl}}{d\phi} - \frac{dH_{exit\ cyl}}{d\phi} \tag{C.2}$$

where the subscript *ent cyl* describes all quantities which enter the cylinder,
whereas *exit cyl* indicates the exiting quantities. The inner energy U_c is calculated according to (2.6) and $x_B(\phi)$, as defined in (3.67), is the already burnt
mass-fraction of fuel and thus its derivative represents the normalized burn

---

[4] Modern Diesel engines use pilot injections before the main injection in order to
increase the premixed part of the combustion.

rate. Multiplied by the total fuel mass injected per cycle and the fuel's heating value, this gives the gross heat-release rate.

The equations for internal energy and enthalpy flows can be found in Section 2.3.1, and the volume of the cylinder is given by (3.66). The mass flows through intake and exhaust valves can be calculated on the basis of valve-lift curves provided by the manufacturer of the engine and using (2.10) for the isothermal orifice. The modeling of the heat losses to the cylinder walls is more complicated, necessitating several approximations. Usually, convective heat losses according to the following equation are assumed

$$\frac{dQ_w}{dt} = \alpha \cdot A_w \cdot (\vartheta_c - \vartheta_w) \tag{C.3}$$

where A_w indicates the area of contact between the hot gases in the cylinder and the walls. This area changes with the crank angle due to the moving piston. An average representative temperature of the cylinder wall has to be chosen. According to Fig. 2.48, an appropriate value for fully warmed up engines lies around $140\,°C$. The heat-transfer coefficient α is often calculated using Woschni's equation

$$\alpha = 130 B \vartheta_c^{-0.53} p_c^{0.8} B^{-0.2} \left[C_1 c_m + C_2 \frac{V_d \vartheta_1}{p_2 V_1} (p_c - p_0) \right] \tag{C.4}$$

where B is the cylinder bore of the engine and c_m, as described in Subsection 2.5.1, is the mean piston speed. Index 1 indicates the state of the gas at intake valve closing, whereas p_0 is the pressure which arised if the engine was motored (not fired). The parameters C_1 and C_2 are determined experimentally. If no swirl is assumed, the following approximations can be used

$$C_1 = 6.18 \tag{C.5}$$
$$C_2 = 0.00324 \quad [m/sK] \tag{C.6}$$

Other approximations can be found in [163].

Complementing the above equations with the ideal gas law (2.5) yields three equations for mass, temperature and pressure. The temperature-dependent gas properties of the mixture, which were first given in [124] and [136], are easily integrated in Matlab^{TM} code using CHEP [63].

There are basically two different ways these equations can be used. First, if the in-cylinder pressure is measured, the heat release and thus the burned mass fraction $x_B(\phi)$ can be derived. Second, given an approximation of $x_B(\phi)$, a complete engine cycle can be simulated. In the subsequent paragraphs, these two applications are outlined.

Cylinder-Pressure Analysis

The mechanical work transferred to the piston during one cycle, i.e. the indicated work, can be written as:

$$W_i = - \int_{cycle} p_c \cdot dV_c$$

Note that for the calculation of the work, a constant offset of the cylinder pressure p_c is irrelevant (which is not the case for the derivation of the heat release, see below). Analogously to (2.93), the mean indicated pressure is then calculated as

$$p_{mi} = \frac{1}{V_d} \cdot W_i = -\frac{1}{V_d} \cdot \int_{cycle} p_c \cdot dV_c \qquad (C.7)$$

In order to obtain the mean effective pressure, the mean friction pressure, which has to be estimated (for example by the model given in (2.112)), has to be subtracted from the indicated one:

$$p_{me} = p_{mi} - p_{mf} \qquad (C.8)$$

Inverting (2.93) finally yields the engine torque.

In order to derive the heat release, only the high pressure part, i.e. the part between intake valve closing (IVC) and exhaust valve opening (EVO) has to be considered. If the low-pressure part between EVO and IVC has to be evaluated as well, the pressures in the intake and exhaust manifold have to be provided (preferably as crank-angle resolved measurements). Using the pressure differences, the gas transferred to and from the cylinder can be calculated and a consistency check with the measured values performed.

Usually, several high-pressure cycles are averaged to gain a representative pressure trace.[5] The pressures at the beginning and at the end of the cycle have to match. The absolute level of each individual pressure trace is usually unknown because the (fast) sensors used provide relative information only and in addition often exhibit a significant drifting behavior. Therefore, the value of the intake pressure at BDC before compression[6] is often used to determine the absolute level of the cylinder pressure.

In addition to the data which can easily be measured, such as engine speed, amounts of fresh air and fuel or spark advance, certain inputs to the cylinder-pressure analysis have to be estimated in a first step, i.e., the relevant cylinder-wall temperature and the amount of residual exhaust-gas. If not enough physical insight is available to assume reasonable values, it is helpful to enlarge the system boundaries and derive the estimations via steady-state

---

[5] The complex gas movements in the cylinder during intake and injection lead to spatial variations of the air-to-fuel ratio and other properties from cycle to cycle. This results in a variability of the (peak) cylinder pressure. The averaged pressure trace therefore represents a "statistical mean" cycle which is useful for the calculation of the average heat release but from which no information on peak pressures and phenomena such as engine knock may be derived.

[6] Since the intake valve is still open and the movement of the piston is minimal at BDC, complete pressure equalization between cylinder and intake manifold can be assumed.

equilibrium considerations. The stationary heat flow from the gas in the cylinder to the cylinder walls, for example, has to be equal to the heat flow from the cylinder walls to the coolant. The increase in temperature of the coolant and its flow rate is easily accessible and thus, by backward calculation, the *representative* cylinder-wall temperature can be estimated.

Cycle Simulation

Using the approximation for the heat release (see Section C.2.2 below) and estimated initial conditions for cylinder charge, amount of residual exhaust-gas and others, the temperature and pressure traces can be calculated. Then, given the pressure trace over the complete engine cycle and (simulated) values for the pressures in the intake and exhaust receivers, the mass transferred to and from the cylinder, indicated engine torque and amount of residual exhaust gas can be derived.

Of course, the state at the end of the cycle has to match the state at the beginning. Using the new, calculated values as refined initial conditions and iterating the calculation eventually leads to a coherent engine cycle, enabling an accurate prediction of mass flows, engine torque and cylinder-charge composition.

C.2.2 Approximations for the Heat Release

Instead of using a heat release which is calculated by means of a pressure analysis as described above, an artificial approximation can be used as input to the full cycle calculation. One widely used approach is the Vibe [208] heat-release function, which is especially convenient for SI engines:

$$x_B(\phi) = \frac{m_{\varphi,burnt}(\phi)}{m_\varphi} = 1 - e^{c \cdot \left(\frac{\phi - \phi_\zeta - \phi_{ign.\,del}}{\Delta \phi_b} \right)^{m_v + 1}} \tag{C.9}$$

In the form presented here, it defines the fraction of the total fuel-mass injected per cycle that has already burnt up to a certain crank angle ϕ.[7] Therefore, it strictly[8] increases from zero to unity. The four adjustable parameters are the ignition angle ϕ_ζ, the ignition delay $\phi_{ign.\,del}$, the combustion duration $\Delta \phi_b$ and the shape parameter m_v. The parameter c is chosen to guarantee a 99.9% conversion at the end of combustion, i.e., $\phi = \phi_\zeta + \phi_{ign.\,del} + \Delta \phi_b$, which usually holds for well-timed combustion.

The Vibe parameters can be obtained by fitting the Vibe function to a heat release calculated by a pressure analysis. The Vibe function is very useful for investigating the influence of varying combustion parameters. An illustrative example is the variation of the spark advance and its effect on

---

[7] Often, the first derivative with respect to ϕ, multiplied by the total mass of fuel m_φ and the fuel's heating value, is used, which yields the gross heat-release rate as it is used in (C.2).

[8] The combustion never intermittently stops.

the brake-specific fuel consumption (bsfc) (all other parameters are kept constant); Varying the spark timing moves the center of combustion ($x_B = 0.5$) and therefore affects the engine's efficiency: If it is chosen too early, the combustion counteracts the compression and if it is late, some of the fuel is burned late in the expansion stroke and can thus only partly be converted into mechanical work (but leaves the cylinder as enthalpy in the exhaust gas). The center of combustion has been varied between 10° crank angle before TDC and 30° crank angle after TDC. The resulting curves of $x_B(\phi)$ are shown in Figure C.2.

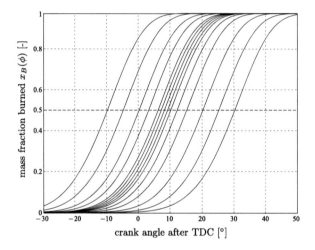

Fig. C.2. Course of the burned mass-fraction for variations of the center of combustion ($x_B = 0.5$) between 10° crank angle before TDC and 30° crank angle after TDC.

Figure C.3 depicts the corresponding variations in the cylinder pressure, plotted over the cylinder volume in linear and in double-logarithmic representation. The advantage of the latter is obvious: The polytropic compression and expansion can be seen clearly (which is not the case for the linear representation).

The resulting brake-specific fuel consumption (bsfc) as a function of the center of combustion is shown in Figure C.4. The curve of the bsfc shows a minimum at 8° crank angle after TDC and confirms the often made statement that the center of combustion has to lie at approximately 10° crank angle after TDC in order to achieve maximum efficiency [97].

C.2.3 Variations of the Vibe Parameters: Csallner Functions

Of course, a change of the combustion, i.e. of the shape of the heat-release curve, affects the boundary conditions of the combustion, i.e., exhaust temperature and pressure, and vice versa. In [49], factors describing the influence

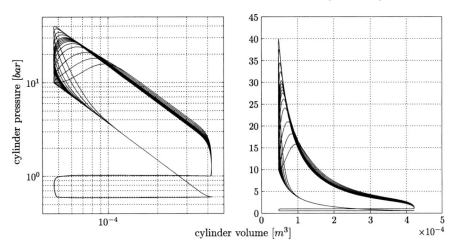

Fig. C.3. Cylinder-pressure variations due to changing combustion-center placement (by spark timing only).

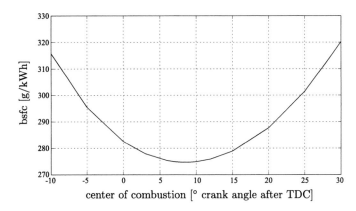

Fig. C.4. Influence of the center of combustion on the brake-specific fuel consumption.

of the most important operating-point parameters on the Vibe function are presented. The three Vibe parameters

$$\Delta\phi_{ign.\,del} = \Delta\phi_{ign.\,del_{ref}} \cdot \prod_i f_i \tag{C.10}$$

$$\Delta\phi_b = \Delta\phi_{b_{ref}} \cdot \prod_i g_i \tag{C.11}$$

$$m_v = m_{v_{ref}} \cdot \prod_i h_i \tag{C.12}$$

are each adjusted by six factors describing the influence of the following operating-point parameters:

- ϕ_ζ spark advance
- λ air-to-fuel ratio
- $\vartheta_{c,300}$ temperature in cylinder at 300°CA after gas exchange TDC
- $p_{c,300}$ pressure in cylinder at 300°CA after gas exchange TDC
- x_{egr} residual gas fraction
- n_e engine speed in rpm

The individual functions are presented in the following; for more details or explanations refer to the original literature.

$$f_\zeta = \frac{70 + \phi_\zeta}{70 + \phi_{\zeta_{ref}}}$$

$$f_\lambda = \frac{\lambda(2.2\lambda - 3.74) + 2.54}{\lambda_{ref}(2.2\lambda_{ref} - 3.74) + 2.54}$$

$$f_\vartheta = 2.16 \cdot \frac{\vartheta_{c,300_{ref}}}{\vartheta_{c,300}} - 1.16 \tag{C.13}$$

$$f_p = \left(\frac{p_{c,300}}{p_{c,300_{ref}}}\right)^{-0.47}$$

$$f_x = 0.088 \cdot \frac{x_{egr}}{x_{egr_{ref}}} + 0.912$$

$$f_n = \frac{1 + \frac{400}{n_e} - \frac{800000}{n_e^2}}{1 + \frac{400}{n_{e_{ref}}} - \frac{800000}{n_{e_{ref}}^2}}$$

$$g_\zeta = 1$$

$$g_\lambda = \frac{y_\lambda \cdot (\lambda(0.8\lambda - 1.36) + 0.56) + 1}{y_\lambda \cdot (\lambda_{ref}(0.8\lambda_{ref} - 1.36) + 0.56) + 1}$$

$$g_\vartheta = 1.33 \cdot \frac{\vartheta_{c,300_{ref}}}{\vartheta_{c,300}} - 0.33 \tag{C.14}$$

$$g_p = \left(\frac{p_{c,300}}{p_{c,300_{ref}}}\right)^{-0.28}$$

$$g_x = 0.237 \cdot \frac{x_{egr}}{x_{egr_{ref}}} + 0.763$$

$$g_n = \frac{1.33 - \frac{660}{n_e}}{1.33 - \frac{660}{n_{e_{ref}}}}$$

$$h_\zeta = 1$$

$$h_\lambda = \frac{F(\lambda)}{F(\lambda_{ref})} \begin{cases} \lambda \le 0.95 : F(\lambda) = 1 + 0.0833(y_\lambda - 2.5)(\lambda - 0.95) \\ \lambda \ge 0.95 : F(\lambda) = 1 - 0.2963(y_\lambda - 2.5)(\lambda - 0.95) \end{cases}$$

$$h_\vartheta = 1 \tag{C.15}$$

$$h_p = 1$$

$$h_x = 1$$

$$h_n = \frac{0.625 + \frac{750}{n_e}}{0.625 + \frac{750}{n_{e_{ref}}}}$$

$$y_\lambda = 7 \tag{C.16}$$

Using the Csallner functions, parameter studies around reference points can be carried out. The operating-point parameters (including the Vibe parameters) are calibrated with measurements in the reference point and, subsequently, parameter studies and optimizations can be performed in the surrounding of the operating point. In [154] and [156] it is shown that variations of up to 20% in all relevant parameters are covered and that therefore many test-bench optimizations can be replaced by computer calculations.

References

1. **Ackermann J.** (1993) *Robuste Regelung.* Springer Verlag, Berlin
2. **Alkidas A.C.** (1997) *Intake-Valve Temperature and the Factors Affecting It.* SAE paper 971729
3. **Alkidas A.C.** (2001) *Intake-Valve Temperature Histories During S.I. Engine Warm-Up.* SAE paper 2001-01-1704
4. **Allmendinger K.**, Guzzella L., Seiler A. and Loffeld O. (2001) *A Method to Reduce the Calculation Time for an Internal Combustion Engine Model.* SAE paper 2001-01-0574
5. **Amstutz A.** and Guzzella L. (1997) *Fuel Cells for Transportation – An Assessment of their Potential for CO_2-Reduction.* Conference on Greenhouse Gas Control, Interlaken, Switzerland
6. **Andersson P.** and Eriksson L. (2004) *Cylinder Air Charge Estimators in Turbocharged SI-Engines.* SAE paper 2004-01-1366
7. **Anonymous** (1997) *Bosch Automotive Handbook.* Robert Bosch GmbH, Stuttgart
8. **Anonymous** (2001) *Emission Measurement Standards of the European Union.* $http : //europa.eu.int/eurlex/en/search/search\_lif.htm$
9. **Aquino C.F.** (1981) *Transient A/F Control Characteristics of a 5 Liter Central Fuel Injection Engine.* SAE paper 810494
10. **Arici O.**, Johnson J.H. and Kulkarni A.J. (1999) *Cooling System Simulation; Part 1 - Model Development.* SAE paper 1999-01-0240
11. **Arsie I.**, Pianese C. and Rizzo G. (1998) *Models for the Prediction of Performance and Emissions in a Spark Ignition Engine.* SAE paper 980779
12. **Aström K.J.** and Hägglund T. (1992) *PID Controllers: Theory, Design and Tuning.* Instrument Society of America, Research Triangle Park, NC
13. **Atkins P.W.** (1988) *Physikalische Chemie.* VCH Verlag, Weinheim
14. **Atkinson C.M.**, Long T.W. and Hanzevack E.L. (1998) *Virtual Sensing: A Neural Network-Based Intelligent Performance and Emissions Prediction System for On-Board Diagnostics and Engine Control.* SAE paper 980516
15. **Auckenthaler T.**, Onder C. and Geering H. (2002) *Modelling of a Solid-Electrolyte Oxygen Sensor.* SAE paper 2002-01-1293
16. **Auckenthaler T.**, Onder C. and Geering H. (2004) *Online Estimation of the Oxygen Storage Level of a Three-Way Catalyst.* SAE paper 2004-01-0525

17. **Auckenthaler T.**, Onder C. and Geering H. (2004) *Aspects of Dynamic Three-Way-Catalytic Converter Behaviour Including Oxygen Storage.* IFAC Symposium "Advances in Automotive Control," Salerno
18. **Balenovic M.** (2002) *Modeling and Model-Based Control of a Three-Way Catalytic Converter.* PhD thesis, Technical University of Eindhoven
19. **Balenovic M.**, Backx T. and de Bie T. (2002) *Development of a Model-Based Controller for a Three-Way Catalytic Converter.* SAE paper 2002-01-0475
20. **Balluchi A.**, Benvenuti L., Di Benedetto M.D., Villa T., Wong-Toi H. and Sangiovanni-Vincentelli A.L. (2000) *Hybrid Controller Synthesis for Idle Speed Management of an Automotive Engine.* Proc. 2000 IEEE American Control Conference, Chicago, vol. 2, pp. 1181–5
21. **Balluchi A.**, Benvenuti L., Di Benedetto M.D. and Sangiovanni-Vincentelli A.L. (2003) *Idle Speed Control Synthesis Using an Assume-Guarantee Approach.* In Nonlinear and Hybrid Systems in Automotive Control, Springer-Verlag, Berlin
22. **Beard A.** and Smith G.J. (1971) *A Method of Calculating the Heat Dissipation from Radiators to Cool Vehicle Engines.* SAE paper 710208
23. **Bejan A.** (1996) *Heat Transfer.* John Wiley and Sons, New York
24. **Bianchi G.M.**, Falfari S., Parotto M. and Osbat G. (2003) *Advanced Modeling of Common Rail Injector Dynamics and Comparison with Experiments.* SAE paper 2003-01-0006
25. **Boehme T.R.**, Onder C., Guzzella L. (2008) *Code-generator based software package for defining and solving one-dimensional, dynamic, catalytic monolith models.* Computers & Chemical Engineering, DOI: 10.1016/j.compchemeng.2008.01.003
26. **Bosch H.** and Janssen F. (1988) *Catalytic Reduction of Nitrogen Oxides - A Review on the Fundamentals and Technology.* Catalysis Today, vol. 2, no. 4, pp. 369–531
27. **Bozek J.W.**, Evans R., Tyree C.D. and Zerafa K.L. (1992) *Operating Characteristics of Zirconia Galvanic Cells (Lambda Sensors) in Automotive Closed-Loop Emission Control Systems.* SAE paper 920289
28. **Brailsford A.D.**, Yussouff M. and Logothetis E.M. (1996) *Theory of Gas Sensors: Response of an Electrochemical Sensor to Multi-Component Gas Mixtures.* International Solid-State Sensors and Actuators Conference, B: Chemical, Stockholm, vol. B34, pp. 407–11
29. **Brand D.** (2004) *Control-Oriented NO Models for SI Engines.* Dissertation ETH Zürich no. 11948
30. **Brandt E.**, Wang Y. and Grizzle J. (2000) *Dynamic Modeling of a Three-Way Catalyst for SI Engine Exhaust Emission Control.* IEEE Transactions on Control Systems Technology, vol. 8, no. 5, pp. 767–776
31. **Brettschneider J.** (1979) *Berechnung des Luftverhältnisses von Luft-Kraftstoff-Gemischen und des Einflusses von Messfehlern auf Lambda.* Bosch Technische Berichte 6, Nr. 4, pp. 177–186
32. **Brettschneider J.** (1997) *Extension of the equation for calculation of the air-fuel equivalence ratio.* SAE paper 972989
33. **Brinkmeier C.**, Büchner S. and Donnerstag A. (2003) *Transient Emissions of a Sulev Catalytic Converter System Dynamic Simulation vs. Dynamometer Measurements.* SAE paper 2003-01-1001

34. **Brussovansky S.**, Heywood J.B. and Keck J.C. (1992) *Predicting the Effects of Air and Coolant Temperature, Deposits, Spark Timing and Speed on Knock in Spark Ignition Engines.* SAE paper 922324

35. **Butts K.**, Sivashankar N. and Sun J. (1995) *Feedforward and Feedback Design for Engine Idle Speed Control Using l_1 Optimization.* Proceedings of the American Control Conference, vol. 4, pp. 2587–2590

36. **Cassidy J.** (1977) *A Computerized On-Line Approach to Calculating Optimum Engine Calibrations.* SAE paper 770078

37. **Cavina N.**, Corti E., Minelli G. and Serra G. (2002) *Misfire Detection Based on Engine Speed Time-frequency Analysis.* SAE paper 2002-01-0480

38. **Chang C.F.**, Fekete N.P. and Powell J.D. (1993) *Engine Air-Fuel Ratio Control Using an Event-Based Observer.* SAE paper 930766

39. **Chatterjee D.**, Deutschmann O. and Warnatz J. (2002) *Detailed surface reaction mechanism in a three-way catalyst.* Faraday Discuss., vol. 119, pp. 371-384

40. **Chin Y.K.** and Coats F.E. (1986) *Engine Dynamics: Time-Based Versus Crank-Angle Based.* SAE paper 860412

41. **Christen U.** (1996) *Engineering Aspects of H_∞-Control.* Dissertation ETH Zürich no. 11433

42. **Christen U.**, Vantine K.J. and Collings N. (2001) *Event-Based Mean-Value Modeling of DI Diesel Engines for Controller Design.* SAE paper 2001-01-1242

43. **Cipollone R.** and Sciarretta A. (2001) *The Quasi-Propagatory Model: A New Approach for Describing Transient Phenomena in Engine Manifolds.* SAE paper 2001-01-0579

44. **Cook J.A.**, Grizzle J.W. and Sun J. (1996) *Engine Control.* In The Controls Handbook, edited by W. S. Levine, CRC Press-Times Mirror Books

45. **Cortona E.**, Onder Ch. H. and Guzzella L. (2002) *Engine Thermal Management with Components for Fuel Consumption Reduction.* International Journal of Engine Research, IMechE, vol. 3, no. 3, pp. 157–170

46. **Cortona E.** (2003) *Engine Thermomanagement For Fuel Consumption Reduction.* Dissertation ETH Zürich no. 13862

47. **Cowart J.S.**, Keck J.C., Heywood J.B., Westbrook C.K. and Pitz W.J. (1990) *Comparison of Engine Knock Predictions Using a Fully-Detailed and a Reduced Chemical Kinetic Mechanism.* Twenty-Third Symposium (Int) on Combustion, pp. 1055-1062, The Combustion Institute, Pittsburgh

48. **Cowart J.** and Cheng W. (1999) *Intake Valve Thermal Behavior During Steady-State and Transient Engine Operation.* SAE paper 1999-01-3643

49. **Csallner P.** (1981) *Eine Methode zur Vorausberechnung der Änderung des Brennverlaufes von Ottomotoren bei geänderten Betriebsbedingungen.* Dissertation TU München, 1981

50. **Curtis E.W.**, Aquino C.F., Trumpy D.K. and Davis G.C. (1996) *A New Port and Cylinder Wall Wetting Model to Predict Transient Air/Fuel Excursions in a Port Fuel Injected Engine.* SAE paper 961186

51. **Curtis E.W.**, Aquino C.F., Plensdorf W.D., Trumpy D.K., Davis G.C. and Lavoie G.A. (1997) *Modeling Intake Valve Warm-Up.* ASME Fall Technical Conference 29-1 (97-ICE-40)

52. **Douaud A.M.** and Eyzat P. (1978) *Four-Octane-Number Method for Predicting the Anti-Knock Behavior of Fuels and Engines.* SAE paper 780080

53. **Doyle J.**, Francis B. and Tannenbaum A. (1992) *Feedback Control Theory.* Macmillan, New York

54. **Dyntar D.**, Onder C. and Guzzella L. (2002) *Modeling and Control of CNG Engines.* SAE paper 2002-01-1295

55. **Eastwood P.** (2000) *Critical Topics in Exhaust Gas Aftertreatment.* Engineering Design Series, Research Studies Press Ltd., Baldock, Hertfordshire, England

56. **Ellman A.** and Piché R. (1999) *A Two Regime Orifice Formula for Numerical Simulation.* ASME J-DSMC, vol. 121, pp. 721–724

57. **Elmqvist C.**, Lindström F., Angström H.E., Grandin B. and Kalghatgi G. (2003) *Optimizing Engine Concepts By Using a Simple Model for Knock Prediction.* SAE paper 2003-01-3123

58. **Emmerling G.** and Zuther F. (1999) *Motorische Verbrennung - aktuelle Probleme und moderne Lösungsansätze.* Berichte zur Energie- und Verfahrenstechnik, Editor A. Leiperz, Heft 99.1, Erlangen

59. **Engh G.T.** and Chiang C. (1970) *Correlation of Convective Heat Transfer for Steady Intake-Flow Through a Poppet Valve.* SAE paper 700501

60. **Eriksson L.**, Nielsen L. and Glavenius M. (1997) *Closed Loop Ignition Control by Ionization Current Interpretation.* SAE paper 970854

61. **Eriksson L.** (2002) *Mean Value Models for Exhaust Temperatures.* SAE paper 2002-01-0374

62. **Eriksson L.**, Frei S., Onder C. and Guzzella L. (2002) *Control and Optimization of Turbocharged Spark Ignited Engines* Proceedings of IFAC 15th World Congress

63. **Eriksson L.** (2004) *CHEPP - A Chemical Equilibrium Program Package for Matlab.* SAE paper 2004-01-1460

64. **Ferguson C.R.** and Kirkpatrick A.T. (2001) *Internal Combustion Engines.* John Wiley & Sons, New York

65. **Franklin G.**, Powell J.D. and Emami-Naeini A. (1986) *Feedback Control of Dynamic Systems.* Addison-Wesley, Reading, MA

66. **Franklin G.**, Powell J.D. and Workman M. (1990) *Digital Control of Dynamic Systems.* Addison-Wesley, Reading (MA)

67. **Franzke D.E.** (1981) *Beitrag zur Ermittlung eines Klopfkriteriums der ottomotorischen Verbrennung und zur Vorausberechnung der Klopfgrenze.* PhD thesis, TU München

68. **Friedrich A.** (1997) *The Assessment of Fuel Cells in Transport from the Environmental Point of View.* Umweltbundesamt, Berlin

69. **Geering H.**, Onder C.H., Roduner C.A., Dyntar D. and Matter D. (2002) *ICX-3 — A Flexible Interface Chip for Research in Engine Control.* Proc. of the FISITA 2002 World Automotive Congress, Helsinki, Paper no. FV02V315

70. **Gerhardt J.**, Hönninger H. and Bischof H. (1998) *A new Approach to Functional and Software Structure for Engine Management Systems – Bosch Me7.* SAE paper 980801

71. **Germann H.**, Tagliaferri S. and Geering H. (1996) *Differences in Pre- and Post-Converter Lambda Sensor Characteristics.* SAE paper 960335

72. **Gizzi W.P.** (2000) *Dynamische Korrekturen für schnelle Strömungssonden in hochfrequent fluktuierenden Strömungen.* Dissertation ETH Zürich no. 13482

73. **Glattfelder A.H.** and Schaufelberger W. (2003) *Control Systems with Input and Output Constraints.* Springer Verlag, Berlin

74. **Göpel W.** and Wiemhöfer H.D. (2000) *Statistische Thermodynamik.* Spektrum Akademischer Verlag, Heidelberg

75. **Gorille I.** (1980) *MOTRONIC – ein neues elektronisches System zur Steuerung von Ottomotoren.* MTZ Motortechnische Zeitschrift 41, vol. 51, pp. 203–215

76. **Gravdahl J.** and Egeland O. (1999) *Centrifugal Compressor Surge and Speed Control.* IEEE Transactions on Control Systems Technology, vol. 7, no. 5, pp. 567–579

77. **Greene A.B.** and Lucas G.G. (1969) *The Testing of Internal Combustion Engines.* English University Press Ltd.

78. **Guzzella L.**, Geering H. and Hirzel H. (1987) *Anwendungsspezifische Chips für die Regelung von Ottomotoren.* Technische Rundschau, vol. 79, no. 1/2, p. 46–49

79. **Guzzella L.** and Amstutz A. (1998) *Control of Diesel Engines.* IEEE Control Systems Magazine, vol. 8, No. 9, pp. 55–71

80. **Guzzella L.** and Amstutz A. (1999) *CAE-Tools for Quasistatic Modeling and Optimization of Hybrid Powertrains.* IEEE Transactions on Vehicular Technology, vol. 48, no. 6, pp. 1762–1769

81. **Guzzella L.** and Sciarretta A. (2007) *Vehicle Propulsion Systems* Springer, Berlin

82. **Guzzella L.**, Wenger U. and Martin, R. (2000) *IC-Engine Downsizing and Pressure-Wave Supercharging for Fuel Economy.* SAE paper 2000-01-1019

83. **Gyarmathy G.** (1983) *How does the COMPREX Pressure Wave Supercharger Work?* SAE paper 830234

84. **Halstead M.P.**, Kirsch L.J. and Quin C.P. (1977) *The Autoignition of Hydrocarbon Fuels at High Temperatures and Pressure – Fitting of a Mathematical Model.* Combustion and Flame, vol. 30, pp. 45–60

85. **Haluska P.** and Guzzella L. (1998) *Control Oriented Modeling of Mixture Formation Phenomena in Multi-Port Injection SI-Gasoline Engines.* SAE paper 980628

86. **Harigaya Y.**, Toda F. and Suzuki M. (1993) *Local Heat Transfer on a Combustion Chamber Wall of a Spark-Ignition Engine.* SAE paper 931130

87. **Harmsen J.** (2001) *Kinetic Modelling of the Dynamic Behaviour of an Automotive Three-Way Catalyst under Cold-Start Conditions.* PhD Thesis, Technical University of Eindhoven

88. **Hasegawa Y.**, Akazaki S., Komoriya I., Maki H., Nishimura Y. and Hirota T. (1994) *Individual Cylinder Air-Fuel Ratio Feedback Control Using an Observer.* SAE paper 940376

89. **Hendricks E.** (1986) *A Compact, Comprehensive Model of Large Turbocharged, Two-Stroke Diesel Engines.* SAE paper 861190

90. **Hendricks E.** and Sorenson S.C. (1990) *Mean-Value Modeling of SI Engines.* SAE paper 900616

91. **Hendricks E.**, Jensen M., Chevalier A. and Vesterholm T. (1994) *Conventional Event Based Engine Control.* SAE paper 940377

92. **Hendricks E.**, Jensen M., Chevalier A. and Vesterholm T. (1994) *Problems in Event Based Engine Control.* American Control Conference, Baltimore, vol. 2, pp. 1585–1587

93. **Hendricks E.**, Chevalier A. and Jensen M. (1995) *Event-Based Engine Control: Practical Problems and Solutions.* SAE paper 950008

94. **Hendricks E.** (2001) *Isothermal vs. Adiabatic SI Engine Mean Value Engine Models.* Proc. 3rd IFAC Workshop on Advances in Automotive Control, Karlsruhe

95. **Herden W.** and Küsell M. (1994) *A New Combustion Pressure Sensor for Advanced Engine Management.* SAE paper 940379

96. **Hertzberg A.** (2001) *Betriebsstrategien für einen Ottomotor mit Direkteinspritzung und NOx-Speicher-Katalysator.* Ph.D. thesis, Universität Karlsruhe

97. **Heywood J.B.** (1988) *Internal Combustion Engine Fundamentals.* McGraw Hill, New York

98. **Ho S.Y.** and Kuo T.W. (1997) *A Hydrocarbon Autoignition Model for Knocking Combustion in SI Engines.* SAE paper 971672

99. **Hoebink J.H.B.J.**, Harmsen J.M.A., Balenovic M., Backx A.C.P.M. and Schouten J.C. (2001) *Automotive exhaust gas conversion: from elementary step kinetics to prediction of emission dynamics.* Topics in Catalysis, vol. 16/17, pp. 319-327

100. **Hrovat D.** and Sun J. (1997) *Models and Control Methodologies for IC Engine Idle Speed Control Design.* Control Engineering Practice, vol. 5, no. 8, pp. 1093–1100

101. **Huber E.** and Koller T. (1977) *Pipe Friction and Heat Transfer in the Exhaust Pipe of a Firing Combustion Engine.* Proceedings of the CIMAC Convention, Tokio

102. **Incropera F.P.** and DeWitt D.P. (1996) *Fundamentals of Heat and Mass Transfer.* John Wiley & Sons, New York

103. **Ingram G.** and Gopichandra S. (2003) *On-line Oxygen Storage Capacity Estimation of a Catalyst.* SAE paper 2003-01-1000

104. **Inhelder J.** (1996) *Verbrauchs- und Schadstoffoptimiertes Ottomotor-Aufladekonzept.* Dissertation ETH Zürich no. 11948

105. **Jensen J.P.**, Kirstensen A.F., Sorenson S.C., Houba N. and Hendricks E. (1991) *Mean Value Modeling of a Small Turbocharged Diesel Engine.* SAE paper 910070

106. **Jobson E.**, Hjortsberg O., Andersson S. and Gottberg I. (1996) *Reactions Over a Double Layer Tri-Metal Three-Way Catalyst.* SAE paper 960801

107. **Johnson T.** (2003) *Diesel Emission Control in Review – The Last 12 Months.* SAE paper 2003-01-0039

108. **Jurgen R.** (1995) *Automotive Electronics Handbook.* McGraw-Hill, New York

109. **Kalghatgi G.T.** (2005) *Auto-ignition quality of practical fuels and implications for fuel requirements of future SI and HCCI engines.* SAE paper 2005-01-0239

110. **Kampelmühler F.T.**, Paulitsch R. and Gschweitl K. (1993) *Automatic ECU-Calibration: An Alternative to Conventional Methods.* SAE paper 930395

111. **Kang J.M.**, Kolmanovsky I. and Grizzle J.W. (2001) *Dynamic Optimization of Lean Burn Engine Aftertreatment.* ASME J-DSMC, vol. 123, no. 2, pp. 153–160

112. **Karim G.A.** and Gao J. (1992) *A Predictive Model for Knock in Dual Fuel Engines.* SAE paper 921550

113. **Kiencke U.** and Nielsen L. (2000) *Automotive Control Systems.* Springer, Berlin

114. **Kiencke, U.** (1999) *Engine Misfire Detection.* IFAC Control Engineering Practice, vol. 7, no.2, pp. 203–208

115. **Kim Y.W.**, Sun J., Kolmanovsky I. and Koncsol J. (2003) *A Phenomenological Control-Oriented Lean NOx Trap Model.* SAE paper 2003-01-1164

116. **Kladopoulou E.A.**, Yang S.L., Johnson J.H., Parker G.G. and Konstandopoulos A.G. (2003) *A Study Describing the Performance of Diesel Particulate Filters During Loading and Regeneration: A Lumped Parameter Model for Control Applications.* SAE paper 2003-01-0842

117. **Klepatsch M.** (1997) *Gemischbildung beim Ottomotor: Neue Möglichkeiten der Simulationsrechnung.* Wiener Motorensymposium, VDI Reihe 12, Nr. 306, pp. 217–233

118. **Koch F.W.** and Haubner F.G. (2000) *Cooling System Development and Optimization for DI Engines.* SAE paper 2000-01-0283

119. **Kolmanowsky I.** and Stefanopoulou A. (2001) *Optimal Control Techniques for Assessing Feasibility and Defining Subsystem Level Requirements: An Automotive Case Study.* IEEE Trans. Control Systems Technology, vol. 9, no. 3, pp. 524–534

120. **Koltsakis G.C.** and Stamatelos A.M. (1999) *Modeling Dynamic Phenomena in 3-Way Catalytic Converters.* Chemical Engineering Science, vol. 54, no. 20, pp. 4567–4578

121. **Konstandopoulos A.G.**, Margaritis K. and Paraskevi H. (2001) *Spatial Non-Uniformities in Diesel Particulate Trap Regeneration.* SAE paper 2001-01-0908

122. **Konstandopoulos A.G.**, Skaperdas E. and Masoudi M. (2001) *Inertial Contributions to the Pressure Drop of Diesel Particulate Filters.* SAE paper 2001-01-0909

123. **Krause W.** and Spies K. H. (1996) *Dynamic Control of the Coolant Temperature for a Reduction of Fuel Consumption and Hydrocarbon Emission.* SAE paper 960271

124. **Krieger R.B.**, Borman G.L. (1966) *The Computation of Apparent Heat Release for Internal Combustion Engines.* ASME paper 66-WA/DGP-4

125. **Krüger S.** (2000) *Advanced Microsystems for Automotive Applications.* Springer Verlag, Berlin

126. **Larsson M.**, Andersson L., Fast O., Litorell M. and Makuie R. (1999) *NOx Trap Control by Physically Based Model.* SAE paper 1999-01-3503

127. **Lee H.K.**, Russell M.F. and Bae C.S. (2002) *Mathematical model of diesel fuel injection equipment incorporating non-linear fuel injection.* Proc. IMechE Part D - Journal of Automobile Engineering, vol. 216, no. D3, pp 191–204

128. **Levenspiel O.** (1972) *Chemical Reaction Engineering.* John Wiley & Sons, New York

129. **Livengood J.C.** and Wu P.C. (1955) *Correlation of Autoignition Phenomena in Internal Combustion Engines and Rapid Compression Machines.* Fifth Symposium (Int) on Combustion, pp. 347–356, Reinhold Publishing Corporation, Pittsburgh, PA

130. **Ljung L.** (1987) *System Identification; Theory for the User.* Prentice Hall, Englewood Cliffs, NJ

131. **Locatelli M.**, Onder C.H. and Geering H. (2003) *A Rapidly Tunable Model of the Droplet Evaporation for the Study of the Wall-Wetting Dynamics.* JSAE paper 20035416

132. **Locatelli M.**, Onder C.H. and Geering H. (2004) *A Rapidly Tunable Wall-Wetting Model for SI Engine Control.* SAE paper 2004-01-1461

133. **Malchow G.L.**, Sorenson S.C. and Buckius R.O. (1979) *Heat Transfer in the Straight Section of an Exhaust Port of a Spark Ignition Engine.* SAE paper 790309

134. **Maloney P.J.** (1999) *An Event-Based Transient Fuel Compensator with Physically Based Parameters.* SAE paper 1999-01-0553

135. **Maroteaux F.** and Le Moine L. (1995) *Modeling of Fuel Droplets Deposition Rate in Port Injected Spark Ignition Engines.* SAE paper 952484

136. **McBride B.J.**, et al. (1963) *Thermodynamic Properties to 6000 K for 210 Substances Involving the First 18 Elements.* NASA SP 3001, Washington, DC

137. **Metzler F.**, Hesse U., Rocklage G. and Schmitt M. (1996) *Thermomanagement.* SAE paper 1999-01-0238

138. **Mhadeshwar A.**, Wang H. and Vlachos D. (2003) *Thermodynamic consistency in microkinetic development of surface reaction mechanisms.* J. Phys. Chem, vol. 107, pp. 12721–12733

139. **Mladek M.**, Onder C.H. and Guzzella L. (2000) *A Model for the Estimation of Inducted Air Mass and the Residual Gas Fraction using Cylinder Pressure Measurements.* SAE paper 2000-01-0958

140. **Mladek M.** (2003) *Cylinder Pressure for Control Purposes of Spark Ignition Engines.* Dissertation ETH Zürich no. 14916

141. **Möller R.** et al., (2009) *Is oxygen storage in three-way catalysts an equilibrium controlled process?* Appl. Catal. B: Environ., doi:10.1016/j.apcatb.2009.05.003

142. **Möller R.** (2009) *Modelling and Control of Three-Way Catalysts.* Dissertation ETH Zürich no. 18426

143. **Morselli R.**, Corti E. and Rizzoni G. (2002) *Energy Based Model of a Common Rail Injector.* Proc. of the 2002 IEEE Conference on Control Applications, Glasgow, vol. 2, pp. 1195–1200

144. **Moore F.K.** and Greitzer E.M. (1986) *A Theory of Post-Stall Transients in Axial Compression Systems.* Journal of Engineering for Gas Turbines and Power, vol. 108, pp. 68–76

145. **Morari M.** and Zafiriou E. (1989) *Robust Process Control.* Prentice Hall, Englewood Cliffs, NJ

146. **Moraal P.** and Kolmanovsky I. (1999) *Turbocharger Modeling for Automotive Control Applications.* SAE paper 1999-01-0908

147. **Moran M.** and Shapiro H.N. (1992) *Fundamentals of Engineering Thermodynamics.* Wiley, New York

148. **Moser F.**, Sams T. and Cartellieri W. (2001) *Impact of Future Exhaust Gas Emission Legislation on Heavy Duty Truck Engine.* SAE paper 2001-01-0186

149. **Müller M.**, Hendricks E. and Sorenson S.C. (1998) *Mean Value Modelling of Turbocharged Spark Ignition Engines.* SAE paper 980784

150. **Müller N.** and Isermann R. (2001) *Control of Mixture Composition Using Cylinder Pressure Sensors.* SAE paper 2001-01-3382

151. **Müller R.**, Hart M., Truscott A., Noble A., Krötz G., Eickhoff M., Cavalloni C. and Gnielka M. (2000) *Combustion-Pressure-Based Engine Management System.* SAE paper 2000-01-0928

152. **Nievergeld A.J.L.** (1998) *Automotive Exhaust Gas Conversion: Reaction Kinetics, Reactor Modelling and Control.* PhD Thesis, Technical University of Eindhoven

153. **Nowak U.**, Frauhammer J. and Nieken U. (1996) *A Fully Adaptive Algorithm for Parabolic Partial Differential Equations in One Space Dimension.* Computers and Chemical Engineering, vol. 20, no. 5, pp. 547–561.

154. **Onder C.H.** (1993) *Modellbasierte Optimierung der Steuerung und Regelung eines Automobilmotors.* Dissertation ETH Zürich no. 10323

155. **Onder C.H.** and Geering H.P. (1993) *Model-Based Multivariable Speed and Air-to-Fuel Ratio Control of an SI Engine.* SAE paper 930859

156. **Onder C.H.** (1995) *Model-Based Engine Calibration for Best Fuel Efficiency.* SAE paper 950983

157. **Onder C.H.**, Roduner C. and Geering H. (1997) *Model Identification for the Air/Fuel Path of an SI Engine.* SAE paper 970612

158. **Pachernegg S.J.** (1069) *A Closer Look at the Willans-Line.* SAE paper 690182

159. **Padeste C.**, Cant N.W. and Trimm D.L. (1993) *The influence of water on the reduction and reoxidation of ceria.* Catalysis Letters, vol. 18, nr. 3, pp. 305-316

160. **Pattas K.**, Stamatelos A., Pistikopoulos P., Koltsakis G., Konstandinidis P., Volpi E. and Leveroni E. (1994) *Transient Modeling of 3-Way Catalytic Converters.* SAE paper 940934

161. **Patton K.J**, Nitschke R.G. and Heywood J.B. (1989) *Development and Evaluation of a Friction Model for Spark Ignition Engine.* SAE paper 890836

162. **Peyton Jones J.**, Jackson R., Roberts J. and Bernard P. (2000) *A Simplified Model for the Dynamics of a Three-Way Catalytic Converter.* SAE paper 2000-01-0652

163. **Pischinger R.**, Krassnig G., Taucar G. and Sams T. (1989) *Thermodynamik der Verbrennungskraftmaschine* Springer-Verlag Wien New York

164. **Plapp G.**, Klenk M. and Moser W. (1989) *Methods of On-board Misfire Detection.* SAE paper 900232

165. **Poulikakos D.** (1994) *Conduction Heat Transfer.* Prentice Hall, Englewood Cliffs, NJ

166. **Powell B.K.** and Cook J.A. (1987) *Nonlinear low frequency phenomenological engine modeling and analysis.* Proc. of the 1987 IEEE American Control Conference, vol. 1, pp. 332–340

167. **Powell J.D.** (1987) *A Review of IC Engine Models for Control System Design.* Proc. of the 10th IFAC World Congress, San Francisco, CA

168. **Powell J.D.** (1995) *IC Engine Models for Control System Design*. Proc. of the 1st IFAC-Workshop on Advances in Automotive Control, Ascona
169. **Ranz W.E.** and Marshall W.R. (1952) *Evaporation from Drops*. Chem. Engrg. Prog., vol. 48, pp. 141–146
170. **Rao H.**, Cohen A., Tennant J. and Van Voorhies K. (1979) *Engine Control Optimization Via Nonlinear Programming*. SAE paper 790177
171. **Rassweiler G.** and Withrow L. (1979) *Motion Pictures of Engine Flames Correlated with Pressure Cards*. SAE Transactions, pp. 297–303
172. **Rausen D.J.**, Stefanopoulou A.G., Kang J.M., Eng J.A. and Kuo T.W. (2004) *A Mean-Value Model for Control of Homogeneous Charge Compression Ignition (HCCI) Engines*. Proc. of the 2004 IEEE American Control Conference, Boston, MA
173. **Ribi B.** (1996) *Radialverdichter im Instabilitätsbereich*. Dissertation ETH Zürich no. 11717
174. **Rizzoni G.** and William B. (1990) *Onboard Diagnosis of Engine Misfires*. SAE paper 901768
175. **Roduner C.A.** (1997) *H∞-Regelung linearer Systeme mit Totzeiten*. Fortschritt-Berichte VDI, Reihe 8, Nr. 708, VDI Verlag GmbH, Düsseldorf
176. **Roduner C.**, Onder C. and Geering H. (1997) *Automated Design of an Air/Fuel Controller for an SI Engine Considering the Three-Way Catalytic Converter in the H∞ Approach*. Proc. 5th IEEE Mediterranean Conference on Control and Systems, Paphos, Cyprus
177. **Schäpertöns H.** and Lee W. (1985) *Multidimensional Modelling of Knocking Combustion in SI Engines*. SAE paper 850502
178. **Schaer C.** (2003) *Control of a Selective Catalytic Reduction Process*. Dissertation ETH Zürich no. 15221
179. **Schmitz G.**, Oligschläger U., Eifler G. and Lechner H. (1994) *Automated System for Optimized Calibration of Engine Management Systems*. SAE paper 940151
180. **Schöggl P.**, Koegeler H.M., Gschweitl K., Kokal H., Williams P. and Hulak K. (2002) *Automated EMS Calibration Using Objective Driveability Assessment and Computer Aided Optimization Methods*. SAE paper 2002-01-0849
181. **Schubiger R.A.**, Boulouchous K. and Eberle M.K. (2002) *Russbildung und Oxidation bei der dieselmotorischen Verbrennung*. Motortechnische Zeitschrift (MTZ), no. 5, pp. 342–353
182. **Schwarz Ch.** (1998) *Theorie und Simulation aufgeladener Verbrennungsmotoren*. Habilitationsschrift, TU Hannover
183. **Sellnau M.C.**, Matekunas F.A., Battiston P.A., Chang C.F. and Lancaster D.R. (2000) *Cylinder-Pressure-Based Engine Control Using Pressure-Ratio Management and Low-Cost Non-Intrusive Cylinder Pressure Sensors*. SAE paper 2000-01-0932
184. **Shamma J.S.** and Athans M. (1992) *Gain scheduling: Potential Hazards and Possible Remedies*. IEEE Control Systems Magazine, vol. 12, no. 3, pp. 101–107
185. **Shafai E.**, Roduner C. and Geering H. (1996) *Indirect Adaptive Control of a Three-Way Catalyst*. SAE paper 961038
186. **Shafai E.**, Roduner Ch. and Geering H. (1997) *On-Line Identification of Time Delay in the Fuel Path of an SI Engine*. SAE paper 970613

187. **Shaver G.M.**, Gerdes J.C., Jain P., Caton P.A. and Edwards C.F. (2003) *Modeling for Control of HCCI Engines.* Proceedings of 2003 American Control Conference, Denver, vol. 1, pp. 749–754

188. **Shaver G.M.** and Gerdes J. (2003) *Cycle-to-Cycle Control of HCCI Engines.* Proceedings of the 2003 ASME International Mechanical Engineering Congress and Exposition, IMECE2003-41966, Washington, DC

189. **Shibata Y.**, Shimonosono H. and Yamai Y. (1993) *New Design of Cooling System with Computer Simulation and Engine Compartment Simulator.* SAE paper 931075

190. **Shirawaka T.**, Miura M., Itoyama H., Aiyoshizawa E. and Kimura S. (2001) *Study of Model-Based Cooperative Control of EGR and VGT for a Low- Temperature, Premixed Combustion Diesel Engine.* SAE paper 2001-01-2006

191. **Spring P.**, Guzzella L. and Onder C. (2003) *Optimal Control Strategy for a Pressure-Wave Supercharged SI Engine.* ASME Spring Technical Conference, Salzburg, paper no. ICES2003-645

192. **Stobart R.K.** (1997) *The Demands of Cylinder Event Control.* SAE paper 970617

193. **Stöckli M.** (1989) *Reibleistung von 4-Takt Verbrennungsmotoren.* LVV Technical Report, ETH Zürich

194. **Stone R.** (1999) *Introduction to Internal Combustion Engines.* SAE Publishing, Warrendale, PA

195. **Streit E.E.** and Borman G.L. (1971) *Mathematical simulation of a large turbocharged two-stroke Diesel engine.* SAE paper 710177

196. **Stuhler H.**, Kruse T., Stuber A., Gschweitl K., Piock W., Pfluegl H. and Lick P. (2002) *Automated Model-Based GDI Engine Calibration Adaptive Online DoE Approach.* SAE paper 2002-01-0708

197. **Sun J.**, Kolmanovsky I., Brehob D., Cook J.A., Buckland J. and Haghgooie M. (1999) *Modeling and Control of Gasoline Direct Injection Stratified Charge (DISC) Engines.* Proc. of the 1999 IEEE Conf. on Control Applications, Kauai, HI, vol. 1, pp. 471–477

198. **Tennant J.**, Giacomazzi R., Powell J. and Rao H. (1979) *Development and Validation of Engine Models Via Automated Dynamometer Tests.* SAE paper 790178

199. **Tounsi M.F.**, Menegazzi P. and Rouchon P. (2003) *NOx Trap Model for Lean-Burn Engine Control.* SAE paper 2003-01-2292

200. **Trella T.** (1979) *Spark Ignition Engine Fuel Economby Control Optimization - Techniques and Procedures.* SAE paper 790179

201. **Treybal R.E.** (1980) *Mass-Transfer Operations.* McGraw-Hill, 3rd Edition, New York, NY

202. **Turin R.C.** and Geering H. (1993) *On-line Identification of Air-to-Fuel Ratio Dynamics in a Sequentially Injected SI Engine.* SAE paper 930857

203. **Turin R.**, Casartelli E. and Geering H. (1994) *A New Model for Fuel Supply Dynamics in an SI Engine.* SAE paper 940208

204. **Turin R.C.** and Geering H. (1995) *Model-Reference Adaptive A/F-Ratio Control in an SI Engine based on Kalman Filtering Techniques.* Proc. of the 1995 American Control Conference, pp. 4082–4090, Seattle, WA

205. **Ulrich O.**, Wlodarczyk R. and Wlodarczyk M.T. (2001) *High-Accuracy Low-Cost Cylinder Pressure Sensor for Advanced Engine Controls.* SAE paper 2001-01-0991

206. **Urlaub A.** (1995) *Verbrennungsmotoren.* Springer Verlag, Berlin
207. **Vantine K.J.**, Christen U., Glover K. and Collings N. (2001) *Analysis of an Event Based Diesel Engine Model for Control Purposes.* Proc. 3rd IFAC Workshop on Advances in Automotive Control, Karlsruhe
208. **Vibe I.I.** (1970) *Brennverlauf und Kreisprozessrechnung.* VEB Verlag Technik Berlin, 1970
209. **Vidyasagar M.** (1978) *Nonlinear Systems Analysis.* Prentice-Hall, Englewood Cliffs, NJ
210. **Wang Y.**, Raman S. and Grizzle J.W. (1999) *Dynamic Modeling of a Lean NOx Trap for Lean Burn Engine Control.* Proc. of the 1999 American Control Conference, San Diego, CA
211. **Warnatz J.**, Maas U. and Dibble R.W. (2006) *Combustion.* Springer, Berlin Heidelberg New York, fourth edition
212. **Watanabe S.**, Machida K., Iijima K. and Tomisawa N. (1996) *A Sophisticated Engine Control System Using Combustion Pressure Detection.* SAE paper 960042
213. **Watson N.** and Janota M. (1982) *Turbocharging the Internal Combustion Engine.* The McMillan Press Ltd., London
214. **Weber F.**, Guzzella L. and Onder C. (2002) *Modeling of a Pressure Wave Supercharger Including External Exhaust Gas Recirculation.* Proc. IMechE, Part D: Journal of Automobile Engineering, vol. 216, no. 3, pp. 217–235
215. **Weeks R.W.** and Moskwa J.J. (1995) *Automotive Engine Modeling for Real-Time Control using MATLAB/SIMULINK.* SAE paper 950417
216. **Wen C.Y.** and Fan L.T. (1975) *Models for Flow Systems and Chemical Reactors.* Marcel Dekker Inc., New York, NY
217. **Woermann R.J.**, Teuerkauf H.J. and Heinrich A. (1999) *A Real-Time Model of a Common-Rail Diesel Engine.* SAE paper 99010862
218. **Worret R.** (2002) *Zylinderdruckbasierte Detektion und Simulation der Klopfgrenze mit einem verbesserten thermodynamischen Ansatz.* PhD thesis, TH Karlsruhe
219. **Yamamoto M.**, Yoneya S., Matsuguchi T. and Kumagai Y. (2002) *Optimization of Heavy-Duty Diesel Engine Parameters for Low Exhaust Emissions Using the Design of Experiments.* SAE paper 2002-01-1148

CPSIA information can be obtained at www.ICGtesting.com
Printed in the USA
LVOW031707041011

249071LV00008B/34/P